Colloid Science

Volume 4

A Specialist Periodical Report

Colloid Science

Volume 4

A Review of the Literature published
1977—1981

Senior Reporter
D. H. Everett, M.B.E., *Department of Physical Chemistry
University of Bristol*

Reporters
J. M. Brown, *University of Oxford*
G. F. Cerofolini, *Polytechnic of Milan, Italy*
J. Davis, *University of Bristol*
E. Dickinson, *University of Leeds*
R. L. Elliot, *Rothamptead Experimental Station, Harpenden*
W. A. House, *Freshwater Biological Assoc., Wareham*
I. A. McLure, *University of Sheffield*
I. L. Pegg, *University of Sheffield*
V. A. Soares, *University of Lisbon, Portugal*

The Royal Society of Chemistry
Burlington House, London, W1V 0BN

NO ANAL

ISBN 0-85186-538-0
ISBN 0305-9723

QD
549
C5781
V.4
CHEM

Typeset and Printed in Northern Ireland at
The Universities Press (Belfast) Ltd.

Preface

The fourth Volume of Reports on Colloid Science opens with a critical review of recent work on adsorption on heterogenous surfaces: this has been an area of intense study during the past decade and the need for an integrated survey has been increasingly apparent. Chapter 2 is closely related to Chapter 1, but provides a quite different historical and philosophical perspective of the relationship between the Dubinin–Radushkevich equation and the heterogeneity of surfaces. Chapter 3 continues for the period 1977–1981 an account of progress in adsorption from solution of (mainly) non-polar molecules. In recent years the applicability of statistical mechanical theory to assemblies of colloidal particles has received increasing attention and promises to be one of the main thrusts forward in Colloid Science: there is after all nothing in the formal development of statistical mechanics which says that the 'particles' with which it is concerned should be single atoms or molecules. Chapter 4 provides an account of the way in which work in this area has developed up to 1981. The continued rapid expansion of work on the structure of micelles and of the effects of micellization on the kinetics and mechanism of reactions justifies a further review. This is an area of research in which organic chemists have played a leading role, while the general topic of chemical reactions in colloidal systems still attracts too little attention from colloid scientists in general. Finally, in a new departure for these Reports, Chapter 6 provides an extensive bibliography of the surface tension of binary liquid mixtures. With increasing interest on the theory of liquid/vapour interfaces it is important to have available a comprehensive bibliography of experimental data that can be called upon to illustrate or to test more critically the theoretical developments.

Bristol, July 1982   D. H. Everett

Contents

                  Suspensions   **150**
                  By E. Dickinson

          1 Introduction   150

          2 Formalism and Methodology   152
              Liquid-state Background   152
              Distribution and Correlation Functions   153
              Computer Simulation   155

          3 Order–Disorder Phase Transitions   157
              Numerical Results and Approximate Theories   159
              Wigner Lattice Model   161
              The Onsanger Transition   162

          4 Structure from Light Scattering   163
              Time-averaged Light Scattering   163
              Light-scattering Studies of Floc Structure   167
              Photon Correlation Spectroscopy   167

          5 Osmotic Equation of State   169

          6 Polydispersity   170
              Thermodynamics of Polydisperse Systems   170
              Light Scattering from Polydisperse Systems   173

          7 Time-dependent Behaviour   174
              Neglect of Hydrodynamic Interactions   174
              Effect of Hydrodynamic Forces between Particles   177

Chapter 5   Micellar Structure and Catalysis   **180**
                  By J. M. Brown and R. L. Elliott

          1 Introduction   180

          2 Micellar Structure and Solubilization   180
              Computational Aspects   180
              Critical Micelle Concentrations   182
              Properties of Micelles   183
              Scattering Techniques   184
              Solubilization   185

          3 Magnetic Resonance   186
              Electron Spin Resonance   186
              Proton Magnetic Resonance   188

1
Adsorption on Heterogeneous Surfaces

BY W. A. HOUSE

1 Introduction

This Report is concerned with the surface heterogeneity of gas–solid and liquid–solid systems. It presents a historical development of the subject with particular emphasis placed on work published in the period 1970 to July 1981. Properties of the solid/gas and solid/liquid interfaces are dependent on the chemical structure and topology of the solid surface. The chemical potential of the adsorbate, whether vapour, gas, or liquid, is dependent upon the nature of the surface and the spatial variation of the adsorption potential energy. Generally the development of physical adsorption theory has tended to concentrate on homogeneous surfaces, although the subsequent applications have been made in a haphazard manner to a wide class of absorbents commonly termed heterogeneous; since the details of surface heterogeneity are invariably unknown, the entire development and results derived therefrom are disputable.

In many instances the macroscopic surface topology may be examined using electron microscopy and such techniques as ultraviolet and X-ray photoelectron spectroscopy, Auger electron spectroscopy, secondary ion mass spectroscopy,[1] and infrared spectroscopy;[2] these are commonly adopted to investigate qualitatively and sometimes quantitatively, the chemical nature of the solid surface. Other methods using ultra-high-vacuum surface techniques such as RHEED and LEED[1,3] are available to probe the microscopic structure of the surface and have proved invaluable in elucidating the structure of metal surfaces with and without adlayers. Unfortunately these methods are not generally applicable to the macroscopic analysis of particulate surfaces.

It is interesting to reflect on the situation described over thirty years ago by Roginskii and Todes:[4] 'The inhomogeneity of the surfaces of most solids is borne out by a number of independent facts. The achievement of electron microscopy justifies the hope that the time is not very far off when the peculiar structure of an active surface will probably be open to direct observation and measurement. At present however, only integral and indirect methods of

[1] M. Prutton, 'Surface Physics', Clarendon Press, Oxford, 1975.
[2] L. H. Little, 'Infrared Spectra of Adsorbed Species', Academic Press, London and New York, 1966.
[3] D. G. Castner and G. A. Somorjai, *Chem. Rev.*, 1979, **79**, 233; G. A. Somorjai, 'Chemistry in Two Dimensions', Cornell Univ. Press, Ithaca, NY, 1981.
[4] S. Roginskii and O. Todes, *Acta Physicochim.*, *URSS*, 1946, **21**, 519.

studying surface inhomogeneity are at our disposal.' As yet no direct and quantitative method of assessing surface heterogeneity is available. The theoretical development and applications of physical adsorption to heterogeneous surfaces has been orientated to deriving information about surface heterogeneity of particular systems (*i.e.* gas/solid combinations) from gas adsorption isotherm data.

The Development and Early Solutions of the Integral Equation of Adsorption.—Langmuir[5] derived the classical isotherm equation which bears his name by assuming the adsorbate to be localized on the surface without lateral interactions and that adsorption was on a homogeneous crystal surface exhibiting only one kind of 'elementary space' (*i.e.* a region of uniform adsorption energy, U). The fractional monolayer coverage of the total surface was expressed as:

$$\theta = p/(K+p), \tag{1}$$

where

$$K = A^0 \exp(-U/RT). \tag{2}$$

Here A^0 is a constant related to the partition function for the adsorbate[6—8] and usually assumed to be temperature dependent and independent of the adsorption energy; this approximation will be discussed in Section 3.

Langmuir himself was the first to generalize his equation:[5] 'Let us assume that the surface contains several different kinds of elementary spaces representing the fractions β_1, β_2, β_3, *etc.*, of the surface so that $\beta_1 + \beta_2 + \beta_3 + \ldots = 1$'. This approach led to an equation describing the total adsorption, θ_T, on a heterogeneous surface:

$$\theta_T = \frac{\beta_1 p}{K_1 + p} + \frac{\beta_2 p}{K_2 + p} + \ldots, \tag{3}$$

where $K_1, K_2 \ldots$, are the constants governing adsorption on sites of types $1, 2, \ldots$ This equation was also written in an integral form to describe adsorption on a surface divided into infinitesimal fractions $d\beta$. Equation (3) found limited application owing to the number of unknowns required to describe a real surface and the uncertainty of the effects of ignoring lateral interactions. In spite of this, the model was developed to describe multilayer adsorption on a surface with one elementary space producing the well known BET equation.[9]

During the same period the analysis of experimental data, particularly at low adsorbate concentrations, led to the popular application of an empirical equation usually attributed to Freundlich:[10]

$$\theta_T = cp^{1/n}, \tag{4}$$

[5] I. Langmuir, *J. Am. Chem. Soc.*, 1918, **40**, 1361.
[6] S. Ross and J. P. Olivier, 'On Physical Adsorption', John Wiley and Sons, New York, 1964.
[7] S. E. Hoory and J. M. Prausnitz, *Surf. Sci.*, 1967, **6**, 377.
[8] W. A. House and M. J. Jaycock, *J. Colloid Interface Sci.*, 1974, **47**, 50.
[9] S. Brunauer, P. H. Emmett, and E. Teller, *J. Am. Chem. Soc.*, 1938, **60**, 309.
[10] H. Freundlich, 'Colloid and Capillary Chemistry', Methuen and Co. Ltd., London, 3rd. Edn. 1926, Ch. 3, p. 111.

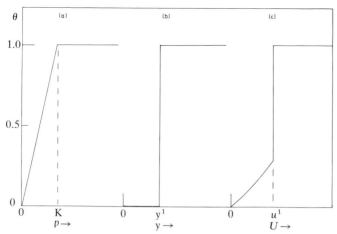

Figure 1 *Approximate function*; (a) *Zeldowich approximation to the Langmuir isotherm*; (b) *condensation approximation.* y *(or* U*) is related to the bulk gas pressure through equation* (91) *i.e.,* $U = -RT \ln(p/A^0)$; (c) *Hobson approximation.* U *is related to the bulk gas pressure through the equation*: $U = -RT \ln(p/A^0) - Q$ [*see the discussion of equation* (96)].
These functions may be regarded either as local isotherms i.e. U *constant and the abscissa as pressure, e.g. Figure* (1a), *or as 'surface filling functions' with pressure constant and the abscissa as the adsorption energy, i.e. Figures* (1b) *and* (1c)

where c is a constant and n an integer $(n > 1)$. The success of this equation for surfaces that were not expected to be homogeneous led a number of researchers to investigate what distribution of adsorption energies the overall isotherm equation (4) represented. Zeldowich[11] in 1935 was perhaps the most successful; he adopted the Langmuir integral equation for adsorption in a form equivalent to:

$$\theta_T(p) = \int_0^\infty \theta(p, K)F'(K)\, \mathrm{d}K = \int_0^\infty pF'(K)/(K + p)\, \mathrm{d}K, \qquad (5)$$

where $F'(K)$ is the distribution function for the elementary spaces on the surface. Zeldowich was not able to solve equation (5) exactly for $F'(K)$ but formulated an approximate method of solution by approximating the Langmuir equation for the individual 'elementary spaces' as shown in Figure 1(a) *i.e.*

$$\theta(p, K) = p/K \qquad 0 < p < K: \text{Henry's Law,}$$
$$\theta(p, K) = 1 \qquad K < p: \text{Surface completely} \atop \text{covered.} \qquad (6)$$

[11] J. Zeldowich, *Acta Physicochim., URSS*, 1935, **1**, 961.

This method produced the result:

$$F'(p) = -p\frac{d^2\theta_T}{dp^2},\tag{7}$$

which when applied to the Freundlich isotherm [equation (4)] gave:

$$F'(K) = A_F K^{(1/n-1)},\tag{8}$$

where

$$A_F = -c/n^2 + c/n.\tag{9}$$

Equation (8) may be rewritten by substituting equation (2);

$$F'(U) = B\exp(-\alpha U/RT),\tag{10}$$

where

$$B = A_F A^{0(1/n-1)},$$
$$\alpha = 1/n - 1\tag{11}$$

When this distribution was adopted to generate $\theta_T(p)$ using equation (5), the Freundlich isotherm at low pressures was obtained.

Although more exact solutions of equation (5) were not forthcoming until later years, much effort using approximate methods was spent, including formidable pioneering work by many Russian investigators (see Tolpin, John, and Field [12] for a review of this work prior to 1952). Roginskii's [4] 'control band method' was particularly enlightening but restricted to distribution functions that do not change rapidly between the limits of the adsorption energy. Once again the Langmuir equation was employed to describe adsorption on the 'elementary spaces' and led to the formulation:

$$F'(K) = \frac{1}{RT}\left[\frac{d\theta_T}{d\ln p}\right]_{p=K}.\tag{12}$$

Roginskii and Todes were fully aware of the problems of determining a more rigorous solution to the integral equation, particularly the demands made upon the quality of experimental adsorption data.

Sips [13] in 1948 was able to solve the integral equation for $F'(U)$ using a Stieltjes transform method (see Section 2). He assumed that the internal partition function for the adsorbed phase (incorporated in K) is independent of the adsorption energy and obtained a solution of the form shown in equation (10). However the normalization integral:

$$\int_{-\infty}^{+\infty} F'(U)\,dU\tag{13}$$

was nonconvergent. This result led Sips to suggest a generalized Freundlich

[12] J. G. Tolpin, G. S. John, and E. Field, *Adv. Catal.*, 1953, **5**, 217.
[13] R. Sips, *J. Chem. Phys.*, 1948, **16**. 490.

isotherm:

$$\theta_T = \frac{cp^{1/n}}{1 + cp^{1/n}},\tag{14}$$

and the subsequent analysis produced an exponential distribution function. At the suggestion of Honig and Hill,[14] Sips[15] later modified his theory by limiting the adsorption energies to positive values.

Extensions of the Adsorption Model to Include Adsorbate Lateral Interactions.—The assumptions implicit in the Langmuir model throw some doubt upon evaluations of the adsorption energy distributions obtained *via* equation (5). It is not surprising that later developments were primarily concerned with adopting more sophisticated local isotherms (sometimes referred to as model isotherms) to describe adsorption on homogeneous surfaces. This however complicates the analysis since once lateral adsorbate interactions are allowed, the spatial distribution of adsorption energies must be specified. Tompkins[16] in 1950 clarified this situation by recognizing two different types of surface characterized by: '(a) sites of equal energy grouped in patches, *i.e.* the adsorbed film is a polyphase system; or (b) sites of different energy distributed individually over the surface such that any small element of area contains a representative array of sites, *i.e.* the adsorbed film is a monophase system.'

The work discussed in this report is concerned with these two groups. The surface of type (a) will be referred to as 'patchwise heterogeneous' after Ross and Olivier[6] and type (b) as 'random heterogeneous' after Hill.[17] The development of the patchwise heterogeneous model followed an extension of the concepts discussed by Langmuir. It was realized that equation (5) could be written in a more general form:[6,18]

$$\theta_T(p, T) = \int_{U_1}^{U_h} \theta(p, T, U) F'(U) \, dU,\tag{15}$$

where $\theta(p, T, U)$ is the local isotherm equation, which for a patchwise heterogeneous model must account for adsorbate lateral interactions within individual patches. U_h and U_1 are the upper and lower limits respectively of the adsorption energy.

Terminology.—It is relevant here to clarify the terminology now associated with surface heterogeneity and adopted throughout this report.

Unisorptic Surface. A hypothetical surface with an adsorption potential energy $U(\tau, z)$ which is independent of τ (where τ is a surface plane vector). This surface is sometimes referred to as a structureless surface.[19,20]

[14] J. M. Honig and E. L. Hill, *J. Chem. Phys.*, 1954, **22**, 851: J. M. Honig, *ibid.*, 1955, **23**, 1557.
[15] R. Sips, *J. Chem. Phys.*, 1950, **18**, 1024.
[16] F. C. Tompkins, *Trans. Faraday Soc.*, 1950, **46**, 569.
[17] T. L. Hill, *J. Chem. Phys.*, 1949, **17**, 762.
[18] W. A. Steele, *J. Phys. Chem.*, 1963, **67**, 2016.
[19] W. A. Steele, *Surf. Sci.*, 1973, **39**, 149.
[20] W. A. House and M. J. Jaycock, *J. Colloid Interface Sci.*, 1977, **59**, 252.

Homotattic Surface.[21] A microscopically uniform and homogeneous surface. This includes ideal crystal surfaces where the adsorption potential energy, $U(\tau, x)$ is a periodic function of τ, *e.g.* the ideal (100) surface of sodium chloride or an ideal basal surface plane of graphitized carbon black.

Intrinsic Heterogeneity. That periodic surface heterogeneity possessed by a single homotattic surface.

Patch. An area of a surface that is either energetically unisorptic or homotattic.

Residual Heterogeneity. A term used when describing a patchwise heterogeneous surface; it is that heterogeneity that cannot be ascribed to homotattic surfaces and is generally caused by impurities and defects on the surface.

2 Solutions of Fredholm Equations of the First Kind

Introduction.—The generalized integral equation for a heterogeneous surface (equation 15) is a linear Fredholm equation of the first kind defined as:

$$\int_a^b K(x, y)f(y)\,\mathrm{d}y = g(x), \qquad c \leqslant x \leqslant d. \tag{16}$$

The solution, $f(y)$, given the kernel, $K(x, y)$ and function $g(x)$, poses special difficulties which have been discussed in an excellent review by Miller.[22] The most obvious difficulty concerns the character and range of the operator $K(x, y)$ as not all functions, $g(x)$, can be written in a form given by equation (16), *e.g.* if $K(x, y) = \cos(x)\cos(y)$ and $f(y)$ is any integrable function, then:

$$g(x) = \cos(x)\int_a^b \cos(y)f(y)\,\mathrm{d}y, \tag{17}$$

i.e. $g(x)$ must be a multiple of $\cos(x)$. In this example even if a solution did exist, it would not be unique as there are an infinity of functions, $\phi(y)$ such that:

$$\leftrightarrow \quad \begin{aligned} &\int_a^b K(x, y)[f(y) + \phi(y)]\,\mathrm{d}y = g(x) \\ &K(x, y)\phi(y) = 0. \end{aligned} \tag{18}$$

The greatest problem however concerns the ill-posed nature of equation (16). A small perturbation in the data function $g(x)$ leads to a large perturbation in the solution, $f(y)$. This may be demonstrated by considering $f(y)$ in equation (16) as an exact and unique solution. Applying a perturbation $\delta g(x)$ to $g(x)$ and examining the subsequent perturbation on the solution $\delta f(y)$ we

[21] C. Sanford and S. Ross, *J. Phys. Chem.*, 1954, **58**, 288.
[22] G. F. Miller, 'Numerical Solutions of Integral Equations', ed. L. M. Delves and J. Walsh, Clarendon Press, Oxford, 1974, Ch. 13, p. 175.

have:

$$\int_a^b K(x, y)[f(y) + \delta f(y)] \, dy = g(x) + \delta g(x), \tag{19}$$

$\leftrightarrow$

$$\int_a^b K(x, y) \, \delta f(y) \, dy = \delta g(x). \tag{20}$$

It is convenient to write the perturbation

$$\delta g(x) = c h_n(x), \tag{21}$$

where

$$h_n(x) = \int_a^b K(x, y) \cos(ny) \, dy, \tag{22}$$

with

$$\lim_{n \to \infty} h_n(x) = 0. \tag{23}$$

Substituting equation (21) into (20) leads to:

$$\delta f(y) = c \cos(ny), \tag{24}$$

$\leftrightarrow$

$$\left| \frac{\delta f(y)}{\delta g(x)} \right| = \left| \frac{\cos(ny)}{h_n(x)} \right|. \tag{25}$$

hence if n is large [the perturbation in $g(x)$ is small], this ratio will be large. The absolute value of the fraction expressed in equation (25) depends on n and the form of the kernel in equation (22). Generally the instability is greater for flat and smooth kernels than for sharply peaked ones. This may be demonstrated by allowing $K(x, y) = \delta(y - y_0)$, where $\delta(y - y_0)$ is a Dirac-delta function; following the argument above, $h_n(x) = \cos(ny_0)$, $\delta g(x) = c \cos(ny_0)$, and $\delta f(y) = c \cos(ny_0)$ thus yielding the fraction $\delta f(y)/\delta g(x) = 1$. Hence the degree of success achieved in solving equation (16) depends to a large extent on the accuracy of $g(x)$ and the shape of the kernel.

General Numerical Methods of Solution.—A comprehensive discussion and comparison of the numerical methods of solution of the Fredholm equation will not be presented here. (See Miller[22] and the references cited by Dormant and Adamson[23]). However, it is pertinent to examine the mathematical basis of some of the methods that have been successful and to discuss their limitation when applied to the solution of equation (15).

The Matrix Regularization Approach. For a data set $\{g(x_1), g(x_2), \ldots, g(x_i), \ldots, g(x_n)\}$ a system of linear equations may be constructed [*cf.* equation (16)].

$$\boldsymbol{Af} = \boldsymbol{g}, \tag{26}$$

[23] L. M. Dormant and A. W. Adamson, *Surf. Sci.*, 1977, **62**, 337.

where **A** is the matrix of quadrature coefficients for the tabular points $\{y_1, y_2, \ldots, y_i, \ldots, y_m\}$ and $\{x_1, x_2, \ldots, x_j, \ldots, x_n\}$ *i.e.* the elements a_{ij} of **A** are $a_{ij} = w_i\, k_{ji}$, where w_i are quadrature weights and k_{ji} elements of the kernel matrix. The ill-conditioned nature of equation (16) means that a physically realistic solution for the matrix equation (26) is impossible unless allowance is made for inherent errors in the data set $g(x)^{24}$ *i.e.*

$$\boldsymbol{Af} = \boldsymbol{g} + \boldsymbol{E}, \tag{27}$$

where the elements $\{e_1, e_2, \ldots, e_i, \ldots, e_n\}$ are the absolute errors for the data set **g**. The values of these errors are of course unknown but a solution to equation (27) is possible if some prior information concerning its general form is available. Twomey[25] has shown that if a 'smooth' solution is required, the minimization of the curvature, *i.e.* $\sum_i (f_{i-1} - 2f_i + f_{i+1})^2$, with the conditions $\boldsymbol{Af} = \boldsymbol{g} + \boldsymbol{E}$ and $\sum_j e_j = $ constant, gives a general solution:

$$\boldsymbol{f} = (\boldsymbol{A}^T\boldsymbol{A} + \gamma\boldsymbol{H})^{-1}\boldsymbol{A}^T\boldsymbol{g}, \tag{28}$$

where γ is a Lagrange multiplier and **H** the expansion coefficient matrix:

$$
\begin{bmatrix}
1 & -2 & 1 & 0 & \cdot & \cdot & \cdot & \cdot & \cdot & \cdot \\
-2 & 5 & -4 & 1 & 0 & \cdot & \cdot & \cdot & \cdot & \\
1 & -4 & 6 & -4 & 1 & 0 & \cdot & \cdot & \cdot & \\
0 & 1 & -4 & 6 & -4 & 1 & 0 & \cdot & \cdot & \\
0 & 0 & 1 & -4 & 6 & -4 & 1 & 0 & \cdot & \\
\cdot & \cdot & \cdot & \cdot & \cdot & \cdot & \cdot & \cdot & \cdot & \\
 & & 0 & 1 & -4 & 6 & -4 & 1 & 0 & 0 \\
 & \cdot & 0 & 1 & -4 & 6 & -4 & 1 & 0 \\
 & & \cdot & \cdot & 0 & 1 & -4 & 6 & -4 & 1 \\
 & & & \cdot & \cdot & & 0 & 1 & -4 & 5 & -2 \\
 & & & \cdot & \cdot & \cdot & \cdot & 0 & 1 & -2 & 1
\end{bmatrix}
$$

Figure 2(a, b) show the results obtained by Phillips[24] and House (unpublished) using this matrix method. The decreasing oscillations in the solution with increasing damping constant, γ, is evident but when the solution itself has a number of peaks, increasing γ tends to reduce their definition and at the same time smooth the shorter-range oscillations. The choice of γ can be guided by a comparison between the anticipated error in the g_i values and the deviations between the values in the dataset **g** and the generated values obtained using the calculated f_i values for a particular γ. Unfortunately when the technique is applied to the solution of equation (15), physically unrealistic negative frequencies may arise as permitted solutions. This problem has been discussed in detail by Merz[26] who postulates a 'concavity criterion', which he claims is a

[24] D. L. Phillips, *J. Assoc. Comp. Mach.*, 1962, **9**, 84.
[25] S. Twomey, *J. Assoc. Comp. Mach.*, 1963, **10**, 97.
[26] P. H. Merz, *J. Comp. Phys.*, 1980, **38**, 64.

necessary condition, but not a sufficient condition, for the use of a particular kernel *i.e.* local isotherm. In other words some adsorption data and local isotherms are incompatible if negative frequencies are to be avoided. This aspect of the problem needs further study. The regularization method is particularly promising but needs further improvement to incorporate a non-negativity constraint and a mathematical criterion for the choice of γ. Merz[26] has already suggested methods for the choice of γ. It is also desirable to limit the analysis to some finite range of adsorption energies, *e.g.* the condensation approximation (see Section below) may be used to define the adsorption energy limits if the adsorption is in the region of the two-dimensional critical temperature.

Approximation of the Kernel. This has been one of the most popular methods of solution of equation (15). In general a schematic of the more exact kernel is

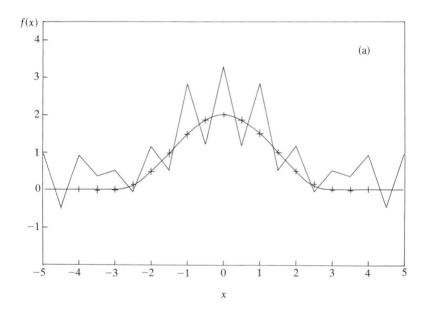

Figure 2 *Results of the matrix regularization method of solution.*
(a) *Solutions of the integral equation* $\int_{-6}^{+6} K(x-\lambda)f(x)\,dx = g(\lambda)$ *with* $K(z) = 1 + \cos(\pi\lambda/3)$ *for* $|z| \leq 3$ *and* $K(z) = 0$, $|z| > 3$ *and* $g(z) = (6+\lambda)[1 - 1/2\cos(\pi\lambda/3) - 9/2\pi\sin(\pi\lambda/3)]$ *for* $|z| \leq 6$ *and* $g(z) = 0$ *for* $|z| > 6$. *A truncation error introduced in g of average absolute value 0.0014 and maximum 0.004.* (i) *Smooth solid line, true solution,* (ii) ——, $\gamma = 0$ *solution,* (iii) + +, $\gamma = 0.01$ *solution*
(Reproduced by permission from *J. Assoc. Comp. Mach.*, 1962, **9**, 84. Copyright 1962, Association of Computing Machinery, Inc.)

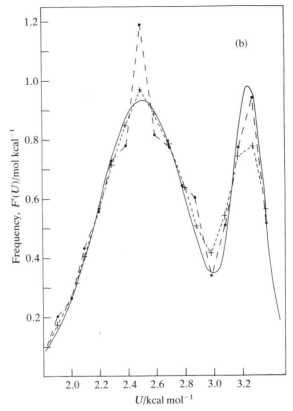

Figure 2 (*cont.*)
(b) *Solutions of the integral equation for* $f(x)$ *a double-peaked Gaussian function and functions g and K determined for the Hill–de Boer equation, see p. 16.* ——, *exact solution*; ● - - - - ●, $\gamma = 0.001$; + - - - - +, $\gamma = 0.01$

devised *e.g.* equation 6. Perhaps the simplest approximation is the step-function suggested by Berenyi[27] and illustrated in Figure 1(b):

$$K(x, y) = 0, \qquad y < y^1;$$
$$K(x, y) = 1 \qquad y > y^1. \qquad (29)$$

Substituting into equation (16):

$$g(x) = \int_{y^1}^{b} f(y) \, dy, \qquad (30)$$

and by Leibniz's rule:

$$\frac{dg(x)}{dy^1} = f(y^1). \qquad (31)$$

[27] L. Berenyi, *Z. Phys. Chem.*, 1920, **94**, 628.

Hence if some relationship between y^1 and x can be formulated $f(y^1)$ is easily computed either by numerical differentiation or by expressing $g(x)$ as a series expansion and determining $f(y^1)$ by analytical methods. Equation (31) forms the basis of the 'condensation' approximation that has been applied to solving equation (15) by many investigators including Cerofolini,[28] Adamson and Ling,[29] and Jaronieć *et al.*[30] Application of equation (31) is very convenient but is only exact in the limit of 0 K when the step-approximation for the local isotherm [equation (29)] becomes exact and the patches are filled in the sequence of their adsorption energies *i.e.* patches are either empty or full. Another kernel approximation investigated by Zeldowich[11] and outlined in Section 1 was extended and applied by Hobson[31] and will be discussed in more detail in Section 4.

Optimization Methods Involving a Generalized Form of $f(y)$. If some generalized two or three parameter function can be postulated for $f_i(y)$ then general optimization methods can be applied to determine the values of the parameters which produce the best agreement, when incorporated in equation (26), with the dataset, g, *e.g.* a criterion for convergence of the 'best-fit' could be the root-mean-square (r.m.s.) relative deviation:

$$\text{r.m.s. deviation} = \left[n^{-1} \sum_{i=1}^{n} \left(\frac{g_i - \tilde{g}_i}{g_i} \right)^2 \right]^{1/2}, \tag{32}$$

where $\tilde{g}_i$ are the values generated using some generalized function.

This method was successful in the first attempts to solve equation (15) using a local isotherm equation that allowed adsorbate lateral interactions within patches. The ingenious graphical method, now referred to as the Ross and Olivier method of analysis, has been described in their monograph[6] and in a condensed account by Ross.[32] The technique used by Ross and Olivier employs a Gaussian probability function for the distribution of adsorption energies, *i.e.* a two-parameter generalized distribution of the form:

$$f_i = F'(U) = \frac{1}{N} \exp[-\gamma(U - \bar{U})^2], \tag{33}$$

where γ is a 'heterogeneity' parameter determining the 'sharpness' of the distribution, $\bar{U}$ the median of the distribution, and N the normalization constant:

$$N = \int_0^5 \exp[-\gamma(U - 2.5)^2] \, dU, \tag{34}$$

with U measured in kcal mol^{-1}. The distribution function was confined to a

[28] G. F. Cerofolini, *Surf. Sci.*, 1971, **24**, 391; *Chem. Phys.*, 1978, **33**, 423.
[29] A. W. Adamson and I. Ling, *Adv. Chem. Ser.*, 1961, **33**, 51.
[30] M. Jaronieć, W. Rudzinski, S. Sokolowski, and R. Smarzewski, *Colloid Polym. Sci.*, 1975, **253**, 164.
[31] J. P. Hobson, *Can. J. Phys.*, 1965, **43**, 1934.
[32] S. Ross, 'The Solid-Gas Interface', ed. E. A. Flood, Marcel Dekker Inc., New York, 1967, Ch. 15.

range of $5 \, kcal \, mol^{-1}$ to avoid negative adsorption energies. Hoory and Prausnitz[33] later suggested a log-normal distribution function of the form:

$$^m f_i = F'(U) = \frac{1}{\sqrt{2\pi}\gamma U} \exp\left[-\frac{(\ln U - \ln \bar{U})^2}{2\gamma^2} \right], \tag{35}$$

which is skewed but only defined for positive values of U.

van Dongen[34] adopted a more flexible generalized distribution function:

$$f_i = F'(U) = \exp(c_0 + c_1 U + c_2 U^2 + \ldots + c_n U^n), \tag{36}$$

where the coefficient c_0 is determined by the usual normalization. The Hill–de Boer multilayer local isotherm equation was adopted since it allowed for both lateral adsorbate interactions and multilayer interactions; this approach will be discussed in Section 6. The parameters in equation (36) were determined by minimizing the sum of the absolute squared deviations between the experimental and calculated adsorption amounts.

Kindle et al.[35] also adopted this method but used another generalized distribution function, a Maxwell–Boltzmann (MB) distribution:

$$f_i = F'(U) = [2\pi/(\pi R T_f)^{3/2}]U^{1/2} \exp[-(U/R T_f)], \tag{37}$$

where T_f is the 'heterogeneity' parameter called the frozen temperature by Kindle et al.[35] They observed that combining the MB distribution and appropriate T_f values, with the Langmuir local isotherm function (the MBL equation), the Langmuir, Temkin,[36] and Freundlich[10] total isotherms could be generated. In the analysis of experimental data they used the Powell optimization algorithm[37] to determine the three parameters corresponding to a minimum in the r.m.s. relative deviation between computed and experimental adsorption data.

Although it is in principle feasible to fit a variety of generalized distribution functions to adsorption data using least-squares methods *e.g.* Powell algorithm[37] or Simplex method,[38] in practice multiple minima may arise; hence considerable care is required in checking the domain of the parameters to obtain initial estimates and the uniqueness of the final solution.

Numerical Iterative Methods of Solution. There are a number of numerical methods that fall into this category but few have been applied to the solution of equation (16). The fact that no prior form of $f(y)$ need be chosen is a distinct advantage.

The regularization method described above may be applied in an iterative manner[39] using the trial function approach described by Twomey.[25] Instead of minimizing the curvature, the solution may be found by minimizing the sum of

[33] S. E. Hoory and J. M. Prausnitz, *Surf. Sci.*, 1967, **6**, 377.
[34] R. H. van Dongen, *Surf. Sci.*, 1973, **39**, 341.
[35] B. Kindl, R. A. Pachovsky, B. A. Spencer, and B. W. Wojciechowski, *J. Chem. Soc., Faraday Trans. 1*, 1973, **69**, 1162.
[36] M. Temkin and V. Levich, *Zh. Fiz. Khim.*, 1946, **20**, 1441.
[37] M. J. Powell, *Comput. J.*, 1965, **7**, 303.
[38] J. A. Nelder and R. Mead, *Comput. J.*, 1965, **22**, 847.
[39] C. B. Shaw, jun., *J. Math. Anal. Appl.*, 1972, **37**, 83.

the squares of the departures from a trial function p, *i.e.*

$$\sum_i (f_i - p_i)^2 \tag{38}$$

is minimized. This method leads to the equation:

$$(\gamma \boldsymbol{I} + \boldsymbol{A}^{\mathrm{T}} \boldsymbol{A})f = \gamma \boldsymbol{p} + \boldsymbol{A}^{\mathrm{T}} \boldsymbol{g}, \tag{39}$$

where $\boldsymbol{I}$ is the identity matrix. If an initial estimate of $\boldsymbol{p}$ is known (otherwise we may let $p_i = $ constant) then equation (39) may be solved in an iterative manner

$$(\gamma \boldsymbol{I} + \boldsymbol{A}^{\mathrm{T}} \boldsymbol{A})f^c = \gamma \boldsymbol{f}^{c-1} + \boldsymbol{A}^{\mathrm{T}} \boldsymbol{g}, \tag{40}$$

where c labels the cycle number. The rate of convergence to the final solution depends on the magnitude of the Lagrange multiplier, γ. This method has been applied to the solution of the random heterogeneous model in Section 5.

An approach that has met with success is the Adamson and Ling iterative method.[40] This method converges to an acceptable solution much faster than any other iterative method although it is not a general iterative solution of the Fredholm equation since the initial approximation to $f(y)$ is obtained by using the condensation approximation. Thus the method utilizes information concerning the physical form of the local isotherm functions (see Section 4 for details and applications).

Another procedure that has been successful is that described by Ross and Morrison[41] in which the adsorption isotherm on a patchwise heterogeneous surface is written as:

$$n_a(p_j) = \sum_{i=1}^{20} s_i f_i(p_j, U_i), \tag{41}$$

where n_a is the number of moles adsorbed at a pressure p_j, and s_i is the specific area of the ith patch. For a set of data points, $j = 1$ to n, a series of equations like (41) may be written in matrix notation as:

$$\boldsymbol{n}_a = \boldsymbol{s} \boldsymbol{f}. \tag{42}$$

In this application equation (42) is ill-conditioned and cannot be solved directly. However one can strive for as close agreement as possible between both sides of the equality. One way to do this is to minimize the sum of the squared relative deviation, D_0 (or r.m.s. deviation):

$$D_0 = \sum_{i=1}^{n} \left(\frac{n_a(p_j) - \sum_{i=1}^{20} s_i f_i(p_j, U_i)}{n_a(p_j)} \right)^2 . \tag{43}$$

This minimum may be determined by any of the methods generally applicable to the solution of several equations with several variables. Ross and

[40] A. W. Adamson and I. Ling, *Adv. Chem. Ser.*, 1961, No. 33, 51: see also A. W. Adamson, 'Physical Chemistry of Surfaces', Interscience Publishers, 2nd Edn., 1967, Ch. XIII, p. 628.
[41] S. Ross and I. D. Morrison, *Surf. Sci.*, 1975, **52**, 103.

Morrison[41] chose the multiparameter Newton–Raphson procedure and Sacher and Morrison[42] compared a variety of optimization techniques (see Section 4) to obtain a solution for **s**.

A completely different approach to the problem has been proposed by Jaroniec and Rudzinski,[43] Sokolowski *et al.*,[44] and Rudzinski *et al.*[45] They essentially transform equation (16) using $U = f(x)$ where $f(x)$ is defined in the interval (a, b) such that $f(a) = U_l$ and $f(b) = U_h$:

$$n_a(p_j) = \int_a^b \theta[p_j, f(x)] \cdot F(x) \, dx, \tag{44}$$

where

$$F(x) = n_m F'[f(x)]f'(x) \tag{45}$$

and n_m is the monolayer capacity. θ and $F(x)$ are then expanded into a series with regard to a complete system of orthonormal functions $\phi_i(x)$ defined in the interval (a, b) *i.e.*

$$\theta[p_j, f(x)] = \sum_{i=0}^{n-1} A_i(p_j)\phi_i(x). \tag{46}$$

$$F(x) = \sum_{i=0}^{n-1} B_i\phi_i(x); \tag{47}$$

e.g. Jacobi polynomials are complete and orthonormal in the interval $(0, 1)$ and Legendre polynomials are complete and orthonormal in the interval $(-1, 1)$.

The coefficients A_i may be obtained from the local isotherm equation and then by substituting equations (46) and (47) into (44) and using the orthonormality condition:

$$n_a(p_j) = \int_a^b \left[\sum_{i=0}^{n-1} A_i(p_j)\phi_i(x)\right]\left[\sum_{i=0}^{n-1} B_i\phi_i(x)\right] dx \tag{48}$$

$\leftrightarrow$

$$n_a(p_j) = \sum_{i=0}^{n-1} A_i(p_j)B_i \quad \text{with} \quad j = 1, 2, \ldots, n. \tag{49}$$

If the linear equation (49) can be solved for the coefficients B_i, then

$$F'(U) = \{n_m f'[\bar{f}(U)]\}^{-1} \sum_{i=0}^{n-1} B_i\phi_i[\bar{f}(U)], \tag{50}$$

where $x = \bar{f}(U)$.

Although this method has only been applied using the Langmuir local isotherm equation, there is no reason why the coefficients A_i could not be obtained for a more sophisticated local isotherm, particularly one that accounts for adsorbate lateral interactions.

[42] R. S. Sacher and I. D. Morrison, *Surf. Sci.*, 1979, **70**, 153.
[43] M. Jaroniec and W. Rudzinski, *Colloid Polym. Sci.*, 1975, **253**, 683.
[44] S. Sokolowski, M. Jaroniec, and G. F. Cerofolini, *Surf. Sci.*, 1975, **47**, 429.
[45] W. Rudzinski, M. Jaroniec, S. Sokolowski, and G. F. Cerofolini, *Czech. J. Phys.*, 1975, **B25**, 891.

General Analytical Methods of Solution.—The solutions obtained using these methods generally depend on finding an analytical form for the dataset $g(x)$ and have been successful for local isotherms that ignore the effects of adsorbate lateral interactions, *i.e.* it must be possible to express the adsorbate coverage per patch in terms of the bulk gas pressure and the adsorption energy of the patch.

Method of the Stieltjes Transform. This is the method originally used by Sips[15] of solving for $F'(U)$ in equation (15) using the local Langmuir equation and assuming the total isotherm equation to be the generalized Freundlich equation. The solution primarily depends on transposing the integral equation to the form:

$$G(s) = \int_0^\infty \frac{F(t)}{s+t} \, dt \tag{51}$$

and then by the theory of Stieltjes transforms,[46]

$$F(t) = \frac{G[t \exp(-\pi i)] - G[t \exp(\pi i)]}{2\pi i} \tag{52}$$

where $F(t)$ is related to $F'(U)$.

Method of the Laplace Transform. In this case the integral equation must be transformed to:

$$G(s) = \int_0^\infty \exp(-ts)F(t) \, dt, \tag{53}$$

and hence the determination of $F(t)$ is reduced to the solution of a Laplace transform:

$$F(t) = \frac{1}{2\pi i} \int_{\beta-i\infty}^{\beta+i\infty} \exp(st)G(s) \, ds, \tag{54}$$

where β is any point of the real positive semi-axis. For many functions $G(s)$, the solution expressed in equation (54) is not simple[47] but in general may be obtained by a numerical method of expanding $F(t)$ as a series of Laguerre functions[48—50] and the data, $G(s)$, by a polynominal expansion.

3 Adsorption on Unisorptic and Homotattic Surfaces

Adsorption Isotherms on Homogeneous Surfaces.—The integral equation describing adsorption on a patchwise heterogeneous surface requires the choice of a local or model isotherm to describe the adsorption on individual patches

[46] D. V. Widdler, 'An Introduction to Transform Theory', Academic Press, New York and London, 1971, p. 128.
[47] M. Jaronieć and S. Sokolowski, *Colloid Polym. Sci.*, 1977, **255**, 374.
[48] A. Papoulis, *Q. Appl. Math.*, 1957, **14**, 405.
[49] W. Rudzinski and M. Jaronieć, *Surf. Sci.*, 1974, **42**, 552.
[50] A. Waksmundzki, S. Sokolowski, J. Rayss, Z. Suprynowicz, and M. Jaronieć, *Sep. Sci.*, 1976, **11**, 29.

of the surface. This choice may be guided by considering submonolayer adsorption on homogeneous surfaces where the degree of residual heterogeneity is very small (typically less than 5% of the total surface area). However assessing the residual heterogeneity is not an easy matter since the interpretation of adsorption isotherms requires prior knowledge concerning the state of the adsorbate. This dilemma has been partially resolved by experimental investigations at adsorption temperatures such that the adsorbate condenses on the surface at a pressure below its saturation vapour pressure. One would expect that if the surface was homogeneous with an appropriately large surface area, then by measuring the isotherm at a temperature below some critical value, a phase transition on the surface would be observed, *i.e.* a 'step' in the isotherm would be evident. This phase transition would occur when the chemical potential of the adsorbate on the homogeneous surface reached a critical value, μ_t; this value depends upon the state of the adsorbate on the surface and in particular the adsorption energy associated with the system. Since the chemical potential of an ideal bulk gas may be written:

$$\mu^g = \mu^{\ominus} + RT \ln p, \tag{55}$$

where $\mu^{\ominus}$ is the standard chemical potential of the gas and p the bulk gas pressure, the phase transition is expected to occur at a critical pressure (when $\mu_t = \mu^g$) depending on the nature of the adsorbent, the state of motion of the adsorbate on the surface, the adsorbate–adsorbate interaction, and the adsorption potential energy field.

de Boer predicted[51] a phase transition by considering the adsorbate to behave as a two-dimensional gas held in dynamic equilibrium with the surface by the adsorption potential energy field. de Boer[51] and Hill[52] developed this approach by deriving an equation analogous to the van der Waals equation for an imperfect gas. This equation, commonly referred to as the Hill–de Boer (HdB) equation, may be written:

$$p = K \frac{\theta}{1-\theta} \exp\left(\frac{\theta}{1-\theta} - \frac{2\alpha\theta}{kT\beta}\right), \tag{56}$$

where K is given in equation (2) and α and β are the two-dimensional analogues of the van der Waals constants a and b, *i.e.* $\alpha/\beta = a/2b$. The HdB equation leads to an ideal two-dimensional critical temperature,[51] $_aT_c = 8\alpha/27R\beta$. Ross and Olivier[53] have shown that a plot of W versus θ [where $W = \ln(\theta/1-\theta) + \theta/(1-\theta) - \ln p = 2\alpha\theta/RT\beta - \ln K$] yields a slope $2\alpha/RT\beta$ from which $_aT_c$ may be calculated.

In spite of the earlier work of Adam and Jessop[54] showing clear and reproducible evidence of two-dimensional condensation of fatty acids on the water surface, reliable data demonstrating two-dimensional condensation on solid surfaces was not forthcoming until later. This was partially due to the

[51] J. H. de Boer, 'The Dynamical Character of Adsorption', The Clarendon Press, Oxford, 1953.
[52] T. L. Hill, *J. Chem. Phys.*, 1946, **14**, 441.
[53] Ref. 6, p. 140.
[54] N. K. Adam and G. Jessop, *Proc. R. Soc. London, Ser. A*, 1926, **110**, 423.

greater technical problems associated with high-vacuum work, particularly the accurate measurement of low gas pressures. Several investigators as early as 1931 reported discontinuities in adsorption isotherms, indeed Allmand and Burrage[55] studied various adsorbates on charcoal and silica showing 'stepped' isotherms. Benton and White[56] obtained 'stepped' isotherms for hydrogen on nickel and copper; Jura *et al.*[57] reported 'steps' on their isotherms of n-heptane on ferric oxide, graphite, and silver, and subsequently Jura and Criddle[58] reported up to four submonolayer first-order transistions for the argon–graphite system at adsorption temperatures below the ideal two-dimensional critical temperature of argon (75.5 K). The validity of this work has been questioned by several researchers including Smith[59] who attempted without success to reproduce the phase transitions for n-heptane on ferric oxide and graphite. However the more recent extensive work using inorganic halides and graphitized carbon blacks has not only demonstrated two-dimensional condensation on homogeneous surfaces, but confirmed the existence of a two-dimensional critical temperature. Ross and Boyd[60] produced data for ethane adsorbed on the (100) face of sodium chloride (at 90 K). Ross and Clark[61] later substantiated their work using methane, ethane, and xenon as adsorbates on sodium chloride and presented data illustrating the existence of a two-dimensional critical temperature. Fisher and McMillan[62] also made extensive measurements for methane and krypton adsorbed on (100) sodium bromide. Other ionic crystals that have been studied include potassium chloride,[60] silver iodide,[63] calcium fluoride,[64,65] and all show 'stepped' isotherms at adsorption temperatures below their ideal two-dimensional critical values. In some instances multiple 'steps' in the submonolayer region arise owing to condensation on different crystal planes, each with its own characteristic adsorption energy. More recently lamellar halides, such as $CdBr_2$, $CdCl_2$, $NiCl$, $MnBr_2$, FeI_2, and PbI_2, have been studied[66,67] and phase transitions observed on the cleavage planes of the crystals. It is probable that future studies on the lamellar halides by conventional techniques and AUGER or LEED will lead to a better understanding of adsorption on homogeneous surfaces.

[55] A. J. Allmand and L. J. Burrage, *Proc. R. Soc. London, Ser. A*, 1931, **130**, 610; *J. Phys. Chem.*, 1931, **35**, 1692.
[56] A. F. Benton and T. A. White, *J. Am. Chem. Soc.*, 1931, **53**, 3301.
[57] G. Jura, E. H. Loeser, P. R. Basford, and W. D. Harkins, *J. Chem. Phys.*, 1945, **13**, 535; 1946, **14**, 117: G. Jura and W. D. Harkins, *J. Am. Chem. Soc.*, 1946, **68**, 1941: G. Jura, W. D. Harkins, and E. H. Loeser, *J. Chem. Phys.*, 1946, **14**, 344.
[58] G. Jura and D. Criddle, *J. Phys. Colloid Chem.*, 1951, **55**, 163.
[59] R. N. Smith, *J. Am. Chem. Soc.*, 1952, **74**, 3477.
[60] S. Ross and G. E. Boyd, 'New Observations on Two-Dimensional Condensation Phenomena', MDDC Rep. 864, USAEC, 1947.
[61] S. Ross and H. Clark, *J. Am. Chem. Soc.*, 1954, **76**, 4291, 4297.
[62] B. B. Fisher and W. G. McMillan, *J. Am. Chem. Soc.*, 1957, **79**, 2969: *J. Chem. Phys.*, 1958, **28**, 549, 555, and 563.
[63] E.W. Sidebottom, W. A. House, and M. J. Jaycock, *J. Chem. Soc. Faraday Trans. 1*, 1976, **72**, 2709.
[64] H. Edelhock and H. S. Taylor, *J. Phys. Chem.*, 1954, **58**, 344.
[65] S. Ross and W. Winkler, *J. Am. Chem. Soc.*, 1954, **76**, 2737.
[66] Y. Larher, *J. Colloid Interface Sci.*, 1971, **37**, 836.
[67] F. Millot, *Cen. Saclay, CEA*, Rep. CEA-N-1865, 1976.

Presently one of the most widely studied and perhaps most homogeneous adsorbents are the graphitized carbon blacks. Polley and Shaeffer[68] showed the surface to be homogeneous and subsequently Singleton and Halsey[69] obtained 'stepped' isotherms for argon and krypton adsorbed on the graphitized carbon black P-33 (2700 K). Many other investigations have been completed using various adsorbates,[70-76] indeed the extensive work of Duval *et al.*[75] and of Larher[76] proposing a 'two-dimensional triple point' temperature, has revived interest in graphitized carbon black systems. Above a 'triple point temperature' the adsorbate condenses to what these authors term a 'liquid structure' and at higher pressures (at constant temperature) this structure transforms to produce a 'solid structure' with a geometry determined by that of the substrate. Recent investigations[77-80] employing AUGER and LEED have confirmed that the first layer of krypton and xenon is in epitaxy with the graphite basal plane.

A characteristic feature of isotherms determined for homogeneous adsorbents should be a linear region at low adsorbate surface coverages. In this region Henry's law is applicable:

$$n_a = A_s K_H p \tag{57}$$

where A_s is the surface area of the adsorbent, n_a the number of moles of adsorbate, and K_H the Henry's constant. This may be expressed as:[81,82]

$$K_H = (1/A_s kT) \int_V \{\exp[-U_s(\mathbf{r})/kT] - 1\} \, d\mathbf{r}, \tag{58}$$

where V is the entire gas volume; hence K_H is dependent on the gas–solid interaction energy, $U_s(\mathbf{r})$. In practice however, an isotherm with a low coverage non-linear 'knee' followed by a linear region is often observed, the 'knee' being attributed to adsorption on high-energy sites.[20,83] The data[84,85] for argon adsorbed on P-33(2700 K) is a good example illustrating the linear Henry's law region without residual heterogeneity effects. Another indication of homogeneity is the variation of the isosteric enthalpy of adsorption with n_a measured at low surface coverages, *i.e.* $\theta < 0.5$. The isosteric enthalpy may be

[68] M. H. Polley, W. D. Schaeffer, and W. R. Smith, *J. Phys. Chem.*, 1953, **57**, 469.
[69] J. H. Singleton and G. D. Halsey, jun., *J. Phys. Chem.*, 1954, **58**, 1011.
[70] C. F. Prenzlow and G. D. Halsey jun., *J. Phys. Chem.*, 1957, **61**, 1158.
[71] W. D. Machin and S. Ross, *Proc. R. Soc. London, Ser. A*, 1962, **265**, 455.
[72] B. W. Davis and C. Pierce, *J. Phys. Chem.*, 1966, **70**, 1051.
[73] H. Cochrane and P. L. Walker, jun., *J. Colloid Interface Sci.*, 1967, **24**, 405; R. A. Pierotti and R. E. Smallwood, *J. Colloid Interface Sci.*, 1966, **22**, 469.
[74] F. Dondi, M. Gonnord, and G. Guiochon, *J. Colloid Interface Sci.*, 1977, **62**, 316.
[75] X. Duval and M. letort, *C. R. Hebd. Seances Acad. Sci.* 1952, **234**, 1363; A. Thomy and X. Duval, *J. Chim. Phys., Phys. Biol.*, 1969, **66**, 1966; *ibid.*, 1970, **67**, 286, 1101.
[76] Y. Larher, *J. Chem. Soc., Faraday Trans. 1*, 1974, **70**, 320.
[77] J. Suzanne, J. P. Coulomb, and M. Bienfait, *Surf. Sci.*, 1973, **40**, 414: *Surf. Sci.*, 1974, **44**, 141.
[78] A. Thomy, J. Regnier, J. Menaucount, and X. Duval, *J. Cryst. Growth*, 1972, **13/14**, 159.
[79] H. M. Kramer and J. Suzanne, *Surf. Sci.*, 1976, **54**, 659.
[80] C. Marti, B. Croset, P. Thonel, and J. P. Coulomb, *Surf. Sci.*, 1977, **65**, 532.
[81] W. A. Steele, *Surf. Sci.*, 1973, **39**, 149.
[82] W. A. House and M. J. Jaycock, *Proc. R. Soc. London, Ser. A*, 1976, **348**, 317.
[83] S. Ross and J. J. Hinchen, 'Clean Surfaces', ed. G. Goldfinger, Marcel Dekker, New York, 1970, p. 115.
[84] J. R. Sams, jun., G. Constabaris, and G. D. Halsey, jun., *J. Chem. Phys.*, 1962, **36**, 1334.
[85] I. D. Morrison and S. Ross, *Surf. Sci.*, 1973, **39**, 21.

expressed as:[81,82]

$$q^{st} = RT\left[\frac{\partial \ln p}{\partial T}\right]_{n_a, A_s}, \tag{59}$$

and generally is not a simple function of the adsorbate concentration. As the adsorbate concentration increases from zero the isosteric enthalpy is expected to increase from its limiting value, q_0^{st}, owing to adsorbate self-attractive interactions. Generally residual heterogeneity manifests itself by a decrease in q^{st} from q_0^{st} at low coverages followed by an increase as the high-energy sites become saturated and adsorbate self-interactions begin to dominate. However, Anderson[86] demonstrated this is not a unique criterion on which to judge the inhomogeneity of an adsorbent. The periodicity and magnitude of $U_s(r)$ may weight a configuration of the adsorbate, which causes net repulsive adsorbate self-interactions; or the adsorbate, because of long-range electrostatic interactions with the surface, may perturb the adsorbent so that the adsorption energy becomes a function of the surface coverage. To the author's knowledge this latter point has not been further investigated, even though for systems such as the lamellar halides and adsorbates with dipole or quadrupole moments, the adsorbent perturbation may be significant.

At adsorption temperatures below the two-dimensional critical temperature, the slope of the 'steps' is an indication of the heterogeneity associated with major exposed crystal planes. The effects of chemical or physical alterations to the surface may be estimated from the changes in the 'step' position and 'step' lengths associated with particular crystal planes.

Intrinsic Heterogeneity.—Ideally the local isotherm should take full account of the intrinsic heterogeneity associated with any real homotattic surface. This is only possible when explicit information concerning $U_s(r)$ for the system under consideration is available. Since $U_s(r)$ is periodic in τ, where τ is the 'in surface plane' vector, there is a barrier to free surface migration of the adsorbate. The magnitude of this translational barrier in comparison to the maximum adsorption energy (*i.e.* $U + {}_aE_0^{vib}$ where ${}_aE_0^{vib}$ is the ground-state vibrational energy of the adsorbate), determines the degree of mobility of the adsorbate (see ref. 6, p. 14). The hindered translation model of Hill[52] and adapted by House and Jaycock[87] has allowed considerable insight into the temperature transition from localized to mobile adsorption. This model fails at moderate adsorbate concentrations because it does not admit adsorbate lateral interactions.

A complete description is possible using the gas–solid virial theorem expounded by Hill and Greenschlag,[88] Steele,[89,90] and House and Jaycock:[82]

$$\frac{n_a}{p} = \frac{B_{AS}}{kT} + \frac{C_{AAS}}{kT}\frac{n_a}{B_{AS}} + \frac{n_a^2}{B_{AS}^2}\left[\frac{D_{AAAS}}{kT} - \frac{C_{AAS}^2}{kTB_{AS}}\right] + \dots, \tag{60}$$

[86] P. R. Anderson, *Surf. Sci.*, 1971, **27**, 60.
[87] W. A. House and M. J. Jaycock, *J. Chem. Soc., Faraday Trans.* 1, 1974, **70**, 1348; *ibid.*, 1975, **71**, 1597.
[88] T. L. Hill and S. Greenschlag, *J. Chem. Phys.*, 1961, **34**, 1538.
[89] W. A. Steele, 'The Solid-Gas Interface', ed. E. A. Flood, Marcel Dekker Inc., NY, 1967, Vol. 1, Ch. 10.
[90] W. A. Steele, 'The Interaction of Gases with Solid Surfaces', Pergamon Press, Oxford, 1975, p. 104.

where B_{AS}, C_{AAS}, and D_{AAAS} are the gas–solid virial coefficients.[90] The evaluation of these coefficients is possible through a series expansion of the adsorption potential energy field. Steele[91] found that the τ dependence of $U_s(\tau)$ for the basal graphitized carbon black/inert gas systems was very small and subsequent investigators[92,93] have taken $U_s(\mathbf{r})$ to be independent of τ. In contrast the results for argon adsorbed on the relaxed face of (100) sodium chloride indicate substantial perturbation of the two-dimensional second virial coefficient,[82] *i.e.* between a 20–47% decrease in value compared to the result for a structureless surface.

The complexity of equation (60) forbids its incorporation as a rigorous and general local isotherm in the integral equation for a patchwise heterogeneous surface. The majority of work published has chosen between equations that allow the adsorbate complete free translational motion over the surface or localized vibrational motion. For some systems, numerical estimates of the ratio of the translational barrier to adsorption energy, can lead to the choice of a realistic local isotherm equation. Alternatively isosteric enthalpy data may offer some guidance. In the absence of this information, there is no justification for the choice of either model and the only course is to employ both extremes in turn and assess the effects on the calculated adsorption energy distribution.

Interrelation Between the Adsorbate Partition Function and the Adsorption Energy.—The majority of local isotherm equations may be written in the general form:

$$p = Kg(\theta, T), \tag{61}$$

where $g(\theta, T)$ is some function which may incorporate terms to account for increasing adsorbate self-interactions with surface coverage. The constant K is given by equation (2) and the value of A^0 is determined from the equation:

$$A^0 = \exp\left\{ -\left[\frac{\partial \ln Q}{\partial n_a}\right]_{A_s, T} - \frac{\mu^\ominus(T)}{kT} \right\}, \tag{62}$$

where $\mu^\ominus(T)$ is the standard chemical potential of the bulk gas [equation (55)] and the total canonical partition function for the adsorbate, Q_T, is:

$$Q_T = Q \exp(U/kT). \tag{63}$$

$\mu^\ominus(T)$ is clearly independent of the adsorption energy, U, although Q is not. Honig and Hill[14] criticized the earlier work of Sips[13,15] for assuming Q to be independent of U, although more recently Cerofolini[94] has argued that for localized adsorption at the 'usual temperatures', the more rigorous integral equation:

$$\theta_T(p, T) = \int_0^\infty \theta \, dU \int_0^\infty \theta(p, T, U, Q) F'(U, Q) \, dQ \tag{64}$$

[91] W. A. Steele, *Surf. Sci.*, 1973, **36**, 317.
[92] F. A. Putman and T. Fort, jun., *J. Phys. Chem.*, 1977, **81**, 2164.
[93] W. A. Steele, *J. Phys. Chem.*, 1978, **82**, 817.
[94] G. F. Cerofolini, *Thin Solid Films*, 1974, **23**, 129.

may be replaced by equation (15). Cerofolini effectively separates from A^0 the partition function for a one-dimensional harmonic oscillator, f_z, for molecular vibration motion perpendicular to the surface plane *i.e.*:

$$A^0 = \left(\frac{2\pi mkT}{h^2}\right)^{3/2} \frac{kT}{f_x f_y} f_z^{-1} = B^0 f_z^{-1}, \qquad (65)$$

where m is the molecular mass and f_x and f_y are the partition functions for vibrational motion parallel to the surface. By assuming B^0 to be independent of U and the perpendicular vibration frequency, $\nu^\perp$, to be proportional to the square-root of the adsorption energy,[95] Cerofolini[94] evaluates the fraction:

$$r_z = \frac{\left|\dfrac{df_z^{-1}}{dU} \bigg/ f_z^{-1}\right|}{\left|\dfrac{d \exp(-U/kT)}{dU} \bigg/ \exp(-U/kT)\right|} \quad \leqslant \frac{h\nu^\perp}{2U} \quad \text{for} \quad \nu^\perp \geqslant \bar{\nu}; \qquad (66)$$

$$\sim \frac{kT}{2U} \quad \text{for} \quad \nu^\perp < \bar{\nu},$$

where $\bar{\nu} \simeq kT/h$. Equation (66) describes the ratio of the relative changes in f_z^{-1} and $\exp(-U/kT)$, which contribute to changes in p [equation (61)] caused by an increment in the adsorption energy. As a typical example, the results of the calculations of House and Jaycock[87] for krypton adsorbed on relaxed (100) sodium chloride yield $\nu^\perp = 0.750 \times 10^{12}$ s^{-1}, which at 77.5 K ($\simeq$liquid nitrogen boiling temperature at atmospheric pressure) is less than $\bar{\nu}$ ($\simeq 1.62 \times 10^{12}$ s^{-1}). Using equation (66), r_z was found to be $\simeq 0.06$, which demonstrates that f_z^{-1} may be considered constant with respect to $\exp(-U/kT)$. In general this is true if $h\nu^\perp \leqslant U$ and $kT \leqslant U$ [equation (66)].

This conclusion has been verified for a more complex potential than the simple harmonic discussed above. Appel[96] considered the adsorbed atom to be oscillating in a Morse potential and from his results for strontium adsorbed on graphite r_z is of the order $10^{-5} T$, where T is the adsorption temperature.

Although the above authors referred to localized models of adsorption, it is not difficult to rewrite equation (65) for mobile adsorption replacing f_x and f_y by the partition function for two-degrees of translational freedom, f_{trans}. In the limit of an ideal two-dimensional gas f_{trans} would be independent of U; however if the translational barrier was significant, the variation of f_{trans} with barrier height and periodicity of $U(\tau)$ would have to be considered. This returns us to the question of correlating the intrinsic heterogeneity with adsorption energy; indeed f_x and f_y [equation (65)] for a particular system, may depend on U. This dependence is much less than for f_z and so Cerofolini's argument is again expected to be valid.

An investigation which has examined the effects of variations in A^0 is that of Hoory and Prausnitz,[33] who used the low-coverage (Henry's law region) data for argon on 'black pearls' (a carbon black) measured at temperatures in the

[95] T. L. Hill, *Adv. Catal.*, 1952, **4**, 211.
[96] J. Appel, *Surf. Sci.*, 1973, **39**, 237.

range 359.3—489.7 K. They calculated the monolayer capacities, n_m, and their standard deviation for the data set by equating Z_H ($Z_H = \lim_{p \to 0} n_a/p$) with $n_m \sum_i F_i' \exp(U_i/RT)/A_i^0$, where i labels the patches of equal adsorption energy. They in fact used three different distribution functions for F_i' but could find no advantage over the Gaussian distribution of Ross and Olivier.[6] Hoory and Prausnitz[33] assumed the relation between $\nu^\perp$ and U suggested by Hill[95] (*i.e.* $\nu^\perp \propto \sqrt{U + {}_a E_0^{vib}}$) and compared the best mean n_m value with that obtained by assuming A^0 a constant. They could not conclude that an allowance for variations in A^0 with changing U led to a better model of adsorption.

Local Isotherm Equations.—To be able to achieve a tractable solution to the integral equation (15), approximate models for local isotherm equations have to be postulated. The previous discussion drew attention to the role of intrinsic heterogeneity although recognizing that a rigorous treatment is difficult for most real systems. Except for helium, hydrogen, and possible neon, classical statistical mechanics ordinarily give an accurate representation of the adsorption system.

The Langmuir Equation. This is perhaps the most approximate but widely adopted equation:

$$g(\theta) = \theta/(1 - \theta). \tag{67}$$

Assuming the adsorbed molecule behaves as a three-dimensional harmonic oscillator and that the internal partition function (for electronic and nuclear energies) remains unchanged upon adsorption, A^0 may be expressed as:

$$A^0 = \prod_i \{1 - \exp(-h\nu_i/kT)\}\exp(-\mu^\Theta/kT), \tag{68}$$

where the subscript i refers to the three separate degrees of vibrational motion of the localized molecule. Unless ν_i are known, an approximate value of A^0 may be obtained by invoking the low-temperature limit *i.e.*:

$$A^0 = \left(\frac{2\pi mkT}{h^2}\right)^{3/2} kT. \tag{69}$$

In fact the value of A^0 is not essential as U is related to $-RT \ln K$ by the equation:

$$U = RT \ln A^0 - RT \ln K. \tag{70}$$

Hence a plot of $-RT \ln K$ is related to U through an axis shift $RT \ln A^0$. The accuracy of equation (69) depends on the system being studied but, as an example, for krypton at 77.5 K adsorbed on sodium chloride, the value of U may be in error by as much as 12% because of the neglect of the vibration-frequency-dependent terms in equation (68).

Other methods of evaluating A^0 have been suggested. de Boer[51] and Hobson[31] utilize the kinetic derivation to obtain a relationship:

$$A^0 = \frac{(MT)^{1/2}\sigma_m}{3.52 \times 10^{22} \times t_0} \text{ Torr,} \tag{71}$$

where σ_m is the monolayer coverage in molecules cm^{-2}, t_0 is the vibration time of an adsorbed molecule in seconds, and M is the chemical molecular weight. For nitrogen, Hobson[31] estimated A^0 as:

$$A^0 = 1.76 \times 10^4 (mT)^{1/2} \text{ Torr.} \tag{72}$$

The adsorption energy calculated by this method is the maximum adsorption energy U^0 (*i.e.* $U + {}_aE_0^{vib}$).

Another method that determines U^0 has been suggested by Dormant and Adamson.[97] This takes the low-pressure limit of the BET equation *i.e.* $p/p^0 \ll 1$ (where p^0 is the saturation vapour pressure of the adsorbate) to give:

$$p = \{\theta p^0/(1-\theta)\}\{\exp[(E_v - U^0)/RT]\}, \tag{73}$$

where E_v is the latent heat of vaporization of the bulk absorbate. Hence:

$$A^0 = p^0 \exp(E_v/RT). \tag{74}$$

Equations (68) and (69) are only applicable to monatomic gases, whereas equations (71) and (74) have been applied to various adsorbates; *e.g.* for nitrogen at its boiling point at atmospheric pressure equations (72) and (74) give values of 8.2×10^5 and 48.9×10^5 Torr, respectively, for A^0. This renders a difference of approximately 12% in the evaluation of U by equation (70).

The Jovanović Equation. In the kinetic derivation of the Langmuir equation the rate of desorption is proportional to the adsorbate surface coverage and independent of the equilibrium bulk gas pressure. Jovanović[98] derived a general equation for monolayer adsorption by taking account of the mechanical contact between the adsorbed and bulk phases. After a complex kinetic derivation he obtained the equation:

$$p = -A^0 \ln(1-\theta)\exp(-U^0/RT), \tag{75}$$

where $A^0 = (\sigma\tau)^{-1}(2\pi mkT)^{1/2}$, with τ denoting the mean residence time of an adsorbed molecule, σ the effective cross-section of interaction equal to $\pi D^2/2$ (where D is the molecular diameter), and m is the molecular mass of an adsorbed molecule. Jovanović found the equation described the adsorption isotherms of several systems, *e.g.* nitrogen and oxygen adsorbed on anatase, n-butane on glass spheres, and n-decane on iron powder. Since the derivation assumes the adsorbent to be homogeneous, it is not clear why its success in describing such systems is presented as a vindication of this local isotherm.

Jaroniec[99] presented a statistical thermodynamic derivation of the Jovanović equation using generalized ensemble theory. The final result may be written in the form:

$$A^0 = \left(\frac{2\pi mkT}{h^3}\right)^{3/2} \frac{kTf_g}{f_{xy}f_z f_a}, \tag{76}$$

where f_g and f_a are the internal partition functions for the bulk gas and

[97] L. M. Dormant and A. W. Adamson, *J. Colloid Interface Sci.*, 1972, **38**, 285.
[98] D. S. Jovanović, *Kolloid Z.*, 1969, **235**, 1203; *ibid.*, 1969, **235**, 1214.
[99] M. Jaroniec, *Colloid Polym. Sci.*, 1976, **254**, 601.

adsorbed molecules, respectively (it is usual to assume $f_g = f_a$). f_z and f_{xy} are the partition functions for the perpendicular vibrational motion of the adsorbate and the two degrees of motion parallel to the surface, respectively. Jovanović argues that his equation is applicable to both localized and mobile adsorption. The difference lies in the evaluation of A^0; in the case of localized adsorption the low-temperature limit of equation (76) yields equation (69) and U replaces U^0 in equation (75). Similarly, in the low-temperature limit, equation (81) is obtained for a mobile model of adsorption. Both the kinetic and statistical thermodynamic derivations have been faulted by Hazlett *et al.*[100] Jaroniec's derivation incorrectly identified the bulk gas pressure for a three-dimensional gas with the spreading pressure of the two-dimensional adsorbed phase. The proposed kinetic derivation of Hazlett, *et al.*[100] produces an isotherm equation of the Langmuir form. In view of the inconsistencies described by Hazlett *et al.* the Jovanović equation must be regarded as an essentially empirical isotherm equation.

The Fowler–Guggenheim Equation.[101] This local isotherm is based on a localized model of adsorption but includes average nearest-neighbour interactions. This is handled on the basis of a random distribution of atoms among the adsorption sites.

The equation may be written in the form:

$$p = [K\theta/(1 - \theta)]\exp(-Z\omega\theta/RT),$$
$$\text{i.e. } g(\theta, T) = [\theta/(1 - \theta)]\exp(-Z\omega\theta/RT), \tag{77}$$

where ω is the nearest-neighbour interaction energy and $Z\theta$ is the average number of occupied nearest-neighbour sites. The value of K is the same as in the Langmuir model *i.e.*

$$K = \left(\frac{2\pi mkT}{h^2}\right)^{3/2} \frac{kT}{f_x f_y f_z} \exp(-U/RT), \tag{78}$$

where $f_i = [1 - \exp(-h\nu_i/kT)]^{-1}$. Alternatively the adsorbed atom can be thought of as confined to a limiting area on the surface, ('site area'), with two degrees of translational freedom, *i.e.*:

$$A^0 = \left(\frac{2\pi mkT}{h^2}\right)^{1/2} \frac{kT}{\sigma f_z}, \tag{79}$$

where σ is the limiting 'site area'. Hence using the low-temperature approximation or knowing the magnitude of $\nu^\perp$, the value of $\ln A^0$ can be determined.

The Fowler–Guggenheim equation exhibits phase-transition loops when $Z\omega/RT \geqslant 4$. The values of the coverage limits of the isotherm step and the corresponding step position, p/K_t, have been tabulated.[102,103]

[100] J. D. Hazlett, C. C. Hsu, and B. W. Wojciechowski, *J. Chem. Soc., Faraday Trans.* 1, 1979, **75**, 602.
[101] R. H. Fowler and E. A. Guggenheim, 'Statistical Thermodynamics', Cambridge University Press, Cambridge, 1949, p. 431.
[102] W. A. House and M. J. Jaycock, *Colloid Polym. Sci.*, 1978, **256**, 52.
[103] J. C. P. Broekhoff and R. H. van Dongen, 'Physical and Chemical Aspects of Adsorbents and Catalysts', ed. B. G. Linsen, Academic Press, London and New York, 1970, Ch. 2.

The Hill–de Boer Equation.[51,52] This local isotherm has been discussed above [see equation (56)]. The value of K is given by:

$$K = \left(\frac{2\pi mkT}{h^2}\right)^{1/2} \frac{kT}{\beta f_z} \exp(-U/RT), \tag{80}$$

where β is the two-dimensional analogue of the excluded volume constant in the van der Waals equation of state, *i.e.* $\theta = \beta n_a/A_s$. Since $f_z = [1 - \exp(-h\nu/kT)]^{-1}$, the low-temperature limit of equation (80) gives:[33,8]

$$A^0 = \left(\frac{2\pi mkT}{h^2}\right)^{1/2} \frac{kT}{\beta}. \tag{81}$$

This local isotherm has been discussed at length by de Boer,[51] Ross and Olivier[6] and also Broekhoff and van Dongen.[103] Phase-transition loops occur when $2\alpha/kT\beta > 6.75$ and the transition step position (*i.e.* coverage limits and corresponding p/K_t) have been tabulated.[102,103]

The Virial Equation for an Imperfect Two-dimensional Gas.[89] The low-temperature limit of equation (60) may be written as:[89]

$$\ln\left(\frac{n_a}{p}\right) = \ln K_H - 2B_{2D}\left(\frac{n_a}{A_s}\right) - (3/2)C_{2D}\left(\frac{n_a}{A_s}\right)^2 \ldots, \tag{82}$$

where K_H is the Henry's constant $[K_H = B_{AS}/A_s kT$, see equation (60)] and B_{2D}, $C_{2D}, \ldots$ the two-dimensional virial coefficients. Steele[81] has presented a method of calculating these coefficients for a structured surface, *i.e.* a surface possessing intrinsic heterogeneity. The same method may be applied to determine the coefficients applicable to an unstructured surface, *i.e.* the two-dimensional virial coefficients, B_{2D}^0, C_{2D}^0, $D_{2D}^0, \ldots$. The first two coefficients, B_{2D}^0 and C_{2D}^0, have been evaluated using the Lennard–Jones 6–12 interatomic potential[81,85] and for argon using the 'exp-six' potential[82] to describe adsorbate self-interactions.

When equation (82) is transformed into the generalized form, $p = Kg(\theta, \tau)$, the value of K is given by:

$$\ln K = \ln(A_s/K_H) - \ln \sigma_m, \tag{83}$$

where σ_m is the limiting area per molecule and K_H is given in equation (58).

Unless the convergence limit of equation (82) is known and the requisite coefficients determinable, some approximation must be made for the infinite series. Ross and Morrison[41] have taken the reduced virial coefficients, D_{2D}/r^{*6}, E_{2D}/r^{*8} and higher terms (where r^* is the Lennard–Jones parameter for the distance of maximum attraction), to be equal to 2, a value supported by the studies of Ree and Hoover.[104] The 'K' used by Ross and Morrison[41] (K_{RM}) may be written as $K_{RM} = A_s/K_H$.

It is not clear what advantages this local isotherm offers over other approximate equations derived for a two-dimensional gas, in particular the simpler Hill–de Boer equation.

[104] F. H. Ree and W. G. Hoover, *J. Chem. Phys.*, 1964, **40**, 931.

Associated Adsorbate Model.[105,106] This model, which leads to the Berezin–
Kiselev equation, accounts for lateral interactions within the adsorbed phase
by proposing the formation of associates in the monolayer. The equation may
be written as:

$$\theta = \theta_1(1 + 2K_2\theta_1 + 3K_2K_3\theta_1^2 + \ldots), \qquad (84)$$

where $\theta_1 = K_1p(1-\theta)$, $K_1 = (1/\alpha^*)\exp(U/RT)$, and $K_i = (1/\beta^*)\exp(Q_i/RT)$, for
$n \geqslant i \geqslant 2$; K_i is a constant for the formation of surface associates of i molecules.
α^* and β^* are temperature-independent entropy factors and Q_i is the pairwise
energy of interaction of adsorbed molecules within the associate. Borowko *et
al.*[107] have discussed two approximations of equation (84), *i.e.* $K_i = K_{i+1}$ for
$i = 2$ to $i = n - 1$, and also $n = 2$ (double associates only). The equations have
been used to demonstrate the relative effects of adsorbate self-interactions and
gas–solid interactions upon the shape of the total isotherm and isosteric
enthalpy curve for both homogeneous and heterogeneous substrates.[107] The
model has also been applied in an investigation of a random distribution of
heterogeneities and will be discussed in Section 5.

Many other local isotherms are available (*e.g.* see ref. 108) but have not
been extensively used in heterogeneity studies.

4 Applications of the Integral Equation of Adsorption

Introduction.—The majority of research concerned with investigating surface
heterogeneity by gas adsorption has assumed the integral equation of adsorption [equation (15)]. The application of this model was popularized by Ross
and Olivier,[6] who applied their analysis method (see Section 2) to a large
number of systems (ref. 6, p. 187), including adsorption on carbon black P-33
after various degrees of graphitization, boron nitride, molecule sieve (Linde
13X), diamond, rutile, and anatase. Since this earlier work the physical model
describing adsorption on a patchwise heterogeneous surface has not changed
but the methods of solution of the integral equation to determine the distribution function, $F'(U)$, have become progressively more sophisticated.

**Generation of Total Isotherms by using a Generalized Distribution Function
and Assuming a Local Isotherm Equation.**—If a local isotherm is chosen, it is a
simple matter to generate a series of total isotherms by selecting different
heterogeneity parameters for a generalized distribution function. The problem
is to evaluate an optimum distribution function consistent with a least-squares
fit to the experimental data. Table 1 presents a number of combinations for
which the total isotherms have been evaluated. Of these the Ross and Olivier
tables,[6] which have been revised and corrected by Jaycock and Waldsax,[109] are
the most usable.

[105] G. I. Berezin and A. V. Kiselev, *J. Colloid Interface Sci.*, 1972, **38**, 227.
[106] J. Garbacz, *J. Colloid Interface Sci.*, 1975, **51**, 352.
[107] M. Borowko, M. Jaroniec, and W. Rudzinski, *Thin Solid Films*, 1977, **46**, 239.
[108] A. Patrykiejew, M. Jaroniec, and W. Rudzinski, *Chem. Eng. J.*, 1978, **15**, 147; A. Patrykiejew
      and M. Jaroniec, *Thin Solid Films*, 1980, **70**, 363.
[109] M. J. Jaycock and J. C. R. Waldsax, *J. Colloid Interface Sci.*, 1971, **37**, 462.

Table 1 *Local isotherms and adsorption energy distributions for which the total isotherm has been calculated (γ is a heterogeneity parameter and N is a normalization constant)*

Local isotherm	Distribution function, $F'(U)$	Ref.
Langmuir	Constant; $1/U_m$ for $U \in [0, U_m]$	a, b, c, d
	Exponential: $\gamma/RT \exp(-\gamma U/RT)$ for $U \in [0, \infty]$	d, e, f
	Exponential: $N \exp(-\gamma U/RT)$ for $U \in [-\infty, +\infty]$	c
Jovanović	Quasi-Gaussian: $2\gamma(U - U_a)\exp[-\gamma(U - U_a)^2]$, where U_a is the minimum adsorptive energy, $U \in [U_a, \infty]$	g
Hill–de Boer	Gaussian: $(1/N)\exp[-\gamma(U - U_m)^2]$ symbols as equation (33)	h, i

[a] A. Frumkin and A. Slygin, *Acta Physicochim.*, *URSS*, 1935, **3**, 791. [b] M. I. Temkin, *Zhur. Fiz. Khim.*, 1940, **14**, 153, *ibid.*, 1941, **15**, 296. [c] G. D. Halsey, and H. S. Taylor, *J. Chem. Phys.*, 1947, **15**, 624. [d] D. N. Misra, *J. Colloid Interface Sci.*, 1975, **43**, 85. [e] J. Zeldowich, *Acta Physicochim.*, *URSS*, 1935, **1**, 961. [f] M. J. Sparnaay, *Surf. Sci.*, 1968, **9**, 100. [g] M. Jaroniec and S. Sokolowski, *Colloid Polym. Sci.*, 1977, **255**, 374. [h] See ref. 6. [i] M. J. Jaycock and J. C. R. Waldsax, *J. Colloid Interface Sci.*, 1971 **37**, 462.

Determination of Distribution Functions for Generalized Total Isotherm Equations.—A number of total isotherm equations have found numerous applications to various adsorption systems. The results of the heterogeneity analysis of some of these isotherms are listed in Table 2.

The Freundlich isotherm [equation (4)] is perhaps the simplest such equation (see Section 1). It may be written in the generalized form as presented by Sips:[15]

$$\theta_T(p, T) = [p/(p + B)]^{1/n}, \tag{85}$$

where B is a temperature dependent constant and n a constant such that $n > 1$. If the Langmuir equation is adopted as the local isotherm, then with $B = A^0$ [see equation (68)], the Stieltjes transform method yields a distribution of the form:[15]

$$F'(U) = \frac{\sin(\pi/n)}{\pi RT}[\exp(U/RT) - 1]^{-1/n}. \tag{86}$$

If $B < A^0$ then the distribution function is translated to lower adsorption energies.[15] A difficulty arises when using the Sips procedure since the distribution function so determined is temperature dependent. Honig and Hill[14] have repeated the analysis but with a constraint requiring the distribution to be independent of temperature thus leading to severe restrictions on the form of the total isotherm function that can be handled by the Stieltjes transform method.

The Dubinin–Radushkevich (DR) isotherm[110] (sometimes referred to as the Dubinin–Radushkevich–Kaganer isotherm) has been discussed at length by

[110] M. M. Dubinin and L. V. Radukshkevich, *Dokl. Akad. Nauk SSSR*, 1947, **55**, 331.

Table 2 *Summary of investigations of generalized total isotherm functions*

Local isotherm	Total isotherm	Method	Ref.
Langmuir	Generalized Freundlich: $\theta_T = \{p/(p + A^0)\}^{1/n}$ where A^0 is the temperature-dependent constant of the Langmuir equation and $1/n < 1$ is a constant	ST	a
Langmuir	Temkin-Pyzhev: $\theta_T = A \log cp$, where A and c are constants	ST	a
Langmuir	Linear combination of Freundlich isotherms: $$\theta_T = \sum_{i=1}^{n} A_i \left(\frac{p}{B+p}\right)^{(i+k)/2(i+k)-1}$$ for $k > 1$ and $B = A^0 \exp(-U_a/RT)$, where A^0 is the Langmuir constant [equation (68)] and U_a the minimum adsorption energy	ST	b
Langmuir	Dubinin–Radushkevich (DR): $$\theta_T = \begin{cases} \exp[-B(RT \ln p/p_m)^2] & \text{for} \quad p \leqslant p_m \\ 1 & \text{for} \quad p > p_m, \end{cases}$$ where B and P_m are constants.	ST CA	c, d e, f
Langmuir	Generalized Dubinin–Radushkevich: $$\theta_T = \begin{cases} \exp\left[\sum_{n=0}^{j} B_n\{\ln(p/p_m)\}^n\right] & \text{for} \quad p \leqslant p_m \\ 1 & \text{for} \quad p > p_m \end{cases}$$	ST CA	g h
Langmuir	Tóth equation: $\theta_T = \dfrac{p}{[B+p^\gamma]^{1/\gamma}}$ where B is a temperature dependent constant and γ a heterogeneity parameter $\gamma \in [0, 1]$	ST CA	i j
Jovanović	Least-squares polynomial	LT	b
Jovanović	Dubinin–Radushkevich	LT	k, l
Jovanović	Generalized Dubinin–Radushkevich	LT	l

Key: ST, Stieltjes Transform; CA, Condensation Approximation; LT, Laplace Transform

[a] R. Sips, *J. Chem. Phys.*, 1950, **18**, 1024. [b] A. Waksmundzki, S. Sokolowski, J. Rayss, Z. Suprynowicz, and M. Jaronieć, *Sep. Sci.*, 1976, **11**, 29. [c] D. N. Misra, *Surf. Sci.*, 1969, **18**, 367. [d] C. V. J. Heer, *J. Chem. Phys.*, 1971, **55**, 4066. [e] W. Rudzinski and M. Jaronieć, *Rocz. Chem.*, 1975, **49**, 165. [f] G. F. Cerofolini, *Surf. Sci.*, 1971, **24**, 391. [g] M. Jaronieć, *Surf. Sci.*, 1975, **50**, 553. [h] M. Jaronieć, W. Rudzinski, S. Sokolowski, and R. Smarzewski, *Colloid Polym. Sci.*, 1975, **253**, 164. [i] J. Tóth, W. Rudzinski, A. Waksmundski, M. Jaronieć, and S. Sokolowski, *Acta Chim. Acad. Sci. Hung.*, 1974, **82**, 11. [j] A. Waksmundzki, J. Tóth, M. Jaronieć, and W. Rudzinski, *Rocz. Chem.*, 1975, **49**, 1003. [k] M. Jaronieć and S. Sokolowski, *Colloid Polym. Sci.*, 1977, **255**, 374. [l] S. Sokolowski, J. Mieczyslaw, and A. Waksmundzki, *Rocz. Chem.*, 1976, **50**, 1149.

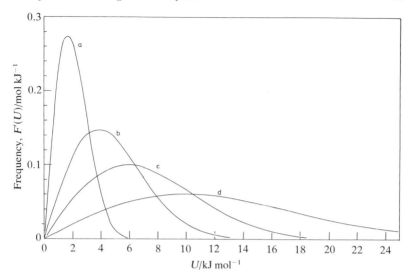

Figure 3 *Distribution functions evaluated from equation (87) at 77.5 K for the DR total isotherm equation and the Langmuir local isotherm. U_m equal to* (a) 2, (b) 4, (c) 6, (d) 10 kJ mol^{-1}

Cerofolini,[94] who also tabulated systems that follow the DR isotherm. Misra[111] analysed the DR* equation using the Langmuir local isotherm equation and obtained the result:

$$F'(U) = (1/\pi RT)\exp(R^2 T^2/2U_m^2)\sin(\pi RTU/U_m^2)$$
$$\times \exp(-U^2/2U_m^2), \tag{87}$$

where $U_m = \sqrt{1/(2B)}$, B is the constant of the DR equation, and U_m the position of the maximum in the distribution when $\tan(2\pi RTBU_m) \simeq 2\pi RTBU_m$. Figure 3 illustrates the form of $F'(U)$ for a selection of values of U_m and clearly shows the 'skew' nature of the distribution. In the limit of low temperatures, *i.e.* $RT \ll U_m$, equation (87) reduces to:[111]

$$F'(U) = (U/U_m^2)\exp[-U/(2U_m^2)] \tag{88}$$

and so becomes independent of temperature.

Tóth *et al.*[112] have noted the wide applicability of another total isotherm equation, the Tóth equation,[113,114] and by using the Stieltjes transform method

---

* Misra[111] wrote the DR equation in the form;

$$\theta_T = \exp\{-B[RT \ln(1 + p_m/p)]^2\}$$

[111] D. N. Misra, *Surf. Sci.*, 1969, **18**, 367.
[112] J. Tóth, W. Rudzinski, A. Waksmundzki, M. Jaroniec, and S. Sokolowski, *Acta Chim. Acad. Sci. Hung.*, 1974, **82**, 11.
[113] J. Tóth, *Acta Chim. Acad. Sci. Hung.*, 1962, **30**, 1.
[114] J. Tóth, *Acta Chim. (Budapest)*, 1971, **69**, 311.

with the Langmuir local isotherm, have investigated the corresponding general adsorption energy distribution:

$$F'(x) = \frac{A^0}{\pi RT} \left\{ \frac{[Bx^\gamma \cos(\gamma\pi) + A^{02\gamma}]^2 + [Bx^\gamma \sin(\gamma\pi)]^2}{[A^{2\gamma} + 2A^{0\gamma}Bx^\gamma \cos(\gamma\pi) + B^2 x^{2\gamma}]^2} \right\}^{1/2\gamma}$$
$$\times \sin\left\{ \arccos \frac{Bx^\gamma \cos(\gamma\pi) + A^{02\gamma}}{([Bx^\gamma \cos(\gamma\pi) + A^{0\gamma}]^2 + [Bx^\gamma \sin(\gamma\pi)]^2)^{1/2}} \right\},$$

(89)

where B and γ are defined in Table 2, A^0 has its usual meaning and $x = \exp(U/RT) - 1$. Figure 4 shows the distribution obtained from the data of Tóth *et al.*[112] for propane adsorbed on Nuxit charcoal. The slight temperature dependence is an inherent feature of the Stieltjes transform method, although Tóth *et al.*[112] attribute the majority of the observed temperature dependence to the neglect of the adsorbate–adsorbate interactions. An important consideration, which may explain why the Tóth equation finds applications when the DR equation fails, is the shape of the distribution function. Figure 4 illustrates the 'tailing' to lower adsorption energies in comparison to the log-normal type distribution obtained from the DR total isotherm.

The condensation approximation discussed in Section 2 has been used by various investigators[28,30,94,115–120] by evaluating the derivative in equation (31) using the step-function:

$$p/A^0 > (p/A^0)_c, \qquad \theta = 1;$$
$$p/A^0 < (p/A^0)_c, \qquad \theta = 0,$$

(90)

where the subscript c refers to the critical pressure at which condensation occurs. In this way a relationship between adsorption energy and the equilibrium gas pressure is established:

$$U = -RT \ln(p/A^0)_c.$$ (91)

Similarly the p_m of the DR equation may be written as:

$$U_a = -RT \ln(p_m/A^0)_c.$$ (92)

By substituting equations (91) and (92), for p and p_m in the total isotherm equation, the derivative in equation (31) may be evaluated:[94]

$$F'(U) = \begin{cases} 2B(U - U_a)\exp[-B(U - U_a)^2] & \text{for } U < U_a, \\ 0 & \text{for } U < U_a. \end{cases}$$ (93)

[115] C. C. Hsu, B. W. Wojciechowski, W. Rudzinski, and J. Narkiewicz, *J. Colloid Interface Sci.*, 1978, **67**, 292.
[116] A. Waksmundzki, M. Jaronieć, and L. Lajtar, *Rocz. Chem.*, 1975, **49**, 1197.
[117] W. Rudzinski and M. Jaronieć, *Rocz. Chem.*, 1975, **49**, 165.
[118] S. Sokolowski, M. Jaronieć, and A. Waksmundski, *Rocz. Chem.*, 1976, **60**, 1149.
[119] A. Waksmundzki, J. Tóth, M. Jaronieć, and W. Rudzinski, *Rocz. Chem.*, 1975, **49**, 1003.
[120] M. Jaronieć, *Surf. Sci.*, 1975, **50**, 553.

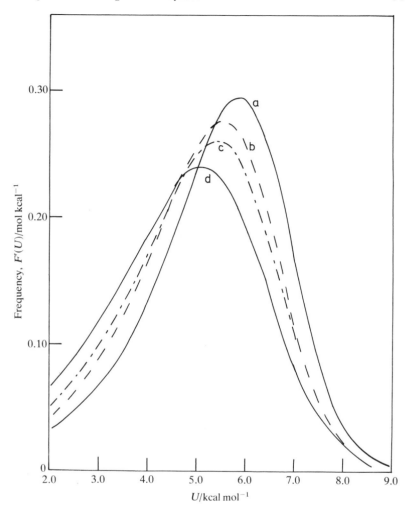

Figure 4 *Distribution functions for propane adsorbed on Nuxit charcoal at temperatures of* (a) 20 °C, (b) 40 °C, (c) 60 °C, and (d) 90 °C. *The isotherm data were fitted to the Tóth equation and analysed using the local Langmuir equation. The distributions have been recalculated using equation (89), constants given by Tóth et al.*[1/2]

$F'(U) = 0$ for $U < U_a$ because the DR equation is expressed so that $\theta_T = 1$ for $p > p_m$.[94] The most frequent patches on the surface have an adsorption energy, $U_m = \sqrt{1/2B} + U_a$, where U_a is the minimum adsorption energy [equation (92)]. If $U_a = 0$, then the distribution curves may be compared to those obtained from the Misra equation.[111] The results of the calculations are presented in Figure 5 for an adsorption temperature of 77.5 K and are in

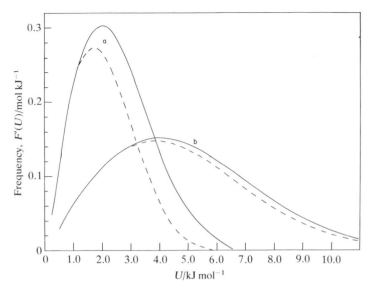

Figure 5 *Examples of distribution functions obtained for DR total isotherm equation at 77.5 K using the local Langmuir function.* ———, *the distribution functions generated from equation* (87) *with* U_m *equal to:* (a) 2, (b) 4 kJ mol^{-1}; ———, *the distribution functions generated by using the condensation approximation* [*equation* (93), *with* $U_a = 0$] *with* U_m *equal to:* (a) 2, (b) 4 kJ mol^{-1}

reasonable agreement, particularly for the broader distribution. With $U_a \neq 0$ the distribution function is merely translated to higher adsorption energies.

In general the relationship between the real distribution, $F'(U)$, and the condensation approximation distribution, $F'_c(U)$ is:[94]

$$F'_c(U) = \int_0^\infty T(U^1, U) F'(U) \, dU^1, \tag{94}$$

where the transform function is,

$$T(U^1, U) = -\left[\frac{\partial \theta(U, U^1)}{\partial U}\right]_{U^1} \text{ and } \theta(U, U^1)$$

is the local isotherm function with U^1 is equivalent to U in equation (91). For the Langmuir equation:[94,121]

$$T(U^1, U) = -\frac{1}{RT} \exp\left(\frac{U - U^1}{RT}\right)\left[1 + \exp\left(\frac{U - U^1}{RT}\right)\right]^{-2}. \tag{95}$$

The transform function acts as a smoothing function upon the real distribution, $F'(U)$, and as demonstrated by Cerofolini[94] and Harris,[121] broadens with increasing temperature; thus the condensation approximation $F'_c(U)$ is less

[121] L. B. Harris, *Surf. Sci.*, 1968, **10**, 129.

accurate at higher adsorption temperatures. The integral equation (94) has been solved by Hsu *et al.*[115] for various function $T(U^1, U)$ including equation (95) and approximate expressions for $F'(U)$ derived using the Fowler–Guggenheim and Hill–de Boer equations for adsorption above the two-dimensional critical temperature.

The condensation approximation has been used to analyse the generalized DR equation (see Table 2). The greater flexibility of this equation allows more complex shaped distributions to be determined, for instance Jaronieć *et al.*[30] obtained a dual peak distribution function for the nitrogen/silica gel system at 78 K.

The high sensitivity of the distribution function to small changes in the experimental adsorption isotherm warrants caution in any curve-fitting procedure, particularly when a limited number of experimental points are available. The extrapolation of data to zero pressure and then evaluating distributions over an adsorption energy range larger than can be experimentally justified is a very doubtful procedure.

Applications of Hobson's Method.—Zeldowich's kernel approximation [equation (6)] discussed in Section 2 has been generalized by Hobson[31] to account for adsorbate lateral interactions. The major problem associated with the Langmuir equation is its inability to predict phase transitions. Rather than use the 'step approximation' to the local isotherm, Hobson chooses a combination of a Henry's law isotherm and a 'step approximation', as shown in Figure 1(c):

$$\begin{aligned} \theta &= (p/A^0)\exp(U/RT) \quad \text{for} \quad U < U^1, \\ \theta &= 1 \quad\quad\quad\quad\quad\quad \text{for} \quad U \geqslant U^1, \end{aligned} \tag{96}$$

where $U^1 = -RT \ln(p/A^0) - Q = U^0 - Q$. When $U^1 = U^0$ *i.e.* $Q = 0$, then equation (96) reduces to equation (6). Hobson showed[31] that by adopting equation (96) as the local isotherm in the integral equation, differentiating twice with respect to pressure at constant T and simplifying with $Q = 0$ yields:

$$F'(U) = -\left(\frac{\partial \theta_T}{\partial U^1}\right)_U - RT\left[\frac{\partial^2 \theta_T}{\partial (U^1)^2}\right]_U. \tag{97}$$

A more complex equation is obtained for $Q \neq 0$.[31] The evaluation of A^0 requires a consideration of the state of the adsorbate, *i.e.* whether it is localized or mobile. Unfortunately the choice of Q is somewhat arbitrary.

Hobson[122] applied the method to analyse the results of the adsorption of argon, nitrogen, and helium on Pyrex at 77.4 K using a range of Q values. A sharply peaked distribution was obtained in each case and the entire distribution was translated to lower adsorption energies by increasing the lateral interaction energy, Q. This result is in agreement with the work of Waksmundzki *et al.*[116]

Cerofolini[28,94] has also examined the Hobson method (which he calls the 'asymptotically correct approximation', with $Q = 0$ and obtains a transform

[122] J. P. Hobson, *Can. J. Phys.*, 1965, **43**, 1941.

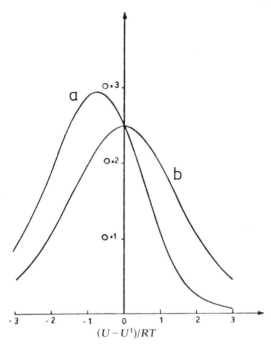

Figure 6 *Comparison of the condensation approximation and the Hobson method (with Q = 0).* (a) $-kT(\delta/\delta U + kT\delta^2/\delta U^2)\theta(U, U^1)$, (b) $-kT\ \delta\theta(U, U^1)/\delta U$, *where $\theta(U, U^1)$ is the Langmuir function*
(Reproduced by permission from *Thin Solid Films*, 1974, **23**, 129).

function [see equation (94)]:

$$T(U^1, U) = \left[\left(\frac{-\partial}{\partial U} - RT\frac{\partial^2}{\partial U^2}\right)\theta(U, U^1)\right]_{U^1}, \tag{98}$$

which he compares with the condensation approximation transform function (Figure 6). The Hobson method leads to a slightly better approximation to the real distribution but the transform function is not centred about $(U - U^1) = 0$ and its application may introduce a small lateral displacement of the distribution's peaks.

An interesting application of Hobson's method has been made by Rudzinski *et al.*[123] by analysing the pressure dependence of chromatographic retention data. They manipulated Hobson's equation, with $Q = 0$, to obtain:

$$F'(U) = -\frac{J}{n_m}\left(\frac{p}{RT}\right)^2\left(\frac{\partial V_N}{\partial p}\right)_T, \tag{99}$$

where J is the James–Martin compressibility factor and V_N the retention volume. The pressure, p, is related to the adsorption energy through equation

[123] W. Rudzinski, A. Waksmundzki, R. Leboda, and Z. Suprynowicz, *J. Chromatogr.*, 1974, **92**, 25.

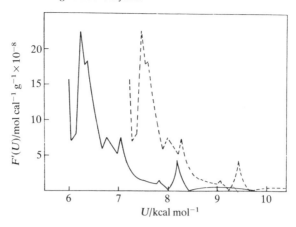

Figure 7 *Distribution of adsorption energies for cyclohexane adsorbed on wide-pore silica gel at 126.3 °C. The solid line denotes the function $F'(U)$ calculated by using the Fowler–Guggenheim local adsorption isotherm; the dashed line denotes $F'(U)$ for the Langmuir isotherm*
(Reproduced by permission from *J. Chromatogr.*, 1975, **110**, 381).

(91). Figure 7 illustrates an example of the results obtained by this method.[124] Equation (97) is convenient to use and the inclusion of Q, as demonstrated by Hobson, only laterally translates the distribution. Waksmundzki *et al.*[116] have investigated this translation using the condensation approximation and assuming different relations between p and U pertaining to a variety of local isotherm functions. With the Fowler–Guggenheim equation, the phase transition is symmetrical about $\theta = 0.5$ (above the ideal two-dimensional critical temperature, the point of inflection in the local isotherm is at $\theta = 0.5$) and hence the step-function determines the relationship:

$$-RT \ln(p/A^0) = U + 0.5Z\omega = U - 2RT_c,\qquad (100)$$

since $Z\omega = -4RT_c$, where T_c is the ideal two-dimensional critical temperature. For the condensation approximation and the Hobson method, this leads to a translation of the distribution function by $2RT_c$ but the shape of the distribution remains unchanged. In general such a simple adsorption energy translation does not apply to the Hill–de Boer equation without making gross assumptions about the phase-transition parameters.[116]

Results of the Numerical Iterative Methods.—The application of more sophisticated local isotherms than the Langmuir or Jovanović dictates the use of numerical iterative methods. The advantages of these methods are they do not require curve fitting of the experimental data to establish a total isotherm equation, nor assumptions regarding the general form of the adsorption energy distribution. However they generally necessitate extensive and reliable experimental data, the measurement of which has been made possible by the more

---

[124] A. Waksmundzki, M. Jaroniec, and Z. Suprynowicz, *J. Chromatogr.*, 1975, **110**, 381.

recent development of low-pressure precision diaphragm gauges and better high-vacuum equipment without mercury burettes.

Applications of the Adamson and Ling Method.[40] This has been one of the most widely used methods of analysis. The first approximation of the iterative procedure is obtained using the condensation approximation. In fact Adamson and Ling[40] used a graphical method and initially calculated the integral adsorption energy distribution, $F(U)$. This curve is subsequently adjusted in accordance with the disagreement between the experimental monolayer coverage and the calculated value obtained by using a realistic isotherm equation $[\theta(p, U)]$, *i.e.*

$$\theta_T(\text{calc.}) = \int_{U_1}^{U_h} \theta(p, U)\, dF, \qquad (101)$$

where U_h and U_1 are the upper and lower limits of the adsorption energy respectively as determined from the step-approximation corresponding to the lowest and highest equilibrium pressures and the form of the local isotherm equation [*i.e.* the adsorption energies are evaluated using equations of the general form of equation (91)]. The iterative improvement consists of adjusting the $F(U)$ values such that:

$$F(U)^c = F(U)^{c-1} \frac{(n_a/n_m)(\text{expt.})}{\theta_T(\text{calc.})}, \qquad (102)$$

where c labels the iteration number. This adjustment ascribes differences between $\theta(\text{expt.})$ and $\theta_T(\text{calc.})$ at a particular pressure, p, to an error in $F(U)$ with U evaluated and related to p through the step-approximation. Adamson and Ling found this to be a good method of optimizing the agreement between experimental and calculated isotherms, although they restricted the majority of their applications by using the Langmuir local isotherm.

The method has been applied (using the Langmuir local isotherm) to the analysis of nitrogen, oxygen, and argon adsorbed on rutile[40] at 78 K and 90 K, nitrogen on boron nitride[125] (77.5 K) using the Langmuir, Fowler–Guggenheim, and Hill–de Boer local isotherms; argon on carbon black[126] and ammonia adsorbed on commercial silica–alumina cracking catalysts[127] (0 °C) using the Langmuir equation, and nitrogen on silica[128,129] at 77 K, again using the Langmuir equation. Jackson and Davis[130] examined argon and nitrogen adsorbed at 77 K on unannealed and annealed sodium chloride by adopting the Fowler–Guggenheim local isotherm. The results of the method have also been compared to those obtained using the Ross and Olivier procedure (using the Hill–de Boer equation) for krypton adsorbed on anatase at 77.5 K.[8] The dual peaked distributions obtained were in good agreement but differed in

[125] A. W. Adamson, I. Ling, L. Dormant, and M. Orem, *J. Colloid Interface Sci.*, 1966, **21**, 445.
[126] P. Y. Hsieh, *J. Phys. Chem.*, 1965, **68**, 1068.
[127] P. Y. Hsieh, *J. Catal.*, 1963, **2**, 211.
[128] J. Whalen, *J. Phys. Chem.*, 1967, **71**, 1557.
[129] J. B. Sorrell and R. Rowan, jun., *Anal. Chem.*, 1970, **42**, 1712.
[130] D. J. Jackson and B. W. Davis, *J. Colloid Interface Sci.*, 1974, **47**, 499.

shape from that distribution derived using the Langmuir local isotherm and the Adamson and Ling method. Waksmundzki *et al.*[131] have applied the method to the evaluation of the adsorption energy distribution from gas-adsorption chromatography data. Their distribution for chloroform adsorbed on Polysorb C at 348.2 K obtained by selecting the local Langmuir equation, shows two distinct peaks. This is in accordance with the later work of Waksmundzki *et al.*[50] who used the Stieltjes transform method of analysis of chromatography data for hexane adsorbed on Polysorb C at 348.2 K.

The major disadvantages of the Adamson and Ling method are its reliance on the BET value of n_m, the lengthy calculations that are involved, and the consequences of applying the condensation approximation. Adamson and Ling[40] drew attention to the extensive pressure range required by any heterogeneity analysis. As the adsorption temperature increases, the condensation approximation becomes poorer and the adsorption energy range so defined becomes more restrictive and less realistic. Morrison and Ross[41,132] have criticized the method as 'analytically unsound as it lacks any provision for leading its approximations on to a unique and optimum solution'. Dormant and Adamson[23] have replied that such a solution is impossible without prior information concerning the implicit errors in the data. As mentioned in the discussion concerning the matrix-inversion method in Section 2, a set of solutions is generally available and it becomes a compromise between physically acceptable forms of the solution and the magnitude of the deviations tolerable between calculated and experimental isotherm data.

The HILDA Algorithm. This is a numerical method developed by House and Jaycock,[102] which uses the iterative scheme of Adamson and Ling.[40] The algorithm is an improvement over the original procedure because:

(*a*) A knowledge of the monolayer capacity is not a prequisite since the algorithm systematically adjusts the normalization constant such that $F(U_1) = 1$ and when U_1 corresponds to the minimum adsorption energy, the normalization capacity becomes the monolayer capacity.

(*b*) The solution is reached when the r.m.s. deviation between the experimental and calculated isotherms is a minimum. The algorithm incorporates an odd-ordered quadratic smoothing routine that allows the isotherm data to be smoothed and if necessary (*i.e.* if large fuctuations in the distribution arise): the individual $F(U)$ curves may also be smoothed.

(*c*) The algorithm permits a choice of local isotherm functions, *i.e.* Langmuir, Fowler–Guggenheim, Hill–de Boer, or the two-dimensional virial.

(*d*) Patches of adsorption energy $>U_h$ are assumed to be full. This approximation avoids the difficulty of proposing a distribution of adsorption energies in the range U_h to ∞.

The algorithm has been tested on isotherm data generated for a dual-peaked Gaussian distribution with the HdB local isotherm function. With isotherm data correct to four significant digits, the original distribution was obtained by the HILDA algorithm without any smoothing procedure.[102] The data were

[131] A. Waksmundzki, S. Sokolowski, M. Jaroniec, and J. Rayss, *VUOTO*, 1975, **8**, 113.
[132] I. D. Morrison and S. Ross, *Surf. Sci.*, 1977, **62**, 331.

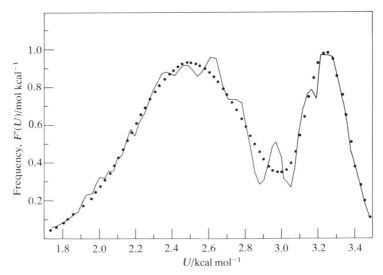

Figure 8 *Results of the analysis of isotherm data with inherent random errors in the adsorption amount (i.e. 10% at 10^{-5} Torr decreasing on a logarithmic pressure scale to 0.4% at 10^{-2} Torr and higher). · · ·, distribution determined by HILDA with the isotherm data correct to 4 significant digits*

processed to include random errors (based on a rectangular probability distribution generator) of a maximum of 10% at 10^{-5} Torr decreasing on a logarithmic scale *i.e.* 6.7% at 10^{-4} Torr, 3.3% at 10^{-3} Torr, to 0.4% at 10^{-2} Torr, and greater. The distribution obtained by HILDA with the smoothing options is shown in Figure 8; in view of the generous random error allowance, the agreement is thought to be good and substantiates the validity of the method. However, the appearance of small details in distributions should not always be interpreted in terms of physical properties of the surface unless confirmed by the analysis of new and accurate experimental data. This problem has also been highlighted by Zolandz and Myers in their recent review article.[133] They introduced a random error of 2% of the value of the isotherm point throughout the pressure range. Without any data smoothing large oscillations in the distribution function were observed (similar to those obtained using the regularization method and shown in Figure 2a). The oscillations decreased in magnitude when the smoothing options were employed. These authors concluded that the highest-energy peak was fairly well determined.

The method has been used by House and Jaycock[20] to assess the effects of annealing high quality sodium chloride particulates (Figure 9) and differences in the surfaces when prepared by two different techniques. HILDA has also

[133] R. R. Zolandz and A. L. Myers, *Prog. Filtr. Sep.*, ed. R. J. Wakeman, 1979, **1**, 1.

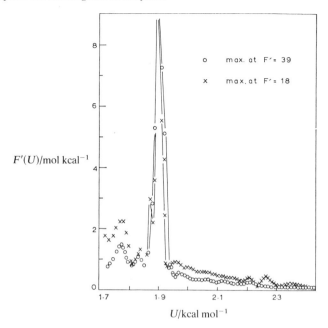

Figure 9 *The adsorption energy distribution function obtained by* HILDA *from the analysis of the adsorption isotherm for krypton on NaCl at 76.1 K. Key:* xxx, *before annealing;* ∘∘∘, *after* 21 h *annealing in a dry nitrogen atmosphere* (600 Torr) *at* 310—315 °C
(Reproduced by permission from *J. Colloid Interface Sci.*, 1977, **59**, 252).

been applied to investigating different preparations of silver iodide by analysing krypton adsorption isotherms measured at different temperatures.[63] Phase-transition 'steps' on the isotherms gave sharp peaks in the distribution functions that were associated with different crystalline faces of the silver iodide. Figure 10 presents an example of this analysis and compares the result with that obtained using CAEDMON[85] (see below).

The algorithm has also been used to analyse the nitrogen/Spheron-6 system at 77 K employing both the Hill–de Boer and the two-dimensional virial local isotherms and the results demonstrate the effects of annealing the carbon black.[134] The absolute nitrogen (78 K) and argon (78 K) isotherms measured by Aristov and Kiselev for adsorption on maximally hydroxylated non-porous and wide-pored silica have been analysed[135] using the Langmuir and Hill–de Boer local isotherms. The argon isotherm has also been analysed using the regularization method and produced excellent agreement with the HILDA distribution.[135]

[134] W. A. House and M. J. Jaycock, *J. Chem. Soc., Faraday Trans. 1*, 1977, **73**, 942.
[135] W. A. House, *J. Chem. Soc., Faraday Trans. 1*, 1978, **74**, 1045.

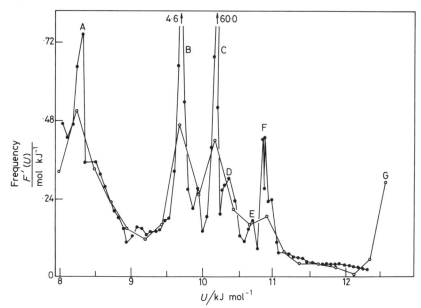

Figure 10 *The adsorption energy distribution functions obtained from the analysis of krypton adsorbed on silver iodide at 77.9 K using the Hill–de Boer local isotherm* (a), ••••, HILDA; ○ ○ ○, CAEDMON
(Reproduced by permission from *J. Chem. Soc., Faraday Trans.* 1, 1977, **73**, 942).

The CAEDMON Algorithm. This method has already been mentioned in Section 2. Ross and Morrison[41] adopted the two-dimensional virial local isotherm and demonstrated the high degree of homogeneity for the nitrogen, argon/P-33 (2700 K) carbon black systems measured at 90.1 K, nitrogen, argon/boron nitride measured at 90.1 K, and the effects of annealing P-33 carbon black from the analysis of nitrogen and argon isotherms at 90.1 K (see Figure 11).

Wesson[136] has used CAEDMON to assess the effects of annealing Spheron-6. The algorithm has also been compared with HILDA[134] and found to give results in reasonable agreement but with somewhat less detail in the distribution for the nitrogen/unannealed Spheron-6 (77 K) system. Another algorithm developed by Sacher and Morrison[42] also called CAEDMON has been used to analyse Wesson's nitrogen/carbon black data.[136] The results show differences with the analysis using the CAEDMON program of Ross and Morrison.[41] In the author's opinion the highest-energy peak obtained using either of the CAEDMON programmes is an artifact of the algorithms and the peak only serves to compensate for adsorption that occurs on 'patches' of energy greater than that value considered in the analysis. The distribution functions shown by

[136] S. Wesson, Ph.D. Thesis, Rensselaer Polytechnic Institute, 1975.

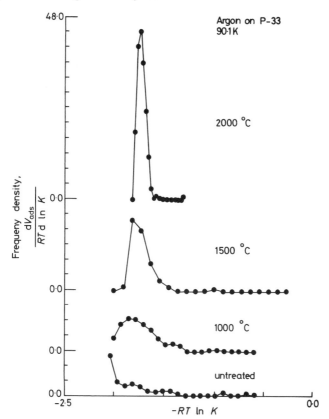

Figure 11 *Adsorption energy distributions obtained by* CAEDMON *showing the effect of thermal conditioning of carbon black P-33 at different temperatures.* V_{ads} *in* cm^3 (STP)g^{-1}, *K in* Torr, *and RT in* kcal mol^{-1}
(Reproduced by permission from *Surf. Sci.*, 1975, **52**, 103).

Sacher and Morrison[42] and determined by the Ross and Morrison CAEDMON program, are not consistent with the distribution presented by Wesson[136] with respect to the existence of the highest-energy peak. The recent results of Sacher and Morrison[42] are in much closer agreement with the original HILDA analysis[134] of Wesson's data. Papanu[137] has analysed the nitrogen/kaolinite (77.15, 89.52 K) and nitrogen/pyrophylite (77.15 K) systems using CAEDMON. The algorithm was modified by Hinman and Halsey[138] and used to analyse data for the adsorption of argon (79,100 K) and nitrogen (79 K) on three progressively sintered tin dioxide samples. The submonolayer data were rather sparse and the analysis achieved little in differentiating between the three samples. Nonetheless differences between the argon and

[137] S. C. Papanu, Ph.D. Thesis, Rensselaer Polytechnic Institute, 1976.
[138] D. C. Hinman and G. D. Halsey, *J. Phys. Chem.*, 1977, **81**, 739.

nitrogen distributions were evident; the former showed single gaussian type peaks, whereas the latter were bimodal.

Results From Specific Adsorption Systems.—A number of specific applications have already been discussed; the most enlightening have been concerned with monitoring changes in adsorption distributions when the surface is subjected to some treatment, *e.g.* heating or annealing. However a number of independent studies have been concerned with a specific system and it is interesting to compare the results.

Argon Adsorbed on Titanium Dioxide (Rutile). The data of Drain and Morrison[139] have been analysed by several methods. Drain and Morrison assessed the heterogeneity of their sample from their enthalpy of adsorption data:

$$F(q_0^{st}) = \frac{dV}{d(q_0^{st})}, \qquad (103)$$

where V is the volume of gas adsorbed at STP and q_0^{st} the isosteric enthalpy at 0 K. The q_0^{st} values were obtained from differential molar heat capacities of the adsorbed phase at constant amounts of adsorbate.[139] The distribution so obtained includes information concerning the lateral interactions and as stressed by Drain and Morrison the relationship (103) is only accurate at high values of q_0^{st}, *i.e.*

$$q_0^{st} = U + {}_a P^{ia} \qquad (104)$$

at absolute zero Kelvin, where ${}_a P^{ia}$ is the lateral interaction potential energy. The distribution they obtained is illustrated in Figure 12(a).

It is pertinent to mention that several authors[140] have unjustifiably tried to evaluate the adsorption energy distributions directly from the relationship (103) but substituting isosteric enthalpies measured at normal adsorption temperatures ($t > 0$). Koubek *et al.*[141] have reiterated the difficulties in separating mutual lateral interaction energies from the effects of surface heterogeneity when analysing isosteric enthalpy data.

The isotherm data of Drain and Morrison[139] for argon adsorbed on rutile (85 K) have been analysed by Dormant and Adamson,[97] and by Rudzinski and Jaroniec[43,117] by employing localized models of adsorption and ignoring the effects of lateral interactions. The Adamson and Ling method gives good agreement with the distribution derived from enthalpies of adsorption (apart from a lateral shift of the adsorption energy spectrum, see Figure 12a). The condensation approximation yields poor agreement and the orthonormal function expansion analysis[43] produces a single peak skewed to higher adsorption energies in the fashion of the DR distribution (Figure 5).

In marked contrast to these results, Rudzinski and Jaroniec[49] obtained a distribution function showing three distinct peaks. They used the Jovanović

[139] L. E. Drain and J. A. Morrison, *Trans. Faraday Soc.*, 1952, **48**, 316, 840.
[140] J. J. Chessick and A. C. Zettlemoyer, *J. Phys. Chem.*, 1958, **62**, 1217; T. Morimoto, M. Nagao, and J. Imai, *Bull. Chem. Soc. Jpn.*, 1974, **47**, 2994.
[141] J. Koubek, J. Pasek, and J. Volf, *J. Colloid Interface Sci.*, 1975, **51**, 491.

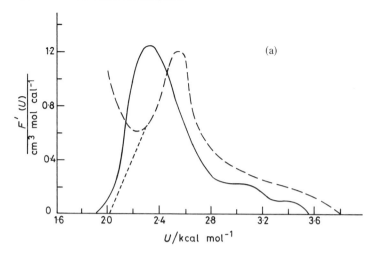

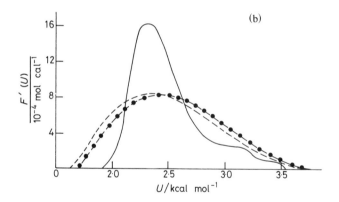

Figure 12 *Adsorption energy distribution for argon on rutile from the data of Drain and Morrison*
(Reproduced by permission from *Trans. Faraday Soc.*, 1952, **48**, 316, 840).

(a) ——, *calculated using the Adamson and Ling method, with the BET local isotherm equation, from the 85 K isotherm of Drain and Morrison;* — —, *obtained by Drain and Morrison from differential enthalpy data (the light dashed line shows their assumed closure at low energies)*
(Reproduced by permission from *J. Colloid Interface Sci.*, 1972, **38**, 285).

(b) ——, *as in (a) above*, –●–●–, *the DR distribution corresponding to the Langmuir local behaviour;* – – –, *DR distribution corresponding to Jovanović local behaviour. The latter two distributions are obtained by using the condensation approximation*
(Reproduced by permission from *Rocz. Chem.*, 1975, **49**, 165).

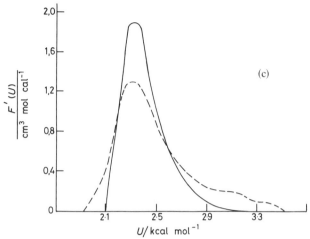

Figure 12 (*cont.*)

(c) – – –, *calculated by the Adamson and Ling method as in* (a) ——, *obtained by using the orthonormal function expansion method with the Langmuir local isotherm*
(Reproduced by permission of *Colloid Polym. Sci.*, 1975, **253**, 683).

local isotherm and by expressing the data as a polynomial expansion, were able to use a Laplace transformation to obtain a distribution function. This distribution does not appear to be constrained to positive values of $F'(U)$ and it is not clear whether the agreement between the experimental isotherm and the calculated isotherm takes this into consideration. The distribution also extends into adsorption energies that are less than the heat of liquefaction of argon. The authors show that the inversion procedure they use does not perform well in a model calculation to determine the heterogeneity function from isotherm data computed for a model heterogeneous system. Oh and Kim[142] concluded that the surface heterogeneity could be described by three different single-adsorption energies. They also use a model that ignores the effects of lateral interactions but does account in a very limited sense for second-layer formation provided that the second-layer adsorption occurs when the first layer is almost complete. The results for the argon/rutile (85 K) system obtained by these authors are compared in Table 3. The comparison is accomplished by translating the adsorption energy scale by 500 cal mol^{-1} so that the lowest-energy peak corresponds to the main peak obtained by Drain and Morrison and the lowest-energy peak of Oh and Kim. Although the translation produces reasonable agreement in the peak positions, the extent of the surface that corresponds to the second peak is in disagreement with the value obtained by Oh and Kim.

Rudziński and Jaroniec explain the differences with the result of Dormant and Adamson and also Drain and Morrison in terms of the Jovanović local isotherm. Since all the treatments ignore lateral interactions, the patchwise

142 B. K. Oh and S. K. Kim, *J. Chem. Phys.*, 1977, **67**, 3416.

Table 3 *Comparison of the two results that show a three-peaked heterogeneity function*

	Peak			
	1	2	3	*Ref.*
Adsorption energy	2.58	3.05	3.50	*a*
/kcal mol⁻¹	2·50	3.00	3.20	*b*†
Fraction of	0.657	0.228	0.115	*a*
surface	0.55	0.42	0.03	*b*

† Adsorption energies increased by $500\,cal\,mol^{-1}$ from the values shown by Rudzinski and Jaronieć (ref. *b*)

a B. K. Oh and S. K. Kim, *J. Chem. Phys.*, 1977 **67**, 3415. *b* W. Rudzinski and M. Jaronieć, *Surf. Sci.*, 1974, **42**, 552

approximation is not involved. However if the three peaks are associated with the cleavage planes of rutile *i.e.* (100), (111), (110) they should give rise to phase-transition steps in the isotherm when adsorption occurs below the two-dimensional critical temperature of the adsorbate. This has been observed for the krypton/silver iodide (77.9 K) system and does lead to adsorption energy distributions showing several well defined peaks.[63] Unfortunately krypton isotherms were not measured on the sample of Drain and Morrison. It is difficult to reconcile the three-peaked distribution with the accurate calorimetry work of Drain and Morrison especially in view of the results obtained by Dormant and Adamson.

Adsorption on Silica. The most widely studied isotherm data are those of Aristov and Kiselev[143] for the adsorption of nitrogen (and argon) on maximally hydroxylated non-porous and wide-pored silica. Identical absolute isotherms for argon and nitrogen (*i.e.* isotherms reduced so that adsorption amounts are presented in units of $mol\,m^{-2}$) were found applicable to a variety of maximally hydroxylated silica adsorbents. The absolute isotherm for argon adsorption was found to be insensitive to the degree of hydroxylation, while that of nitrogen was significantly sensitive. This behaviour is not surprising considering the different modes of interaction of the argon and nitrogen with the silica surface. However more extensive and accurate argon adsorption data should indicate some differences between adsorbents having various degrees of surface hydroxylation. The absolute nitrogen isotherm has been analysed by several investigators[30,118,120,135] all using local isotherms that do not allow for lateral interactions, *i.e.* Langmuir and Jovanović. All the results clearly demonstrate two peaks, although the position of the lower-energy peak is slightly uncertain.[118,120] In the unlikely event that the Hill–de Boer equation is applicable as the local isotherm for this system, House[135] as shown that two peaks are still obtained. It has been suggested[119,135] that the peaks correspond to distinct kinds of hydroxy-groups on the silica surface. This is certainly supported by the infrared spectroscopy[144] studies of these silica samples.

[143] B. G. Aristov and A. V. Kiselev, *Colloid J.*, 1965, **27**, 246.
[144] V. Ya Davydov, A. V. Kiselev, and L. T. Zhuravlev, *Trans. Faraday Soc.*, 1964, **60**, 2254.

The absolute argon isotherm has been studied by House[135] and leads to an exponential distribution function, similar to that obtained by van Dongen[34] for the krypton/aeroal system (at 77 K). The HILDA analysis and the regularization analysis (see Section 2) both indicate structure in the adsoption energy distribution, which may be associated with the surface properties that give rise to the two peaks in the nitrogen distribution; more extensive experimental data are required to substantiate this point. Hsu *et al.*[145] used the condensation approximation (and correction function[145]) to determine a single-peaked distribution, for the argon/aerosil system (75 K), which 'tailed' to higher adsorption energies. However the distribution was obtained by using the BET equation as the local isotherm and automatically corrected for multilayer adsorption effects. The problems associated with multilayer adsorption will be discussed in Section 6.

Other investigations of the silica surface have been completed including systems such as nitrogen/Fransil (77.5 K),[116] nitrogen, krypton/aerosil (77 K),[34] nitrogen, krypton/aeroal (77 K),[34] benzene/silica gel (303 K),[120] cyclohexane, cyclohexene/silica gel, and esterified silica gel,[146] cyclohexane and cyclohexene/wide-pore and narrow-pore silica gel,[123] nitrogen/silica gel (77.3 K),[119] nitrogen/silica (77 K),[128] and nitrogen/modified silica gels.[129] The gas-adsorption chromatography methods give results showing considerable detail, *e.g.* five adsorption energy peaks have been distinguished for the cyclohexane, cyclohexene/wide- and narrow-pore silica gels.[123]

5 The Random Heterogeneous Model of Adsorption

If a model of monolayer physical adsorption incorporates an account of adsorbate lateral interactions, the location as well as the distribution of the adsorption energy sites becomes important. This situation has been discussed so far in terms of the patchwise heterogeneity model of adsorption. In some circumstances this representation may not be a realistic model of the surface heterogeneity, indeed a completely random distribution of adsorption energies may be applicable, *e.g.* heterogeneity caused by impurity atoms in the surface. A correlation function, analogous to that used in liquid theory, could be formulated to describe any correlation in the position of the adsorption energy sites and so incorporate the patchwise and the random heterogeneity models as extremes.

Hill[17] developed a canonical ensemble approach to investigate the case of a random distribution of adsorption energies. He assumed that only interactions between adsorbed molecules on nearest-neighbour sites needed to be accounted for and developed an equation for the partition function:

$$\ln Q = \sum_i B_i \left[-\ln\left(\frac{\gamma}{\gamma+1}\right) + \frac{\alpha}{\gamma+1} \right] + \frac{n^1 \omega}{kT}, \qquad (105)$$

[145] C. C. Hsu, W. Rudzinski, and B. W. Wojciechowski, *Phys. Lett. A*, 1975, **54**, 365.
[146] R. Leboda, A. Waksmundzki, and S. Sokolowski, *Rocz. Chem.*, 1976, **50**, 1719; R. Leboda and S. Sokolowski, *J. Colloid Interface Sci.*, 1977, **61**, 365.

where B_i is the number of sites of adsorption energy U_i.

$$\alpha = -1/kT\left[\mu + \omega\frac{\left(\partial n^1\right)}{\left(\partial n_a\right)_{B,T}}\right]$$

$$\gamma = \exp[\alpha - (U_i/kT)]/f_i$$

μ is the chemical potential of the bulk gas and ω the nearest-neighbour interaction energy. n^1 is a configuration constant defined by Hill,[17] and f_i is the adsorbate partition function on the ith site.

The isotherm equation of Hill may be written in a more general form:

$$\theta_T(p, T) = \int_{U_l}^{U_h} \frac{F'(U)p/K}{\exp\left[-\dfrac{\omega}{kT}\left(\dfrac{\partial n^1}{\partial n_a}\right)_{B,T}\right] + p/K}, \qquad (106)$$

where K is defined in equation (78).

Hill proposed two approximations to obtain $(\partial n^1/\partial n_a)_{B,T}$:

(i) The Bragg–Williams approximation (molecules distributed as if $\omega = 0$) $(\partial n^1/\partial n_a)_{B,T} = Z\theta_T(p, T)$ where Z is the number of nearest neighbours.

(ii) The quasi-chemical approximations:*

$$\left(\frac{\partial n^1}{\partial n_a}\right)_{B,T} = \frac{Z}{2}\left\{1 - \frac{kT}{\omega}\ln\left[\frac{(\beta - 1 + 2\theta_T)(1 - \theta_T)}{(\beta + 1 - 2\theta)\theta_T}\right]\right\}, \qquad (107)$$

with $\beta = \{1 - 4\theta_T(1 - \theta_T)[1 - \exp(\omega/kT)]\}^{1/2}$. By assuming a rectangular distribution for $F'(U)$ and the quasi-chemical approximation, Hill[17] demonstrated isotherms exhibiting two phase changes; firstly a partial condensation on the high-energy sites followed by a second condensation near the completion of the monolayer. This later result has been substantiated by Gordon,[147] who used a computer-simulation approach to generate a graph of μ/kT *versus* θ and obtained 'steps' that were larger and in slightly different positions than those found by Hill.[17] This is an important consequence since it suggests first-order phase transitions may be possible on a surface with a wide and random distribution of adsorption energies. To the author's knowledge, this effect has not been associated with any experimental data so far published.

The question now arises as to whether the distribution derived by inverting equation (106) will be significantly different from that distribution evaluated assuming the patchwise model, *i.e.* can the two models be differentiated by analysing isotherm data? It is possible that both models will yield equally valid distributions leaving no criterion on which to choose between them. This has indeed been the case found by Rudzinski *et al.*[148] when the condensation approximation was applied to the solution of the patchwise integral equation

---

* Equation (106) is simplified using the relationship:

$$\exp\left[-\frac{\omega}{kT}\left(\frac{\partial n^1}{\partial n_a}\right)_{B,T}\right] = \left(\frac{2 - 2\theta}{\beta + 1 - 2\theta}\right)^z.$$

[147] R. Gordon, *J. Chem. Phys.*, 1968, **48**, 1408.
[148] W. Rudzinski, L. Lajtar, and A. Patrykiejew, *Surf. Sci.*, 1977, **67**, 195; W. Rudzinski, A. Patrykiejew, and L. Lajtar, *Surf. Sci.*, 1978, **77**, L655-L657.

and equation (106) using the Bragg–Williams approximation. This is equivalent to using the Fowler–Guggenheim equation to describe adsorption on patchwise and random heterogeneous surfaces. The ingenious method applied by Rudzinski *et al.*[148] leads to a general relationship between the random adsorption energy distribution, $F'_R(U)$, and the patchwise adsorption energy distribution, $F'_P(U)$ (within the limitations of the condensation approximation):*

$$F'_R(U_c^R) = F'_R(U_c^P)/[1 + Z\omega F'_P(U_c^P)], \qquad (108)$$

where

$$U_c^R = RT \ln A^0 - Z\omega\theta_T - RT \ln p \text{ and}$$
$$U_c^P = U_c^R - Z\omega(1 - 2\theta_T)/2.$$

The theory was demonstrated using the Dubinin–Radushkevich total isotherm for nitrogen/graphite (78 K) data and as anticipated, because of the form of equation (108), the patchwise distribution was narrower with a larger maximum frequency. This behaviour was also observed using different local isotherms in the patchwise model; nevertheless the adsorption energy range for the random distribution obtained by Rudzinski *et al.* was wider by an amount $Z\omega$ compared with the distribution obtained using the patchwise model. House[149] has used the regularization method (see Section 2) to solve equation (106) without the condensation approximation and using the Bragg–Williams and quasi-chemical equations. The results obtained by using the Bragg–Williams approximation were in good agreement with the distribution derived using the transform equation (108). The quasi-chemical approximation makes only a small difference to the distribution by sharpening the peaks with a change in the peak height increasing with decreasing adsorption energy.

The patchwise and random heterogeneity models have also been compared[150] by generating isotherms for various distributions of adsorption energy using a simple lattice model originally suggested by Nicholson.[150] The results indicate that a random distribution of adsorption energies has less of a 'smoothing' effect on the overall isotherm than the patchwise model with the same adsorption energy distribution. Recently Rudzinski and Lajtar[151] have made a systematic study of the behaviour of isotherms, isosteric enthalpies, and heat capacities calculated for the random heterogeneity model by assuming various exponential distributions of adsorption energy and accounting for adsorbate self-interactions by using the quasi-chemical approximation.

Another approach has been taken by Steele[18] and Pierotti and Thomas.[152] They use the gas–solid virial expansion, a general theory which is not restricted to localized adsorption; Pierotti and Thomas included corrections for the

---

*Rudzinski *et al.*[148] wrote $Z\omega = 4RT_c$, where T_c is the two-dimensional critical temperature of the adsorbate.

[149] W. A. House, *J. Colloid Interface Sci.*, 1978, **67**, 166.
[150] D. Nicholson and R. G. Silvester, *J. Colloid Interface Sci.*, 1977, **62**, 447; D. Nicholson, *J. Chem. Soc., Faraday Trans. 1*, 1975, **71**, 238.
[151] W. Rudzinski and L. Lajtar, *J. Chem. Soc., Faraday Trans. 2*, 1981, **77**, 153.
[152] R. A. Pierotti and H. E. Thomas, *J. Chem. Soc., Faraday Trans. 1*, 1974, **9**, 1725.

gas-phase imperfections. The derivation and treatment of the analysis is particularly lengthy and the reader is referred to the original papers for the details. The results of Steele for the argon/carbon black P-33(2700 K) and 'Black Pearls No. 71', obtained by using the random and patchwise distributions combined with a Gaussian adsorption energy distribution, are inconclusive in distinguishing between the two models. However Pierotti and Thomas found that for the argon/carbon black (Black Pearls) system a Gaussian and random-site energy distribution gave a more realistic result, although there was some question whether the data were sufficiently extensive to warrant such a detailed analysis. Ripa and Zrablich [153] have also used the gas-virial theorem to develop a model of adsorption that incorporates a Gaussian distribution of adsorption energies and a multivariant Gaussian function to describe the spatial distribution of adsorption energies. By incorporating a stochastic model to describe the heterogeneity, they postulate the validity of the statistical homogeneity hypothesis (*i.e.* that any macroscopic portion of the surface has all the information about the total surface) and were able to introduce a correlation function for the adsorption energy distribution. Again the additional variable of the correlation length severely limits the application of such a model to real systems.

A recent study [154] that adopts the associated adsorbate model of Berezin and Kiselev [105,106] may prove particularly useful in elucidating heterogeneity topology. In their theoretical treatise, Jaroniec and Borowko [154] assumed a dual adsorbent surface and the possibility of double associates. Thus by formulating quasi-chemical reactions:

$$A + S_i \overset{K_i}{\rightleftharpoons} A_{i(1)}.$$

$$A_{i(1)} + A_{i(1)} \overset{L_i}{\rightleftharpoons} A_{i(2)},$$

$$A_{i(1)} + A_{j(1)} \overset{L_{1,2}}{\rightleftharpoons} A_{i,j(2)},$$

where A denotes an adsorbate molecule and S_i ($i = 1, 2$) denotes an adsorption site of the *i*th type, they were able to express the total surface coverage in terms of the equilibrium constants K_i, L_i, and $L_{1,2}$ that correspond to the three equilibrium processes. When L_1, L_2, $L_{1,2} \neq 0$, the surface has a random configuration of adsorption heterogeneities. Two other cases were considered: (*a*) $L_1 = L_2 = 0$, $L_{1,2} \neq 0$, a 'chessboard' configuration and (*b*) L_1, $L_2 \neq 0$ and $L_{1,2} = 0$, a patchwise distribution. Unfortunately the development has not reached a point where the adsorption energy distribution can be determined from isotherm data. However the method does give an insight into the effect of heterogeneity topology on the adsorption isotherm shape. [154]

[153] P. Ripa and G. Zrablich, *J. Phys. Chem.*, 1975, **79**, 2118.
[154] M. Jaroniec and M. Borowko, *J. Colloid Interface Sci.*, 1978, **63**, 362; J. K. Garbacz, J. Siedkewski, and M. Jaroniec, *ibid.*, 1979, **72**, 344.

6 The Role of Surface Heterogeneity in Multilayer Adsorption

Very little work has been published on this subject, although it is well known that multilayer adsorption may begin at low adsorbate surface coverages. The majority of treatments have adopted the patchwise heterogeneity model or have not accounted for adsorbate self-interactions and extended the analysis by using a multilayer adsorption isotherm as the local isotherm in the integral equation. The effects of multilayer adsorption in the patchwise heterogeneity analysis with a monolayer local isotherm have been demonstrated by van Dongen[34] and House.[102,135] In general adsorption on the highest-energy patches merely presents a 'new' lower adsorption energy surface on which multilayer adsorption can occur and which contributes to the adsorption process at higher bulk gas pressures. Hence if mixed monolayer–multilayer adsorption is likely, the final distribution contains information concerning both the gas–surface interactions and the adsorption energies associated with multilayer adsorption. This has been illustrated by House and Jaycock[102] by analysing the BET equation by substituting different values of c and using HILDA with the Hill–de Boer local isotherm. At values of $c > 50$ a sharp increase in the distribution frequency occurs at the onset of multilayer formation and gives a ratio: $n_m^{HILDA}/n_m^{BET} \approx 1.09$, which is in agreement with the results of Ross and Olivier[6] and Waldsax.[155] With decreasing c the beginning of multilayer formation becomes less distinct and eventually the monolayer capacity is indeterminable.

An admirable attempt to account for monolayer–multilayer adsorption on heterogeneous surfaces was made by Halsey[156] in 1951. He assumed a patchwise model of adsorption with an exponential distribution of adsorption energies decaying with the third power of the distance from the surface. By using a 'step' approximation to determine the minimum adsorption energy at a particular pressure, he was able to formulate a series expansion for the adsorbate surface coverage and hence generate various adsorption isotherms. Unfortunately the model is severely limited because it ignores the effects of lateral interactions between adsorbate molecules and in assuming a third-power decay equation, is only applicable to adsorption systems where gas–solid dispersion interactions determine the adsorption energy.

A similar approach to Halsey's was also devised by D'Arcy and Watt[157] by extending Langmuir's equation [*i.e.* equation (3)] but including an extra term to account for multilayer formation (assumed to occur at a constant adsorption energy independent of the surface heterogeneity), *i.e.*

$$\theta_T(p) = \sum_{i=1}^{m} \frac{N_i p}{K_i + p} + \frac{Dp}{K_m - p}, \tag{109}$$

where N_i is the number of patches with adsorption energy characterized by K_i and D is the number of patches available for multilayer adsorption. An optimization procedure was used to determine N_i, K_i, K_m, and D with i usually

[155] J. C. R. Waldsax, Ph.D. Thesis, Loughborough University of Technology, 1970, p. 71.
[156] G. D. Halsey, jun., *J. Am. Chem. Soc.*, 1951, **73**, 2693.
[157] R. L. D'Arcy and I. C. Watt, *Trans. Faraday Soc.*, 1970, **66**, 1236.

small, *i.e.* $i = 1$ or 2. The method has been extended by Rudzinski *et al.*[158] leading to a simple graphical technique whereby a plot of $(\partial \ln p/\partial \ln \theta_T)_T$ *versus* p leads to an estimate of the adsorption energies and the sequence of filling of monolayer and multilayer patches.

Again adsorbate lateral interactions are ignored and the simplifying assumption that at any pressure either monolayer or multilayer adsorption takes place leads to some uncertainity in the validity of the results. However the model may be realized for systems possessing distinct strongly and weakly binding patches and is worth considering as a supplement to other methods of analysis.

The most popular method of analysing multilayer adsorption isotherms has undoubtably been through the BET equation. It is not surprising that this model of multilayer adsorption, originally postulated for a homogeneous surface, should have been adopted to investigate multilayer adsorption on a heterogeneous surface. The BET equation provides poor agreement with experimental data at higher relative vapour pressures even on homogeneous surfaces, and a number of modifications have been suggested.[159,160] Dormant and Adamson[97] however accepted the basic BET formulation as the local isotherm and used their numerical iterative method to analyse the argon/rutile (85 K) isotherm data of Drain and Morrison [see Figure 12(a)] and obtained good agreement with the calorimetric distribution of these authors (also derived by assuming adsorbate lateral interactions to be negligible). Dormant and Adamson[97] point out that the value of the adsorption energy, U_c, obtained from the c of the BET analysis is lower than the average, $\bar{U}$, calculated from:

$$\bar{U} = (1/A_s) \int_{U_1}^{U_h} F'(U)U \, dU, \tag{110}$$

where A_s is the surface area. This reflects the range of 'fit' of the BET equation (*i.e.* U_c represents the average adsorption energy of the patches filled in the BET pressure range).

The results of Dormant and Adamson have also been discussed by Jaroniec and Rudzinski.[161] They present an elegant procedure for determining the adsorption energy distribution by employing a modified BET equation (Anderson's equation[160]) in which a finite number of adsorbed layers are allowed:

$$\theta_m(p) = \frac{ck(p/p^0)}{[1 - k(p/p^0)][1 + (c-1)kp/p^0]}. \tag{111}$$

The method is essentially numerical and leads to a series form for $F'(U)$ in terms of orthogonal Legendre polynomials and coefficients obtained by a least-squares 'fit' to the experimental data. Hsu *et al.*[162] have also adopted the modified BET equation (111) and by combining it with a Maxwell–Boltzmann

[158] W. Rudzinski, J. Tóth, and M. Jaroniec, *Phys. Lett. A*, 1972, **41**, 449.
[159] G. L. Pickett, *J. Am. Chem. Soc.*, 1945, **67**, 1958; T. L. Hill, *Adv. Catal.*, 1952, **4**, 232.
[160] R. B. Anderson, *J. Am. Chem. Soc.*, 1946, **68**, 686; S. Brunauer, J. Skalny, and E. E. Bodnor, *J. Colloid Interface Sci.*, 1969, **30**, 546; M. Dole, *J. Chem. Phys.*, 1948, **16**, 25.
[161] M. Jaroniec and W. Rudzinski, *Acta Chim. Acad. Sci. Hung.*, 1976, **88**, 351.
[162] C. C. Hsu, W. Rudzinski, and B. W. Wojciechowski, *J. Chem. Soc., Faraday Trans 1*, 1976, **72**, 453.

distribution [equation (37)] have evaluated 'best fit' parameters[35,162] to characterize the surface heterogeneity. They analysed the argon/rutile (85 K) data of
Drain and Morrison, argon/aerosil (77.5 K), nitrogen/silica gel—Fransil (77 K),
and nitrogen/graphite (78 K) using the integral equation of adsorption with the
Langmuir and modified BET as local isotherms as well as the modified BET
and BET as overall isotherms. In the case of the more homogeneous surfaces,
graphite and aerosil, both the modified BET and the heterogeneity analysis
with the modified BET resulted in equally poor agreement in reproducing the
total experimental isotherm. This disagreement was attributed to the nature of
the model employed and indicated that the adsorbate was perhaps mobile or
that lateral interactions were important. For the argon/rutile and
nitrogen/silica gel systems, the heterogeneity analysis with the modified BET
equation gave good agreement with the experimental isotherm and indicated
the surface to be very heterogeneous.

Cerofolini[163] has also incorporated a form of the BET equation into the
integral equation describing multilayer adsorption on a heterogeneous surface.

$$\theta_{Tm}(p) = \int_0^{-\infty} \theta_m(p, U) F'(U)\, dU, \tag{112}$$

where θ_{Tm} is the total isotherm describing monolayer and multilayer adsorption and θ_m is the local multilayer isotherm. When:

$$\left.\begin{aligned}
\theta_m(p, U) &= f(p)\theta(p, U), &\text{(a)}\\
\theta_m(p, U) &= g(p) + \theta(p, U); &\text{(b)}
\end{aligned}\right\} \tag{113}$$

then:

$$\left.\begin{aligned}
\theta_{Tm}(p) &= f(p)\theta_T(p), &\text{(a)}\\
\theta_{Tm}(p) &= g(p) + \theta_T(p). &\text{(b)}
\end{aligned}\right\} \tag{114}$$

and

[where $\theta_T(p)$ is the total isotherm describing monolayer adsorption and $\theta(p, U)$
the monolayer local isotherm].

If the BET constant $c \gg 1$, then $\theta_m(p, U)$ may be written in the form of
equation (113a) with $\theta(p, U)$ as the Langmuir function. Cerofolini[163] showed
that if the Dubinin–Radushevich equation describes monolayer adsorption,
$\theta_T(p)$, then the total isotherm is given by equation (114a). Adamson and
Ling[40] originally applied equation (114a) using the same function as Cerofolini, *i.e.*; $f(p) = 1/(1 - p/p^0)$ to apply a multilayer correction to adsorption data.
The function suggested by Cerofolini does not lead to a new adsorption energy
distribution; 'fitting' the Dubinin–Radushkevich equation to the lower
region of the experimental isotherm to determine the monolayer adsorption
only ignores the problem of mixed monolayer–multilayer adsorption.

The method of Cerofolini has been extended by Rudzinski *et al.*[164] and
tested using the Dubinin–Radushkevich and the Tóth equations to describe the
monolayer adsorption. The function $f(p)$ was derived for the BET and modified BET equations and used to compute the multilayer isotherms [equation

[163] G. F. Cerofolini, *J. Low Temp. Phys.*, 1972, **6**, 473.
[164] W. Rudzinski, S. Sokolowski, M. Jaronieć, and J. Tóth, *Z. Phys. Chem. (Leipzig)*, 1975, **256**, 273.

(114a)]. The agreement between these isotherms and the experimental data was satisfactory for the hexane/graphite MT1 (at 293 K) system obeying the Dubinin–Radushkevich equation in the submonolayer region and for benzene/graphite MT1 (293 K) obeying the Tóth equation.

The problem of analysing adsorption data for systems where mixed monolayer–multilayer adsorption is important remains a challenge for future research. In cases where the surface heterogeneity is determinable from monolayer data, the effects of heterogeneity on multilayer adsorption is of considerable interest. However it is unlikely that a satisfactory account of multilayer adsorption on heterogeneous surfaces will be possible without including terms for adsorbate self-interactions. Even for homogeneous surfaces, the additional difficulties of allowing for adsorbate lateral interactions have stifled progress. One tractable solution that has met with some success[34] is an extension of the Hill–de Boer equation to an n-layer two-dimensional mobile model discussed by Broekhoff and van Dongen.[103] However the extent of the decrease in the adsorption energy with adsorbate layer thickness introduces another uncertainty. The Fowler–Guggenheim equation may also be generalized to account for multilayer formation[103] but so far has not been included in a heterogeneity analysis. The lattice model of Nicholson[150] also deserves further evaluation.

7 Heterogeneity Effects in Adsorption from Solution

It is not the objective of this discussion to present a comprehensive view of adsorption from solution. The theory is not as advanced as that of gas adsorption and consequently the magnitude and effects of substrate heterogeneity are less certain. No mention will be made concerning adsorption of solutes from pure liquids or binary mixtures since these generally involve quite complex molecular species and adsorbate configurations become a further problem. Neither will adsorption from electrolyte solutions be discussed, for which although of paramount importance in colloid theory, the surface heterogeneity effects are extremely difficult to disentangle from other physical properties that determine the behaviour of these systems.

Following the example set by gas-adsorption studies, a useful approach is to examine the more ideal systems such as simple binary organic liquids in equilibrium with inert homogeneous solids. If these systems can be understood, then it is probable that the effects of surface heterogeneity in more complex systems will be more easily discerned. However as a precursor to this, it is worthwhile considering adsorption from binary gas mixtures as this serves as a link with pure gas adsorption theory.

Comparisons with Adsorption from Binary Gas Mixtures.—Compared to pure gas adsorption, the amount of research concerning adsorption of binary gas mixtures on heterogeneous surfaces is sparse. This is not surprising as the latter requires an understanding of the simpler pure gas systems; in fact a popular approach in verifying mixed gas adsorption theories is to predict mixed gas isotherms from pure gas isotherms measured at the same temperature.

The analogy with adsorption from solution is readily demonstrated by considering as a first approximation Langmuir's theory applied to the adsorption of a gas component 1 from a mixture of gases (1 and 2) at a total pressure p chosen so that one of the components (say 2), when adsorbed from the pure gas phase, forms a monolayer.[165]

$$n^0_{1(2)} = n_m(1 + K_{12}p_{21})^{-1}, \tag{115}$$

where n_m is the monolayer capacity for the adsorption of pure component 1 (or if the molecules of the two gases are of equal size then $n^0_{1(2)} + n^0_{2(1)} = n_m$); $n^0_{1(2)}$ is the number of moles of component 1 adsorbed from the mixture at a total pressure p ($p = p_1 + p_2$); $K_{12} = K_1/K_2$ with $K_1 = A^0_1 \exp(-U_1/RT)$ [see equation (2)] and $p_{21} = p_2/p_1$. Equation (115) may be transformed:

$$n^0_{1(2)} = n_m\{1 + [A^0_1/(xA^0_2)]\exp[(U_2 - U_1)/RT]\}^{-1}, \tag{116}$$

where $x = x_1/(1 - x_1)$ and x_1 is the mole fraction of component 1 in the bulk gas phase. This may be compared to the solution adsorption equation obtained by Everett[166] for an ideal bulk liquid and adsorbed monolayer: (surface phase quantities are shown with a superscript σ and where necessary the bulk liquid phase is signified with a superscript l)

$$x^\sigma_1 = \left[1 + \frac{1}{cx}\exp\left(\frac{\varepsilon_2 - \varepsilon_1}{RT}\right)\right]^{-1}, \tag{117}$$

where $c = \exp[(S^*_1 - S^*_2)/R]$ with $S^*_i = S^l_i - S^\sigma_i$ and S^*_i denotes the thermal entropy of adsorption;[166] $\varepsilon_i = \varepsilon^l_i - \varepsilon^\sigma_i$, where ε^l_i and ε^σ_i are the molar internal energies of the molecules in the bulk liquid and surface phases, respectively.

Both equations (116) and (117) may be generalized for a patchwise heterogeneous surface. In the case of equation (116) this becomes:[165]

$$n_{T,1(2)}(p_{12}) = n_m\int_{-\infty}^{+\infty}\left\{1 + \frac{A^0_1}{p_{12}A^0_2}\exp[-\tilde{U}/(RT)]\right\}^{-1}F'(\tilde{U})\,d\tilde{U}, \tag{118}$$

where $n_{T,1(2)}(p_{12})$ is the total amount of component 1 adsorbed from the mixture, and $\tilde{U} = U_1 - U_2$. Therefore with isotherm data measured at a constant pressure, [and in the case of equation (118), at full monolayer coverage], the total adsorption of component 1 from 2 is a function of $\tilde{U}$, the adsorption energy difference, rather than the individual adsorption energies of the pure components.

Equations (116) and (118) are special cases for mixed gas adsorption. The more general adsorption isotherm for a patchwise heterogeneous surface may be written as:[167]

$$\theta_T(p_1, p_2, T) = \int_{\Delta_1}\int_{\Delta_2}\theta(p_1, p_2, U_1, U_2)F'(U_1, U_2)\,dU_1\,dU_2, \tag{119}$$

where Δ_i determine the range of adsorption energies applicable to each

[165] M. Jaroniec, *J. Colloid Interface Sci.*, 1977, **59**, 230.
[166] D. H. Everett, *Trans. Faraday Soc.*, 1964, **60**, 1803.
[167] M. Jaroniec and W. Rudzinski, *Phys. Lett. A*, 1975, **53**, 59.

adsorbate. Although each patch is characterized by two adsorption energies, it is possible that a patch that is homotattic with respect to one adsorbate may be heterogeneous with respect to another. The inversion of equation (119) to determine the bivariant distribution is extremely difficult and has not been achieved. However some progress has been made by expressing $F'(U_1, U_2)$ as a generalized bivariant Gaussian distribution[168] or log-normal distribution.[169] In the former case:

$$F'(U_1, U_2) = [2\pi\gamma_1\gamma_2(1-\rho^2)^{1/2}]^{-1} \exp\left[\frac{U_1^{*2} - 2\rho U_1^* U_2^* + U_2^{*2}}{-2(1-\rho^2)}\right], \quad (120)$$

where $U_i^* = (U_i - \bar{U}_i)/\gamma_i$ with $i = 1$ or 2 and $\bar{U}_1$, $\bar{U}_2$ are the median values of U_1 and U_2, respectively; γ_1, γ_2 are the variances of the distribution (heterogeneity parameters) and ρ is the correlation coefficient. When $\rho = 0$ there is no correlation between the adsorption energies U_1 and U_2; for chemically similar adsorbates a strong correlation is expected, *i.e.* $\rho \approx 1$. Hoory and Prausnitz[168] found $\rho = 1$ for (ethene + ethane)/activated carbon (293 K, 751 Torr). Jaronieć and Rudzinski[167] determined $\rho = 0.95$ for (ethene + ethane)/charcoal (333 K). This indicates that the two single-component adsorption energy distributions are related through a simple translation along the adsorption energy axis. Indeed if equation (118) is written for any constant total pressure, p, the function $n_{T,1(2)}(p_{12})$ is still dependent on $\bar{U}$: for a simple adsorption energy translation $F'(U_1)$ to $F'(U_2)$, the function $F'(\bar{U})$ is a Dirac-delta function and $n_{T,1(2)}(p_{12})$ becomes independent of the shape of $F'(U_1)$ or $F'(U_2)$ and so $n_{T,1(2)}(p_{1,2})$ is evaluated as if the surface was homogeneous. This explains the success of the analysis of Jaronieć[165] for adsorbed hydrocarbon mixtures on Nuxit charcoal (see propane/Nuxit distribution function, Figure 4). The relationship between the distributions for these systems is a simple linear translation along the energy scale, the distribution shapes remaining the same.[165,170—172] Jaronieć found that the experimental results agreed well with the theory for adsorption on a homogeneous surface.

A number of attempts have been made to solve equation (119) by making the approximation:[173—175]

$$F'(U_1, U_2) = F'(U_1)F'(U_2), \quad (121)$$

i.e. $\rho = 0$ in equation (120). To assume no correlation between the adsorption energies seems a gross approximation. Nonetheless, it does lead to simplifications if the mixed gas Jovanović[176] equation is adopted as $\theta(p_1, p_2, U_1, U_2)$ in equation (119).[174] A number of other approaches to the solution of equation

[168] S. E. Hoory and J. M. Prausnitz, *Chem. Eng. Sci.*, 1967, **22**, 1025.
[169] M. Jaronieć and M. Borowko, *Surf. Sci.*, 1977, **66**, 652.
[170] M. Jaronieć and J. Tóth, *Colloid Polym. Sci.*, 1976, **254**, 643.
[171] M. Jaronieć, *Colloid Polym. Sci.*, 1977, **255**, 32.
[172] J. Tóth, *Acta Chim. Acad. Sci. Hung.*, 1971, **69**, 311.
[173] M. Jaronieć and J. Tóth, *Acta Chim. Acad. Sci. Hung.*, 1976, **91**, 153.
[174] M. Jaronieć, *J. Colloid Interface Sci.*, 1975, **53**, 422.
[175] M. Jaronieć, *J. Colloid Interface Sci.*, 1977, **59**, 371.
[176] M. Jaronieć, *Chem. Zvesti.*, 1975, **29**, 512.

(119) have been investigated by Jaroniec;[177—179] in certain cases the results allow mixed gas adsorption isotherms to be predicted from a knowledge of the pure gas adsorption characteristics.

Adsorption from Binary Liquid Mixtures.—There is no doubt that surface heterogeneity can be important when considering adsorption from solution,[180—182] however very little progress has been made in accounting for the effects in a quantitative manner. Following the research of Schuchowitzky[183] and Sisková and Erdös,[184] the integral equation of adsorption [*e.g.* equation (5)] has been the only model applied to surface heterogeneity investigations by adsorption from solution. The evaluation of the local excess isotherm function may only be achieved by comparing theory with adsorption data obtained for a homogeneous surface, *e.g.* Graphon.

Many experimental investigations of adsorption from binary liquids have been completed, although few have been satisfactorily interpreted using molecular models. Nagy and Schay[185] demonstrate the variety of shapes of the combined or surface excess isotherms obtained for different systems. Several adsorption theories have been postulated to explain some of these isotherms[186,187] but the most successful and unified are those derived on a thermodynamic basis by Everett.[166,188,189] The simplest model is for 'ideal' adsorption from a perfect binary mixture of molecules of equal size in equilibrium with homogeneous surfaces. Further developments described adsorption from regular mixtures of equal size molecules (1:1, regular solution), from athermal mixtures of molecules of different sizes (1:r, athermal solution where r is the ratio of the molecular sizes) and from non-athermal mixtures of molecules of different sizes (1:r, non-athermal solution).

The equation for 'ideal' adsorption has been widely applied in the form:[166,190-193]

$$\frac{x_1 x_2}{n^0 \Delta x_1/m} = \frac{m}{n^\sigma}\left(x_1 - \frac{1}{K-1}\right), \qquad (122)$$

where n^0 is the total number of moles of solution, x_1 and x_2 are the mole fractions of components 1 and 2 in the solution in equilibrium with the

[177] M. Jaroniec, *Z. Phys. Chem. (Leipzig)*, 1976, **257**, 449.
[178] M. Jaroniec and J. Tóth, *Colloid Polym. Sci.*, 1976, **254**, 643; *ibid.*, 1977, **255**, 32.
[179] M. Jaroniec, *Colloid Polym. Sci.*, 1977, **255**, 176.
[180] G. Delmas and D. Patterson, *J. Phys. Chem.*, 1960, **64**, 1829.
[181] M. T. Coltharp and N. Hackermann, *J. Colloid Interface Sci.*, 1973, **43**, 176.
[182] B. R. Puri, D. D. Singh, and B. C. Kaistha, *Carbon*, 1972, **10**, 481.
[183] A. Schuchowitzky, *Acta Physicochim., URSS*, 1938, **8**, 531.
[184] M. Sisková and E. Erdös, *Collect. Czech. Chem. Commun.*, 1960, **25**, 1729, 2599.
[185] L. G. Nagy and G. Schay, *Acta Chim. Acad. Sci. Hung.*, 1963, **39**, 365.
[186] G. Schay, in 'Surface Colloid Science', ed. E. Matijevic, John Wiley and Sons, 1969, p. 155.
[187] G. Schay, L. G. Nagy, and T. Szekrenyesy, *Period. Polytech., Univ. Budapest*, 1962, **6**, 91.
[188] D. H. Everett, *Trans. Faraday Soc.*, 1965, **61**, 2476.
[189] D. H. Everett, in 'Colloid Science' ed. D. H. Everett (Specialist Periodical Reports), The Chemical Society, London, 1978, Vol. 1, Ch. 2.
[190] S. Sircar and A. L. Myers, *J. Phys. Chem.*, 1970, **74**, 2828.
[191] T. Kaguja, Y. Sumida, and T. Tachin, *Bull. Chem. Soc. Jpn.*, 1972, **45**, 1643.
[192] G. Schay and L. G. Nagy, *J. Colloid Interface Sci.*, 1972, **38**, 302.
[193] M. T. Coltharp, *J. Colloid Interface Sci.*, 1976, **54**, 311.

adsorbent, m is the mass of the adsorbent, n^σ the number of moles of the adsorbed phase, Δx_1 is the change in the solution composition after adsorption, and K is the 'separation factor', $K = x_1^\sigma x_2 / x_1 x_2^\sigma$. Equation (122) may be rewritten to give equation (117) $[x_1^\sigma = (n^0 \Delta x_1 / n^\sigma) + x_1$ and assuming c to be constant and independent of the adsorption energy]. Equation (117) has also been derived by Sisková and Erdös.[184] If equation (117) is now incorporated in the integral equation and the distributions $F'(\varepsilon_1)$ and $F'(\varepsilon_2)$ are related through a simple translation of the energy axis, the $F'(\bar{\varepsilon})$, (where $\bar{\varepsilon} = \varepsilon_2 - \varepsilon_1$), becomes a Dirac-delta function, $\delta(\bar{\varepsilon}^1)$ so that the excess isotherm is determined by the local isotherm with a $\bar{\varepsilon}$ value of $\bar{\varepsilon}^1$. The same argument applies if the regular solution monolayer model of Everett is used as the local isotherm.[188]

Unfortunately the systems which are most ideal, because of the close similarity of the components physical and chemical properties, are the most difficult to investigate experimentally owing to the small difference in their adsorption. One of the most thoroughly investigated mixtures is cyclohexane + benzene. This has been classified as a near-regular (1:1) solution by Ash *et al.*,[194] indeed these authors report good agreement of their data for (cyclohexane + benzene)/Graphon (298–328 K), with the theory of Everett for regular solutions but with an α^σ parameter,[188] [defined as $\alpha = N_0 Z(Q_{12} + 1/2(Q_{11} + Q_{22})$, where Q_{ij} is the energy of interaction between a pair of molecules (i, j) on nearest-neighbour lattice sites and Z is the number of nearest neighbours], equal to zero. This is interpreted in terms of the relative orientation of the adsorbed benzene.[194] It is interesting to compare this result with that obtained by Ościk *et al.*[195] for (benzene + cyclohexane)/silica gel (293 K) and active carbon (293 K). These authors use equation (117) as the local excess isotherm function in the integral equation [analogous to equation (118)] for adsorption on a heterogeneous surface and represent the total excess isotherm by the Dubinin–Radushkevich equation. Using the same techniques employed in gas adsorption, they invert the integral equation using a Stieltjes transform method to obtain an expression for $F'(\bar{\varepsilon})$. The condensation method applied by Cerofolini may also be used to obtain an approximate solution to $F'(\bar{\varepsilon})$. For the (benzene + cyclohexane)/silica gel system, a dual peaked distribution is obtained with a majority of the distribution indicating the preferential adsorption of benzene. One would expect that if the surface of the silica gel was relatively homogeneous with respect to cyclohexane adsorption, then $F'(\bar{\varepsilon})$ should reflect the benzene/silica gel distribution function. Because of the π-electron donating character of benzene, this function is expected to be similar to that of nitrogen/silica gel (with a specific quadrupole interaction with the surface hydroxyls). Indeed both distributions show two peaks. The same system was also analysed by Ościk *et al.*[196,197] using the Stieltjes transform method but writing the total surface excess isotherm as a linear combination of Freundlich isotherms. In this instance a single-peaked distribution was obtained. The difficulty with these results is that the authors are attributing a

[194] S. G. Ash, R. Bown, and D. H. Everett, *J. Chem. Soc., Faraday Trans. 1*, 1975, **71**, 123.
[195] J. Ościk, A. Dabrowski, and M. Jaroniec, *J. Colloid Interface Sci.*, 1976, **56**, 403.
[196] J. Ościk, W. Rudzinski, and A. Dabrowski, *Rocz. Chem.*, 1974, **48**, 1991.
[197] W. Rudzinski, J. Ościk, and A. Dabrowski, *Chem. Phys. Lett.*, 1973, **20**, 444.

small difference between the experimental data and the theoretical equations, obtained for a perfect monolayer by Everett's theory, to the effects of surface heterogeneity. This difference could perhaps equally well be explained in terms of the non-ideal character of the bulk liquid or adsorbed phase. In fact the results of Ash *et al.*[194] suggest that the regular solution (1;1) model would be more satisfactory as the local excess isotherm.

The experimental results of Bown[198] and Ash *et al.*[194] for adsorption of (benzene + cyclohexane)/Graphon (298–328 K) have been analysed by House[199] by assuming an ideal bulk and adsorbed phase and also assuming regular solution behaviour for the two phases. The regular solution theory provides the best description of the data if the adsorbed phase is assumed ideal. The behaviour is in agreement with the conclusions of Ash *et al.*[194] House[199] went on to use the same models of adsorption to analyse the data of Sircar and Myers[200] for the adsorption on silica gel (30 °C) and found that the disagreement between the experimental and theoretical values of the specific adsorption excess decreased by 35% if heterogeneity was taken into account. Although a numerical method (the regularization method) was used to solve the integral equation, no satisfactory criteria could be found to estimate the range of $\tilde{\varepsilon}$. The study did show that if heterogeneity effects are ignored and with α and n^σ as adjustable parameters, the optimum value of α was much lower than the value calculated from the (benzene + cyclohexane)/Graphon data.

These difficulties could perhaps be resolved by a systematic study of adsorption from a binary liquid mixture on graphitized carbon blacks and various partially annealed carbon blacks. If the surface heterogeneity changes are also investigated by gas adsorption of the pure components, this would vastly aid the interpretation of the $F'(\tilde{\varepsilon})$ distributions derived from liquid-adsorption studies. In spite of these reservations there seems to be no doubt about the important role of heterogeneity in some adsorption systems. The recent work by Dabrowski and Jaroniec[201] illustrates the problem of evaluating even the adsorption capacity of the surface phase from isotherm data. The agreement between theory and experiment is always better if heterogeneity is taken into account. However this is not sufficient a criterion to judge the validity of the heterogeneity model.

[198] R. Bown, Ph.D. Thesis, University of Bristol, 1973.
[199] W. A. House, *Chem. Phys. Lett.*, 1978, **60**, 169.
[200] S. Sircar and A. L. Myers, *AICHE J.*, 1971, **17**, 186.
[201] A. Dabrowski and M. Jaroniec, *Z. Phys. Chem.*, (*Leipzig*), 1980, **261**, 359; *J. Colloid Interface Sci.*, 1980, **77**, 571.

2

The Dubinin–Radushkevich (DR) Equation: History of a Problem and Perspectives for a Theory

BY G. F. CEROFOLINI

1 Introduction

When in 1947 Dubinin and Radushkevich proposed their adsorption isotherm, they were probably not aware of the crucial role that their equation, the DR equation, would later play in the physical chemistry of surfaces. Indeed, the DR equation was originally proposed on empirical grounds to describe the adsorption of some vapours on porous solids, with no mention of possible extensions or modifications. Only later, thanks to the work of many theoretical and experimental researchers, the study of the DR equation revealed the vital role of this empirical law in physical chemistry.

Though the DR problem does not involve any fundamental change in the description of Nature, it does however appear particularly interesting from the methodological point of view because it shows how a casual discovery became a theory which rapidly was endowed with appropriate language and mathematical tools.

2 The Proposal of Dubinin and Radushkevich

Dubinin and Radushkevich[1] observed that the volume V of vapour (at standard temperature and pressure) retained by a microporous solid at temperature T often obeys the following empirical law:

$$\ln V = A - B\varepsilon^2, \tag{1}$$

where A and B are two suitable constants, ε is the Polanyi potential, *i.e.* the work required to transfer a molecule from the gas phase at pressure p to the liquid phase at saturation pressure p^0,

$$\varepsilon = -k_B T \ln(p/p^0) \tag{2}$$

and k_B is the Boltzmann constant. The DR equation, together with its extension,[2] is still used to describe pore filling as pressure increases.

[1] M. M. Dubinin and L. V. Radushkevich, *Dokl. Akad. Nauk SSSR*, 1947, **55**, 331.
[2] M. M. Dubinin, in 'Adsorption-Desorption Phenomena,' ed. F. Ricca, Academic Press, London, 1972, p. 3.

Later Kaganer[3] observed that the constant A can be identified with $\ln V_m$, V_m being the monolayer capacity measured using the standard Brunauer–Emmett–Teller (BET) method. Bearing in mind this result, equation (1) can be written in the following way

$$\ln \vartheta = -B\varepsilon^2, \tag{3}$$

where ϑ is the fractional coverage. Equation (3) describes submonolayer adsorption: in fact, ϑ ranges from 0 for $p = 0$ ($\varepsilon = +\infty$) to 1 for $p = p^0$ ($\varepsilon = 0$). Relationship (3) is known, and will hereafter be referred to as the DR isotherm or DR equation.

The DR isotherm remained known to a relatively small group of researchers until Hobson called attention to its applicability to non porous surfaces. In a letter to the *Journal of Chemical Physics*,[4] Hobson, not without wonder, observed that 'the appearance of the Dubinin–Radushkevich equation in the present context (submonolayer adsorption of nitrogen on Pyrex) is surprising for two reasons. First, it is a particular equation within the Polanyi potential theory, which is a theory of condensation and might not be expected to apply to physical adsorption at very low coverage. Second, most of the adsorbents to which the Dubinin–Radushkevich equation have been applied[3] have been porous, whereas our conclusion suggests that Pyrex is non-porous for nitrogen. Thus, unless and until a basic derivation for this equation is provided, it can only be considered as a useful empirical relation'.

After the work of Hobson, in the sixties many others dedicated themselves to the task of testing the validity of the DR equation. These included Hobson himself and Armstrong,[5] Hansen,[6] Haul and Gottwald,[7] Endow and Pasternak,[8] Ricca and his school,[9] Schram,[10] Wightman and co-workers,[11] to name only a few.

This task was pursued with such success that Schram[12] in 1967 wrote that, although 'no satisfactory theoretical basis has been developed for the equation first given by Dubinin and co-workers for physical adsorption on microporous substrates, the general applicability of this isotherm equation to represent almost all available experimental data has definitely introduced its use in the last few years. Not only is the surface area easily computed, but also isosteric heats have been calculated from this isotherm by several authors'.

However, great difficulties are linked with equation (3): (i) It is strange that the adsorption isotherm is expressed in the submonolayer region in terms of Polanyi potential. (ii) By applying the Clausius–Clapeyron equation to equations

[3] M. G. Kaganer, *Dokl. Akad. Nauk SSSR*, 1957, **116**, 251.
[4] J. P. Hobson, *J. Chem. Phys.*, 1961, **34**, 1850.
[5] J. P. Hobson and R. A. Armstrong, *J. Phys. Chem.*, 1963, **67**, 2000.
[6] N. Hansen, *Vakuum Technik.*, 1962, **11**, 70.
[7] R. Haul and B. A. Gottwald, *Surf. Sci.*, 1966, **4**, 334.
[8] N. Endow and R. A. Pasternak, *J. Vacuum Sci. Technol.*, 1966, **3**, 196.
[9] F. Ricca, R. Medana, and A. Bellardo, *Z. Phys. Chem.*, (*Frankfurt*), 1967, **52**, 276; F. Ricca and A. Bellardo, *Z. Phys. Chem.*, (*Frankfurt*), 1967, **52**, 318.
[10] A. Schram, *Suppl. Nuovo Cimento*, 1967, **5**, 291.
[11] M. Troy and J. P. Wightman, *J. Vacuum Sci. Technol.*, 1970, **7**, 429; *ibid.*, 1971, **8**, 84; *ibid.*, 1971, **8**, 743. J. S. Chung and J. P. Wightman, *J. Vacuum Sci. Technol.*, 1972, **9**, 1470.
[12] A. Schram, *Suppl. Nuovo Cimento*, 1967, **5**, 309.

(3) and (2) the isosteric heat of adsorption can be computed; this quantity becomes infinite as p vanishes. (iii) The DR equation does not, in the low-pressure limit, converge on the Henry isotherm, to which any realistic adsorption isotherm must reduce as $p \rightarrow 0$.

The fact that the DR isotherm is expressed in terms of Polanyi's potential suggests that the structure of the adphase may, in some way, resemble that of the liquid phase. Working in this framework, *i.e.* assuming the adphase to be liquidlike, the behaviour of the isosteric heat of adsorption should be explained in terms of heterogeneity induced by adsorption.[12] However, feature (ii) cannot be ascribed to any known interaction between adsorbed molecules. The opposite possibility, that the DR equation was in reality due to fixed surface heterogeneity, was often considered in the past,[12] though feature (i) made it seem unlikely. Another point against this possibility was the great number of systems obeying the DR equation; in fact, (iv) if the DR equation is indeed due to surface heterogeneity, and taking into account all the possible ways in which heterogeneity can be manifested, then it is difficult to realize why so many vapour/solid systems are described by the same equation.

Each of features (i) to (iv) involved a remarkable amount of work, and the solutions to these problems led to the 'theory of adsorption on equilibrium surfaces'.

3 Surface Heterogeneity

The simplest way to consider surface heterogeneity is to assume that the surface consists of a collection of many patches, each being characterized by a constant adsorption energy q for a given gas. Assuming that the adsorption energy is the only parameter which varies from one patch to another, the heterogeneous surface can be parametrized by a distribution function $\varphi(q)$, where $\varphi(q)\,dq$ represents the fraction of surface with adsorption energy between q and $q + dq$.

If $\theta(p, q)$ is the local isotherm, *i.e.* the adsorption isotherm holding true on each energetically homogeneous patch supposed to be indefinitely extended, the overall adsorption isotherm is given by

$$\vartheta(p) = \int_{0}^{+\infty} \theta(p, q)\varphi(q)\,dq, \qquad (4)$$

where lower and upper integration limits have conventionally been put equal to 0 and $+\infty$, respectively, though $\varphi(q)$ differs from zero only in a narrow set Ω of energy values — the energy spectrum.

The idea leading to the integral representation (4), *i.e.* the assumption that the surface is patchwise heterogeneous, goes as far back as Langmuir's work[13] and obviously does not represent the only way to treat heterogeneity. For instance, patchwise heterogeneity can be seen in an intermediate position in this sort of space ordering: homogeneity $\rightarrow$ patchwise heterogeneity $\rightarrow$ random heterogeneity. Obviously, this classification is not exhaustive; for instance, it

[13] I. Langmuir, *J. Am. Chem. Soc.*, 1918, **40**, 1361.

considers neither induced heterogeneity nor correlated heterogeneity,[14] both of which can play an important role in adsorption phenomena. In spite of this, patchwise heterogeneity remains the most useful model in view of the possibilities of choosing the local isotherm and providing a wide class of adsorbing surfaces.

Furthermore, relationship (4) permits problems to be stated in a simple language. In fact, assuming $\vartheta(p)$ to be experimentally known and taking a particularly meaningful local isotherm (such as the Langmuir or the Hill–de Boer isotherm) as $\theta(p, q)$, relation (4) can be considered as a Fredholm integral equation of the first kind.* In connection with equation (4), together with the obvious problems of existence and uniqueness of the solution, another question must be considered. Assuming that equation (4) can be solved and yields a unique solution, if $\vartheta(p)$ is exact, equation (4) gives the correct distribution $\varphi(q)$. However, due to inherent unavoidable experimental errors, the overall isotherm is known only in an approximate form $\vartheta_\sigma(p)$, where $\|\vartheta_\sigma - \vartheta\| < \sigma$. By carrying out the experimental observations with ever increasing precision, the error σ can be reduced to the minimum desired, *i.e.* the limit for $\sigma \to 0$ can be considered. By so doing the corresponding solution $\varphi_\sigma(q)$ converges only in the mean to $\varphi(q)$ and, for a Fredholm integral equation of the first kind, there are *a priori* non-indications that $\varphi_\sigma(q)$ converges uniformly to $\varphi(q)$ as $\sigma \to 0$. The problem of solving equation (4) for $\varphi(q)$ belongs to the class of improperly-posed problems, and therefore particular care must be taken.†[15] Fortunately, the researchers engaged in this problem were not aware of this conceptual difficulty and pursued, with naive diligence, the objective of solving equation (4).

The first solution to this problem was given by Sips[16] at the end of the forties. Sips observed that if the local isotherm is the Langmuir one,

$$\theta(p, q) = p/[p + p_L \exp(-q/k_B T)], \tag{5}$$

(p_L being a characteristic pressure), then equation (4) can be transformed in such a way that $\vartheta(p)$ is related to a Stieltjes transform of $\varphi(q)$. Thus, if $\vartheta(p)$ satisfies some conditions of regularity, the reverse transform can be computed so yielding $\varphi(q)$.

Though the suggestions of Sips were received well, by Sips himself[16] who proposed a 'generalized Freundlich isotherm,' by Misra[17] who considered a kind of 'generalized' DR equation, by Rudziński and co-workers[18] who extended the method to multilayer adsorption, and by Tóth *et al.*[19] who

---

* A concise description of the theory of integral equations is given in A. E. Taylor, 'Introduction to Functional Analysis,' Wiley-Interscience, New York, 1958.
† Problem and care are fully discussed in reference 15; the physical analysis given here remains valid in spite of a coarse mathematical error.

[14] J. G. Dash, 'Films on Solid Surfaces,' Academic Press, New York, 1975.
[15] G. F. Cerofolini, *Chem. Phys.*, 1978, **33**, 423.
[16] R. Sips, *J. Chem. Phys.*, 1948, **16**, 490; *ibid.*, 1950, **18**, 1024.
[17] D. N. Misra, *Surf. Sci.*, 1969, **18**, 367; *J. Chem. Phys.*, 1970, **52**, 5499.
[18] W. Rudziński, M. Jaronieć, and S. SokoJowski, *Phys. Lett. A*, 1974, **48**, 171.
[19] J. Toth, W. Rudziński, A. Waksmundzki, M. Jaronieć, and S. SokoJowski, *Acta Chim. Acad. Sci. Hung.*, 1974, **82**, 11.

applied the Sips method to another empirical isotherm (the Tóth equation), the present reviewer remains of Adamson's opinion,[20] namely that the features of regularity required by the Sips method are such that if $\vartheta(p)$ is approximated to a form which allows an inverse transform to be found, then the errors so introduced become too important to be neglected.

Before proceeding further a few words should be spent on the choice of the Langmuir isotherm. Obviously, the choice of this isotherm as local isotherm restricts the applications only to submonolayer localized adsorption with weak lateral interactions; however, this seems the case where DR behaviour is observed.[21]

The method most widely applied after that of Sips is the condensation approximation, first proposed by Roginsky[22] but developed to a reliable method only after the detailed discussion by Harris.[23] In the condensation approximation no unrealistic prerequisite of the overall isotherm is required and the correct local isotherm (about which, I wish to stress, there are only theoretical inferences) is replaced by the 'condensation isotherm', *i.e.*, by the step function (6)

$$\theta_c(p, q) = \begin{cases} 0 & \text{for} \quad 0 \leqslant p < p_c(q); \\ 1 & \text{for} \quad p_c(q) \leqslant p, \end{cases} \tag{6}$$

where $p_c(q)$ is a 'condensation pressure,' which may be evaluated for any local isotherm either by the Harris rule[23] or by variational considerations.[24,25] Both rules, however, give the same qualitative results.

Although it is true that the condensation approximation introduces *a priori* errors in the computation of the distribution function (but these errors become negligible when the absolute temperature vanishes[23,25]) at the same time this method has two great advantages: it does not involve complicated numerical computations in any step, and it can be applied to all local isotherms, the particular isotherm influencing only the expression of the condensation pressure.[23,26]

The solution of equation (4), when the local isotherm is the condensation isotherm, is given by

$$\varphi_c(q) = -[\partial \tilde{\vartheta}(Q)/\partial Q]_{Q=q}, \tag{7}$$

where $Q(p)$ is the reverse function of $p_c(q)$ and $\tilde{\vartheta}(Q) = \vartheta[p(Q)]$. The local isotherm influences only the function $Q(p)$, which, in the case of the Langmuir isotherm (5), is given by

$$Q(p) = -k_B T \ln(p/p_L). \tag{8}$$

Note that the second feature of the condensation approximation allows the

[20] A. W. Adamson, *Physical Chemistry of Surfaces*, Interscience, New York, 1967.
[21] G. F. Cerofolini, *J. Low Temp. Phys.*, 1972, **6**, 473.
[22] S. Z. Roginsky, *C.R. Acad. Sci. URSS*, 1944, **45**, 61; ibid., p. 194.
[23] L. B. Harris, *Surf. Sci.*, 1968, **10**, 128; ibid., 1969, **13**, 377; ibid., 1969, **15**, 182.
[24] G. F. Cerofolini, *Surf. Sci.*, 1971, **24**, 391.
[25] G. F. Cerofolini, *Surf. Sci.*, 1975, **47**, 469.
[26] M. Jaroniec, S. Sokołowski, and G. F. Cerofolini, *Thin Solid Films*, 1976, **31**, 321.

attention to be focused upon a particular isotherm, the extension to others being quite obvious.

The first application of the condensation approximation to the DR equation was made by Harris to calculate the heat of adsorption from experimental data;[21] only later was the present author attracted by the simplicity of the method for evaluating the adsorption energy distribution.[24] The application of the condensation approximation allows difficulties (i) and (ii) to be overcome by an extremely simple argument.[21]

If, with the risk of introducing more complex symbols, the DR equation is taken in the following modified form (for which the name 'modified DR isotherm' was proposed[21])

$$\vartheta(p) = \begin{cases} \exp\{-B[k_B T \ln(p/p_m)]^2\} & \text{for} \quad 0 \leqslant p < p_m; \\ 1 & \text{for} \quad p_m \leqslant p, \end{cases} \tag{9}$$

where p_m is now a free parameter, then an overall isotherm is obtained, which reduces to equation (3) when p_m is taken equal to p^0, but which is obeyed by a number of vapour/solid systems. The behaviour, $\vartheta(p) = 1$ for $p \geqslant p_m$, has been chosen to permit the monolayer to be completed at sufficiently high pressures. Up to this point everything is obvious, because a quantity with a well defined physical meaning (p^0) has been replaced by an empirical parameter (p_m) whose value is determined only by a best fit procedure. However, applying equations (8) and (7) to the modified DR isotherm (9),

$$\varphi_c(q) = \begin{cases} 2B(q - q_m)\exp[-B(q - q_m)^2] & \text{for} \quad q_m \leqslant q \\ 0 & \text{for} \quad 0 \leqslant q < q_m \end{cases} \tag{10}$$

which shows that q_m,

$$q_m = -k_B T \ln(p_m/p_L) \tag{11}$$

represents the lowest adsorption energy of the surface. Observing that q_m, because of its physical meaning, must be independent of temperature, p_m can be written in the form $p_m = p_L \exp(-q_m/k_B T)$, which shows that p_m is characteristic of the vapour/solid system (while p^0 depends on the vapour only). Difficulty (i) is overcome by denying its contents: it is only fortuitous that, because in some cases $p_m \simeq p^0$, the DR equation can be expressed in terms of the Polanyi potential.

If the above interpretation is true, by making an Arrhenius plot, i.e. by plotting $\ln p_m$ (determined with best fits at various temperatures) versus $1/k_B T$, a straight line should be obtained: its slope should give $-q_m$ and the intercept at $1/k_B T = 0$ should give $\ln p_L$. After the proposal to consider p_m as a free parameter, already-known adsorption data were re-analysed. The results confirmed the new interpretation: for hexane on graphite MT-1 at 293 K it is[27] $p_m = 2.4$ Torr, $p^0 \simeq 120$ Torr; for argon on graphite at 78 K it is[28] $p_m = 1.8$ Torr, $p^0 \simeq 220$ Torr; and for nitrogen on graphite at 78 K it is[28] $p_m = 1.6$ Torr, $p^0 = 860$ Torr.

[27] W. Rudziński, S. Sokołowski, M. Jaronieć, and J. Tóth, Z. Phys. Chem., (Leipzig), 1975, **256**, 273.
[28] S. Sokołowski, Vuoto, 1975, **8**, 45.

At this point is is interesting to mention that some researchers were close to finding the actual meaning of p_m; for instance, Ricca, Medana, and Bellardo[9] observed that noble gases are adsorbed on Pyrex glass according to equation (3); however a better fit was obtained choosing the vapour pressure of the adsorbent in the solid phase (instead of the liquid phase) as p^0. The same fact was observed by Kindl *et al.*[29] The lack of a theoretical model prevented these researchers from taking the small step separating them from the actual meaning of p_m.

The condensation approximation furnishes other features of the overall isotherm; from equations (7) and (8) it is immediately seen that $\varphi_c(q)$ is independent of T (as its physical meaning requires) only if ϑ does not depend on p and T separately but through quantity (8). Regarding feature (ii), the difficulty which derives from it is simply surpassed by observing that the singularity of the isosteric heat of adsorption is related to an energy spectrum with an infinite upper bound, $\Omega = (q_m, +\infty)$. But the physical meaning of an infinite adsorption energy remains to be explained.

4 The Onset of the Henry Law

Any local isotherm must reduce to the Henry law in the low pressure limit. The Langmuir isotherm is not an exception to this rule:

$$p \rightarrow 0 \Rightarrow \theta(p, q) \sim \frac{p}{p_L} \exp\left(\frac{q}{k_B T}\right). \tag{12}$$

The insertion of the asymptotic behaviour of the local isotherm into equation (4) gives the low-pressure behaviour of the overall isotherm

$$p \rightarrow 0 \Rightarrow \vartheta(p) \sim kp/p_L, \tag{13}$$

where

$$k = \int_0^{+\infty} \exp\left(\frac{q}{k_B T}\right) \varphi(q) \, dq. \tag{14}$$

Since the integrand is positive, the coefficient k satisfies either condition $k > 0$ or $k = +\infty$. If the energy spectrum is bounded, $\Omega \subseteq (q_m, q_M)$, the second possibility is however excluded.

The DR isotherm, both when expressed by equation (3) and in the modified form (9), does not satisfy condition (13), because $d\vartheta/dp = 0$ for $p = 0$. This difficulty has been summarized in feature (iii) and was clearly recognized in early works too. For instance, Hobson and Armstrong[5] proposed the following empirical rule to modify the DR equation: 'Plot equation (3) in a log–log graph; a parabola is obtained. Plot a sheaf of straight lines with slope $\pi/4$; from this sheaf choose the tangent to the parabola. The actual adsorption isotherm is given by the straight line at pressures below the intersection one, and by the parabola above this pressure'. This construction allows an isotherm to be obtained that coincides with the Henry law at low pressures and with the DR isotherm at higher pressures. The change of behaviour takes place with

[29] B. Kindl, E. Negri, and G. F. Cerofolini, *Surf. Sci.*, 1970, **23**, 299.

continuity and with continuous first derivative: this avoids invoking a (strange and unknown) phase transition to explain the change of behaviour (DR → Henry).

In spite of the merits of the Hobson–Armstrong rule, it should be remembered that this construction is not based upon any well established model. Bearing this in mind, Hobson[30] developed a method (the 'asymptotically-correct approximation') which solves equation (4) taking into account the correct asymptotic behaviour of the local isotherm at very low pressure. This method was developed for both the Langmuir and the Hill–de Boer isotherms (and later extended to the Fowler–Guggenheim isotherm[31]) but, for simplicity, here the attention will be concentrated on the Langmuir isotherm only. In this case, the asymptotically correct approximation consists in replacing the true kernel (5) by the following function:

$$
\theta_a(p, q) = \begin{cases} \dfrac{p}{p_L} \exp\left(\dfrac{q}{k_B T}\right) & \text{for} \quad 0 \leqslant p < p_L \exp\left(\dfrac{-q}{k_B T}\right); \\[3mm] 1 & \text{for} \quad p_L \exp\left(\dfrac{-q}{k_B T}\right) \leqslant p. \end{cases} \tag{15}
$$

If kernel (15) is inserted in equation (4), an easily soluble equation is obtained, and its solution is given by

$$
\varphi_a(q) = - \left[\frac{\partial \bar{\vartheta}(Q)}{\partial Q} + k_B T \frac{\partial^2 \bar{\vartheta}(Q)}{\partial Q^2} \right]_{Q=q}, \tag{16}
$$

where $Q(p) = -k_B T \ln(p/p_L)$ and $\bar{\vartheta}(Q) = \vartheta[p(Q)]$ as in the previous Section.

At this point the ideas of Hobson are recalled to find the transition from the Henry to the DR region. He solved equation (4) in the asymptotically correct approximation choosing the DR equation as overall isotherm. The distribution function so computed was then cut off at a value q_M and the integral (4) was recomputed using the asymptotically correct isotherm (15). This method is largely arbitrary, because the assumed value of q_M strongly influences the zone of transition, and only later did the author succeed in finding a self-consistent way of justifying the Hobson–Armstrong rule.[24] The idea is very simple: it is accepted that, at higher pressures, the true overall isotherm behaves indeed as the modified DR isotherm (9), but this expression must be modified in the low-pressure limit to take behaviour (13) into account.

The simplest way of considering both features consists of assuming an overall isotherm of the following form

$$
\vartheta(p) = \begin{cases} kp/p_L & \text{for} \quad 0 \leqslant p < p_M; \\ \exp\{-B[k_B T \ln(p/p_m)]^2\} & \text{for} \quad p_M \leqslant p < p_m; \\ 1 & \text{for} \quad p_m \leqslant p, \end{cases} \tag{17}
$$

where p_M is a suitable pressure. Applying equations (16) to (17) the following

[30] J. P. Hobson, *Can. J. Phys.*, 1965, **43**, 1941.
[31] G. F. Cerofolini, *Surf. Sci.*, 1975, **52**, 195.

distribution function is obtained explaining this overall behaviour in the asymptotically correct approximation:

$$
\varphi_a(q) = \begin{cases} 0 \\ \{2Bk_BT[1-2B(q-q_m)^2]+ \\ 0 \end{cases}
$$

$$
\begin{array}{ll}
& \text{for} \quad 0 \leqslant q < q_m; \\
2B(q-q_m)\} \exp[-B(q-q_m)^2] & \text{for} \quad q_m \leqslant q < q_M; \quad (18) \\
& \text{for} \quad q_M \leqslant q,
\end{array}
$$

where $q_M = -k_BT \ln(p_M/p_L)$.

Distribution (18) confirms the meaning of p_m as given by the condensation approximation and shows that the existence of an upper limit to integral (4) is a consequence of the Henry behaviour at very low pressures.

The value of q_M is found by imposing the normalization condition to $\varphi_a(q)$:

$$
\int_{q_m}^{q_M} \varphi_a(q)\, dq = 1, \tag{19}
$$

which is an equation for q_M. This equation has two solutions: either $q_M = +\infty$ (corresponding to $p_M = 0$) or

$$
q_M = q_m + 1/(2Bk_BT), \tag{20}
$$

(corresponding to $p_M = p_m \exp[-1/2B(k_BT)^2]$). It is easy to show that if the trivial solution, $q_M = +\infty$, is taken, then $\varphi_a(q)$ becomes negative for high values of q. Rejecting this non-physical solution, (20) is accepted, for which $\varphi_a(q)$ remains positive in the whole interval. Condition (20) is equivalent to the result obtained with the Hobson–Armstrong condition (if p_m is identified with p^0), so that the above discussion can be considered a demonstration of that rule. Because of its physical meaning, the upper adsorption energy should be independent of temperature. Hence the difference $q_M - q_m$ should be independent of temperature and characteristic of the vapour/surface system. Therefore (20) suggests that Bk_BT should be independent of T, *i.e.* B should depend on T through the relation[32]

$$
B = 1/[2(q_M - q_m)k_BT]. \tag{21}
$$

In his experimental study of the adsorption of argon on nickel in the temperature range 70—120 K, Schram[10] in fact observed temperature variations of B (this quantity decreases as T increases) but no tests of equation (21) have yet been performed. Whether B is considered as a constant or as a function of T [according to equation (21)], the distribution function (20) is temperature dependent. This non-physical behaviour is however overcome by observing that even great variations of T imply small changes of $\varphi_a(q)$ and of all observable properties related to it (see Figures 1 and 2).

Concluding this Section, it can be stated that the above discussion has shown that difficulty (iii) can be avoided by modifying the DR isotherm in the

---

[32] G. F. Cerofolini, *Thin Solid Films*, 1974, **23**, 129.

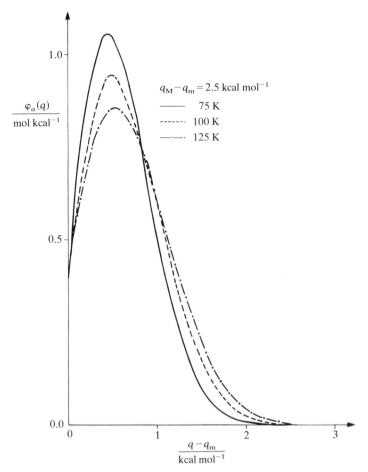

Figure 1 *Distribution (18) is plotted for three different values of temperature [75 K (full line), 100 K (dashed line), and 125 K (dashed and dotted line)] for an assumed energy range (q_m, q_M) of 2.5 kcal mol^{-1}. Though $\psi_a(q)$ depends upon T, its gross observable properties [such as minimum, most probable, average and maximum adsorption energy as well as the shape of $\psi_a(q)$] either are constant or show only a weak dependence on temperature*

low-pressure range; this change leads to an isotherm [the Henry-modified DR isotherm, equation (17)] for which the isosteric heat of adsorption remains finite for $p \to 0$.

5 Equilibrium Surfaces

The methods described in the two previous Sections, *i.e.* the condensation approximation and the asymptotically correct approximation, relate, through

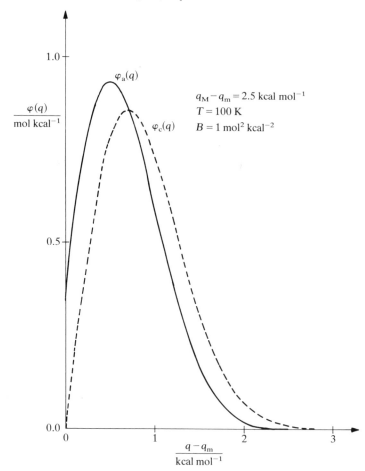

Figure 2 *The distribution function $\varphi_a(q)$ computed at $T = 100$ K for $q_M - q_m = 2.5$ kcal mol^{-1} is compared with the distribution function $\varphi_c(q)$ computed, in the framework of the condensation approximation, from the modified DR isotherm for which the value of B is given by (21) ($B = 1$ mol^2 kcal^{-2}). Note how the two distributions are very similar*

equations (7) and (16), respectively, the distribution function in q to the overall isotherm at $Q(p) = q$. Therefore the range of pressure (p_1, p_2) where the overall isotherm is known with precision identifies an energy range $[Q(p_2), Q(p_1)]$ where one has, within the validity of the approximate methods, reliable indications about the adsorption energy distribution. This feature is lost when exact methods to solve equation (4) are considered. Exact methods are non-local, as they relate the value of the distribution function in q to the value of the overall isotherm not only at $Q(p) = q$ but even out of the physically accessible pressure region $(0, +\infty)$.[15] In the ultimate analysis this is the reason for instability of the solution of equation (4).

The behaviour of the overall isotherm outside the accessible region influences the computed distribution function, and no criterion exists for extrapolating the experimental behaviour; the mixture of these features suffices for the birth of unstable solutions. The situation is well described by the words of Edgar Lee Masters;

> "And as many voices called to me in life
> Marvel not that I could not tell
> The true from the false,
> Nor even, at last, the voice that I should have known".

In the present case, however, the physical and mathematical conditions that can be imposed on the distribution function computed by the approximate local methods lead to meaningful extensions of the overall isotherm. In addition, these extensions, when used as known data in the exact methods,[17,28] lead to distribution functions which do not differ appreciably from the ones computed by the approximate methods.[15] This circumstance is a clear indication of the validity of the results previously obtained.

At this point the essential features of problems (i) to (iii) have been explained. However, the most important problem, *i.e.* (iv), still remains to be solved. Indeed, the DR isotherm is observed with such a frequency that not only is it meaningful but also mandatory to look for the reasons of this behaviour.

It is perhaps interesting to note how difficult it is to recognize the original DR equation in the Henry-modified DR isotherm, equation (17). In fact, equation (3) has been modified to such an extent, both in the analytical expression and in the meaning of its parameters (recall that B was initially supposed to be independent of temperature) that only historical reasons allow the name DR to be kept.*

Furthermore, equation (17) has been modified in order to describe multilayer formation.[21,32] In the BET model (*i.e.*, an adsorption model which describes multilayer formation and which reduces to the Langmuir model in the submonolayer region), if the least adsorption energy, q_m, is greater than the adsorption energy in the second and higher layers, q_s, *i.e.* if $p_m \ll p^0$ (for the experimental consistency of this inequality see the discussion in Section 3), then the overall behaviour is still described by equation (17) in the submonolayer range, while the heterogeneous character of the surface is gradually lost as the coverage becomes larger than the monolayer. In the light of this behaviour BET and DR surface areas are easily demonstrated to be nearly equal; this fact justifies the proposal of Kaganer[3] (see Section 1) and the

---

* About this point, though I do not directly know Dubinin's thought, the following referee's remark to one of my papers is perhaps of interest: 'If it is true that a formalism similar to that described in Dubinin paper applies to low coverage adsorption, this has been shown by Kaganer. Dubinin has not considered the problem of submonolayer adsorption; consequently, he does not agree when one speaks of Dubinin equation in describing adsorption in this range; I have reached this opinion when I discussed the matter with Dubinin in person' (translated from French).

subsequent measures by Kaganer himself[33] and by Hobson.[34] The application of the above model to a particular system (argon/aerosil) by Hsu *et al.*[35] was satisfactory and several other works consolidated and extended that result.[18,27,36,37]

At this point, and in spite of a rather outdated interpretation, let us to recall the experimental work of Hobson,[34] who was the first to undertake the task of studying the adsorption isotherm in the whole range, from the Henry behaviour to the formation of the liquid layer.

Coming back to the explanation of feature (iv), giving rise to this problem the author observed[21] that 'according to the analysis of Marsh and Rand[38] the DR isotherm is not related to the (presumed) fact that a log–log[2] plot linearizes almost all adsorption isotherms. This suggests that the DR behaviour is due to a particular (equilibrium?) structure of the adsorbing surface.' This line was followed up in an attempt to clarify what is meant by the notion of equilibrium surface. Obviously, as the DR isotherm is usually observed in the adsorption of gases on solid surfaces, the surface cannot be thought of as being in equilibrium at the adsorption temperature. It is however possible to retain the concept of equilibrium surface without demanding the equilibrium be conserved even at the adsorption temperature: in fact, it is sufficient to assume that the surface was grown in equilibrium at a given temperature T_g and then suddenly quenched to the adsorption temperature so keeping the old structure.

This idea allows a formal treatment. Characterizing each surface atom by means of the depth of the potential energy well within which it is moving, then under equilibrium conditions the number N_i of surface atoms with energy u_i (<0) does not vary with time. Atoms with energy u_i are formed due to the trapping of atoms (either from the vapour phase or from the mobile pool along the surface) on atoms with energy u_j, or when atoms with energy u_k leave the bound state. A similar but opposite reasoning holds for the death of atoms with energy u_i: they can evaporate or migrate along the surface, or be buried by other atoms.

The problem of studying the evolution of the population $\{N_i\}$ is a very difficult problem of random walk and lattice statistics. However, some simplifying hypotheses are possible[39] and lead to the following result

$$\frac{N_i}{N_0} = \exp\left(-\frac{u_i - u_0}{k_B T_f}\right), \tag{22}$$

where u_0 is the ground-state energy and the temperature T_f is proportional to, and greater than, T_g. Assuming that the energy spectrum $\{u_i\}$ is densely contained in the interval (u_0, u_M), $u_M < 0$, and defining the extra energy,

[33] M. G. Kaganer, *Dokl. Akad. Nauk SSSR*, 1961, **138**, 405.

[34] J. P. Hobson, *J. Phys. Chem.*, 1969, **73**, 2720.

[35] C. C. Hsu, W. Rudziński, and B. W. Wojciechowski, *Phys. Lett. A.*, 1975, **54**, 365.

[36] C. C. Hsu, W. Rudziński, and B. W. Wojciechowski, *J. Chem. Soc., Faraday Trans. I*, 1976, **72**, 453.

[37] G. F. Cerofolini, *Z. Phys. Chem.*, (*Leipzig*), 1977, **258**, 937.

[38] H. Marsh and B. Rand, *J. Colloid Interface Sci.*, 1970, **33**, 101.

[39] G. F. Cerofolini, *Surf. Sci.*, 1976, **61**, 678.

$E = u - u_0$, the following distribution function is obtained:

$$\frac{\mathrm{d}\phi}{\mathrm{d}E} = \begin{cases} \dfrac{r}{k_B T_f} \exp\left(\dfrac{-E}{k_B T_f}\right) & \text{for} \quad 0 \leqslant E < E_M; \\ 0 & \text{otherwise,} \end{cases} \tag{23}$$

where $E_M = u_M - u_0$ and $r = [1 - \exp(-E_M/k_B T_f)]^{-1}$ is a normalization factor.

In order to connect the above result with adsorption properties, one observes that high values of E correspond to highly unsaturated surface atoms. Hence, atoms with high extra energy have bonds available for adsorption, *i.e.*, adsorption takes place preferentially on surface atoms with high extra energy. A measure of the ease of adsorption is given by the 'vertical' adsorption energy q_v. In view of the previous arguments q_v must be an increasing function of E. Tentatively one can assume a relation of the following kind

$$q_v = q_v^0 + \beta E, \tag{24}$$

where β is a suitable constant and $q_v^0 = q_v(E)_{E=0}$.

On the contrary, for patchwise heterogeneous surfaces the lateral binding energy q_l is a decreasing function of E. In fact, the adlayer density decreases as the extra energy increases (note that this property holds true only for patchwise heterogeneous surfaces; for random heterogeneous surfaces the adlayer density should be rather independent of the extra energy); since q_l is proportional to the density, the lateral binding energy must be a decreasing function of the extra energy. A tentative expression with correct asymptotic behaviour ($E \to 0 \Rightarrow q_l = q_l^0$, $E \to +\infty \Rightarrow q_l \to 0$) is given by $q_l = q_l^0[1 - (E/E_0)^2]$ for $E \leqslant E_0$ and $q_l = 0$ for $E > E_0$, where E_0 is a characteristic extra energy. Combining this expression with (24), the total adsorption energy is given by

$$q = q_v + q_l = q_v^0 + \beta E + q_l^0[1 - (E/E_0)^2]. \tag{25}$$

This function can be reversed to give $E(q)$.

The following expansion is algebraically possible[40]

$$E = \alpha_1(q - q_m) + \alpha_2(q - q_m)^2, \tag{26}$$

where $\alpha_1 = 1/\beta$, $\alpha_2 = q_l^0/4\beta^3 \varepsilon_0^2$, and $q_m = q_v^0 + q_l^0$. Expression (26) was first proposed on erroneous grounds,[41] and only later was a correct derivation given.[40] The first term of (26) takes into account mainly the vertical interaction, and the second term is the correction due to the lateral interactions. This term is important for patchwise heterogeneous surfaces, but it is negligible for random heterogeneous surfaces.

As $\varphi(q) = (\mathrm{d}\phi/\mathrm{d}E)(\mathrm{d}E/\mathrm{d}q)$, equation (26) combined with equation (23) allows the adsorption energy distribution to be found. In fact, with q_M denoting the physically meaningful solution of the equation

$$E_M = \alpha_1(q - q_m) + \alpha_2(q - q_m)^2, \tag{27}$$

[40] G. F. Cerofolini, *J. Colloid Interface Sci.*, 1982, **86**, 204.
[41] G. F. Cerofolini, *Surface Sci.*, 1975, **51**, 333; *Z. Phys. Chem.*, (*Leipzig*), 1981, **262**, 289.

the distribution function is given by

$$\varphi(q) = r \frac{\alpha_1 + 2\alpha_2(q - q_m)}{k_B T_f} \exp\left[-\frac{\alpha_1(q - q_m) + \alpha_2(q - q_m)^2}{k_B T_f} \right] \qquad (28)$$

for $q_m \leqslant q < q_M$, and $\varphi(q) = 0$ otherwise.

6 Adsorption on Equilibrium Surfaces

Function (28) inserted into integral (4) allows the computation of the overall adsorption isotherm on equilibrium surfaces. Two cases are conveniently distinguished: weak heterogeneity $(q_M - q_m \gg k_B T_f / \alpha_1)$ and strong heterogeneity $(q_M - q_m \lesssim k_B T_f / \alpha_1)$.

Weak Heterogeneity.—In this case $(q_M - q_m \gg k_B T_f / \alpha_1)$ the upper bound of integration can be ignored, $r \approx 1$, and distribution (28) can reasonably be supposed to hold in the whole range $(q_m, +\infty)$. Putting $T_F = T_f / \alpha_1$, $\chi_0 = \alpha_1 / \alpha_2$, equation (28) becomes

$$\varphi(q) = \frac{1 + 2(q - q_m)/\chi_0}{k_B T_F} \exp\left[-\frac{(q - q_m) + (q - q_m)^2/\chi_0}{k_B T_F} \right], \qquad (29)$$

for $q_m \leqslant q < +\infty$, and $\varphi(q) = 0$ otherwise. Two energy ranges can be considered. In the *low-energy range* $(q - q_m \ll \chi_0)$ distribution (29) is approximately given by

$$\varphi(q) \simeq \frac{1}{k_B T_F} \exp\left(-\frac{q - q_m}{k_B T_F} \right); \qquad (30)$$

while in the *high-energy range* $(q - q_m \gg \chi_0)$ distribution (29) is approximately given by

$$\varphi(q) \simeq \frac{2(q - q_m)}{k_B T_F \chi_0} \exp\left[-\frac{(q - q_m)^2}{k_B T_F \chi_0} \right]. \qquad (31)$$

Distribution (30) is known to be the distribution function computed, in the condensation approximation, from the Freundlich isotherm,[42]

$$\vartheta_F(p) = \begin{cases} (p/p_m)^{T/T_F} & \text{for} \quad 0 \leqslant p < p_m, \\ 1 & \text{for} \quad p_m \leqslant p. \end{cases} \qquad (32)$$

Since the work of Zeldovich[43] it has been established that the above behaviour is due to an exponential distribution of adsorption energy. Various expressions of the Freundlich isotherm are known: for instance, Sips[16] modified the original expression to show saturation at high pressure and to permit the exact evaluation of the distribution function; equation (32) is suggested by the use of the condensation approximation[44] and is in good agreement with rigorous

[42] H. Freundlich, *Trans. Faraday Soc.*, 1932, **28**, 195.
[43] J. B. Zeldovich, *Acta Physicochim. URSS*, 1935, **1**, 961.
[44] G. F. Cerofolini, *Vuoto*, 1975, **8**, 178.

statistical-mechanical computations,[45] at least in the low-temperature range where a 'compensation factor' neglected in equation (32) can be ignored.

Distribution (31), compared with distribution (10), shows that in the high-energy range the distribution function of equilibrium surfaces strictly resembles the distribution function explaining, in the condensation approximation, the modified DR isotherm if B is identified with $1/k_B T_F \chi_0$.

Concluding, if $2k_B T_f \leqslant \alpha_1^2/\alpha_2$ (and this is surely to be expected in the hypotheses that E_0 is very high and $k_B T_f$ not too high), the distribution function (29) has the following extreme behaviours

$$\varphi(q) \simeq \begin{cases} \dfrac{1}{k_B T_F} \exp\left(-\dfrac{q-q_m}{k_B T_F}\right) & \text{for} \quad q-q_m \ll \chi_0 \\[3mm] \dfrac{2(q-q_m)}{k_B T_F \chi_0} \exp\left[-\dfrac{(q-q_m)^2}{k_B T_F \chi_0}\right] & \text{for} \quad q-q_m \gg \chi_0, \end{cases} \tag{33}$$

which in the condensation approximation should indicate an overall isotherm behaving as the modified DR isotherm at low pressures [*i.e.*, for $p \ll p_m \exp(-\chi_0/k_B T)$] and as the Freundlich isotherm at high pressures [*i.e.*, for $p_m \exp(-\chi_0/k_B T) \ll p \leqslant p_m$].

More refined computations,[40] performed inserting distribution (29) into integral (4) and calculating this integral numerically, show that the resulting overall isotherm has the following behaviours;

$$\vartheta(p) \simeq \begin{cases} \text{Henry isotherm, for very low pressures} \\ \text{modified DR isotherm, for low pressures} \\ \text{Freundlich isotherm, for higher pressures} \\ \text{full monolayer, for } p \geqslant p_L \exp(-q_m/k_B T) \end{cases}$$

In a log-log plot the Freundlich zone tends to vanish to the advantage of the DR zone as χ_0 becomes smaller, and both the Freundlich and the DR zones cannot be recognized, when T_F tends to zero, *i.e.* when the surface becomes homogeneous. Examples of these behaviours are sketched in Figures 3 and 4.

In conclusion, the DR isotherm is observed at lower pressures on patchwise surfaces while the Freundlich isotherm is observed at high pressures, and the former behaviour is due to the relevance of lateral interactions.

At this point it is worthwhile mentioning the pioneering work of Sparnaay,[46] who first arrived at similar results by taking a discrete exponential distribution of adsorption energy, without however introducing the notion of equilibrium surface.

Strong Heterogeneity.—The case of strong heterogeneity, $q_M - q_m \leqslant k_B T_F$, leads to different results. In fact, in this case the factor r may be very great

[45] J. Appel, *Surf. Sci.*, 1973, **39**, 237.
[46] M. J. Sparnaay, *Surf. Sci.*, 1968, **9**, 100; M. J. Sparnaay, in 'Clean Surfaces: Their preparation and characterization for interfacial studies', ed. G. Goldfinger, Marcel Dekker, New York, 1970, p. 153.

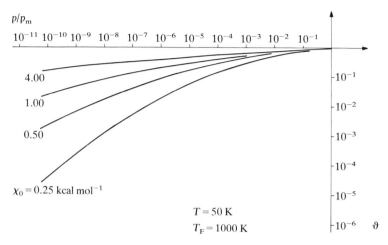

Figure 3 Log-log plot of the adsorption isotherms which should be observable on weakly heterogeneous surfaces grown at $T_F = 1000$ K and frozen to $T = 50$ K. The figure shows that the Freundlich range gradually extends as χ_0 increases

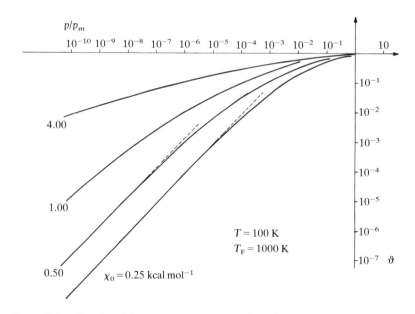

Figure 4 Log-log plot of the adsorption isotherms which should be observable on weakly heterogeneous surfaces grown at $T_F = 1000$ K and frozen to $T = 100$ K. The Henry behaviour is extrapolated by the dashed lines

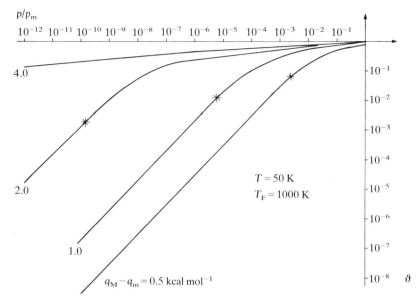

Figure 5 *Log–log plot of the adsorption isotherms which should be observable on strongly heterogeneous surfaces grown at $T_F = 1000$ K and frozen to 50 K for different energy spectra (from 0.5 to 4.0 kcal mol^{-1}). The stars roughly indicate the pressures beyond which the Henry behaviour is no longer observed*

and the distribution function is approximately flat in the interval $(q_m, q_M) = (q_m, q_m + \Delta)$:

$$\varphi(q) \simeq 1/\Delta \qquad (34)$$

for $q_m \leqslant q < q_m + \Delta$, and $\varphi(q) = 0$ otherwise.

Inserting (34) in (4) gives an integral which can be computed in closed form:

$$\vartheta(p) = \frac{k_B T}{\Delta} \ln \left[\frac{1 + \dfrac{p}{p_m} \exp\left(\dfrac{\Delta}{k_B T}\right)}{1 + \dfrac{p}{p_m}} \right]. \qquad (35)$$

This isotherm (35) behaves as the Henry isotherm in the low-pressure range,

$$p \to 0 \Rightarrow \vartheta(p) \simeq \frac{k_B T}{\Delta} \exp\left(\frac{\Delta}{k_B T}\right) \frac{p}{p_m}, \qquad (36)$$

and as the Temkin isotherm[47] in the intermediate range

$$p_m \exp\left(\frac{-\Delta}{k_B T}\right) \lesssim p \lesssim p_m \Rightarrow \vartheta(p) \simeq 1 + \frac{k_B T}{\Delta} \ln \frac{p}{p_m}. \qquad (37)$$

[47] M. I. Temkin, *J. Phys. Chem. USSR*, 1940, **14**, 1153.

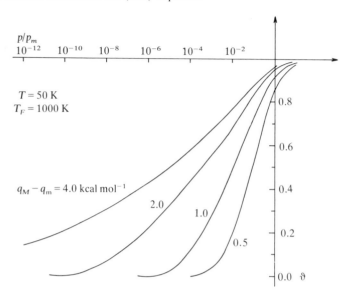

Figure 6 *Linear-log plot of the same isotherm sketched in Figure 5. The graph shows that the Temkin isotherm (rectilinear behaviour in this plot) becomes sharper as the width of the energy spectrum decreases*

That the Temkin isotherm could be ascribed to adsorption on equilibrium surfaces was first guessed by Ionescu,[48] but his derivation was based on rather weak arguments. The above consideration, on the contrary, is already satisfactory and can be made more precise by inserting distribution function (28) in equation (4) and by computing this integral numerically. The results in a log-log plot are shown in Figure 5, and confirm the Henry behaviour at low pressures and the saturation at high pressures.

To verify that the Temkin behaviour is indeed obtained in the intermediate range, the computed overall isotherms are shown in Figure 6 as a linear-log plot: they display a rectilinear behaviour in the intermediate range as required by (37); the rectilinear range is more extended the more $k_B T_F$ exceeds $q_M - q_m$. Thus in the case of strong heterogeneity, the overall behaviour is the following:

$$\vartheta(p) \simeq \begin{cases} \text{Henry isotherm, for low pressures,} \\ \text{Temkin isotherm, for intermediate pressures,} \\ \text{full monolayer, for } p \gtrsim p_m. \end{cases}$$

7   An Extension: Desorption Kinetics

The previous Sections started from the experimental validity of the DR equation and showed how this behaviour can be ascribed to a suitable distribution (the Rayleigh distribution) of adsorption energy.

[48] N. I. Ionescu, *Surf. Sci.*, 1976, **61**, 294; *Rev. Roum. Chim.* 1979, **24**, 83.

To explain why so many systems obey the DR isotherm, it was then supposed that this behaviour is due to a suitable equilibrium structure of the adsorbing surface. Not only does this hypothesis explain the reason for the DR behaviour, but it also solves two other classic problems; the reasons for the Freundlich and the Temkin behaviours.

In view of positive results obtained using the notion of equilibrium surface, it is interesting to apply this notion beyond the original field of application. The first obvious extension is towards kinetic instead of equilbrium properties.

The application of the previous conclusions to adsorption kinetics is quite difficult, because the typical kinetic parameter, the sticking probability Σ, is scarcely correlated (if correlated at all) with the adsorption energy q. As the surface has been parametrized only by q, the previous theory is insufficient to describe adsorption kinetics on heterogeneous surfaces.

The situation is different in desorption kinetics. In this case the activation energy of desorption, $\xi^{\ddagger}$, may be assumed to be equal to the adsorption energy q. For consistency reasons, surfaces formed by non-interacting sites, each with its own activation energy of desorption q, should be considered. If $R(t, q)$ is the desorption rate at time t from the sites with activation energy q, the overall desorption rate is given by

$$\rho(t) = \int_0^{+\infty} R(t, q)\varphi(q)\,\mathrm{d}q. \tag{38}$$

Consider now desorption phenomena on surfaces which, in the equilibrium condition, obey the Langmuir isotherm locally. As the Langmuir isotherm can be obtained assuming desorption kinetics of the first order and adsorption kinetics of zero order (Langmuir arrived at his expression in this way), consistency reasons imply that $R(t, q)$ is given by first-order kinetics

$$R(t, q) = \frac{1}{\tau_0} \exp\left[-\frac{t}{\tau_0} \exp\left(\frac{-q}{k_{\mathrm{B}}T}\right) \right] \exp\left(\frac{-q}{k_{\mathrm{B}}T}\right), \tag{39}$$

where τ_0 is a suitable time (the mean sojourn time on the sites at $T = +\infty$). By inserting expression (39) and the value of $\varphi(q)$ given by equation (28) into integral (38), the overall desorption rate can be evaluated.

Two extreme cases are considered.

Weakly Heterogeneous Surfaces.—In this case, $q_{\mathrm{M}} - q_{\mathrm{m}} \gg k_{\mathrm{B}}T_{\mathrm{F}}$, the expression of the energy distribution function is given by equation (29). In addition, and in order to simplify the computations, it is supposed here that lateral interactions are negligible (χ_0 very high). The energy distribution function reduces to the exponential distribution (30), which under equilibrium conditions leads to Freundlich behaviour.

Inserting equations (30) and (39) into (38) gives

$$\rho(t) = \frac{1}{k_{\mathrm{B}}T_{\mathrm{F}}\tau_{\mathrm{m}}} \int_0^{+\infty} \exp\left[-\frac{t}{\tau_{\mathrm{m}}} \exp\left(\frac{-\chi}{k_{\mathrm{B}}T}\right) \right] \exp\left[\frac{-\chi}{k_{\mathrm{B}}T}\left(1 + \frac{T}{T_{\mathrm{F}}}\right) \right] \mathrm{d}\chi, \tag{40}$$

where $\tau_m = \tau_0 \exp(q_m/_B T)$ and $\chi = q - q_m$. Putting $s = T/T_F$, if $s \ll 1$ (this choice restricts the description to desorption phenomena at low temperature) then the above integral can be manipulated to assume the following expression

$$\rho(t) = \frac{s}{t} \int_0^{+\infty} \exp\left[-\frac{t}{\tau_m} \exp\left(\frac{-\chi}{k_B T}\right) \right] \frac{t}{\tau_m} \exp\left(\frac{-\chi}{k_B T}\right) \frac{d\chi}{k_B T}, \qquad (41)$$

which can be computed in closed form:

$$\rho(t) = \frac{s}{t}\left[1 - \exp\left(\frac{-t}{\tau_m}\right) \right]. \qquad (42)$$

As soon as $t \gg \tau_m$, the kinetics of (42) become $\rho(t) \simeq s/t$. This behaviour is well known as Elovich behaviour:[49] it takes place at low temperature on the same heterogeneous surfaces which, under equilibrium condition, obey the Freundlich isotherm.* The kinetics of (42) hide a difficulty: indeed, if it held true exactly, the total desorbed quantity would diverge at $t \to \infty$. This difficulty can be overcome by introducing a finite upper adsorption energy.

Strongly Heterogeneous Surfaces.—In this case, $q_M - q_m \lesssim k_B T_F$, so that $\varphi(q) = 1/\Delta$ for $q_m \leqslant q < q_m + \Delta$, and $\varphi(q) = 0$ otherwise.

Inserting this distribution into integral (38) leads to

$$\rho(t) = \int_0^\Delta \frac{1}{\tau_m} \exp\left[-\frac{t}{\tau_m} \exp\left(\frac{-\chi}{k_B T}\right) \right] \exp\left(\frac{-\chi}{k_B T}\right) \frac{d\chi}{\Delta}, \qquad (43)$$

where the symbols have the same meaning as in the previous sub-section. A few manipulations give

$$\rho(t) = \frac{k_B T}{\Delta} \frac{1}{t} \left\{ \exp\left[-\frac{t}{\tau_m} \exp\left(\frac{-\Delta}{k_B T}\right) \right] - \exp\left(-\frac{t}{\tau_m}\right) \right\}.$$

In the time range $\tau_m \lesssim t \lesssim \tau_m \exp(\Delta/k_B T)$ and for all temperatures the above expression reduces to

$$\rho(t) \simeq \frac{k_B T}{\Delta} \frac{1}{t},$$

which is again the Elovich behaviour.

As the Elovich behaviour is observed both for weakly and strongly heterogeneous surfaces grown under equilibrium conditions, the reasons for

---

* A rigorous computation gives

$$\rho(t) = \frac{\tau_m^s s}{t^{1+s}} \int_0^{t/\tau_m} \exp(w) w^s \, dw \sim \frac{\tau_m^s s}{t^{1+s}} \Gamma(1+s) \quad \text{for} \quad t \to +\infty,$$

where $\Gamma(1+s)$ is the factorial function. The last result reduces to the Elovich behaviour for $s \ll 1$.

[49] S. Yu. Elovich and G. M. Zhabrova, *J. Phys. Chem. USSR*, 1939, **13**, 1761.

the frequent occurrence of this desorption kinetics isotherm are easily under-stood.[50]

8 Perspectives

In the previous Sections it has been shown how the notion of equilibrium surface can usefully be introduced in the description of some phenomena.

According to a well known principle of strategy, for which an objective, once conquered, must be kept, consolidated, and extended, I believe that the notion of equilibrium surface should be applied, at least tentatively, beyond the original fields of application. Some suggestions are already possible.

Films.—Adsorption data permit both overall geometric properties (through the roughness factor) and detailed energetic properties (through the distribution function) to be obtained. Data of this kind, obtained by studying films prepared in various conditions, could be an almost ideal test of the theory since films deposited on substrates kept at low temperatures cannot undergo structural rearrangements to assume the crystal structure of the bulk phase.

In these deposition conditions a reasonable model of growth is the vertical stack growth, for which the number of atoms in each stack is a random variable distributed according to Poisson's law.[51] The roughness factor of these films is often very high (say ≥ 10)[51,52] and increases in proportion to the thickness.

After annealing at constant temperature, the vertical stack films rearrange themselves to assume first a columnar structure and then an island structure.[53] Consequently, the rougness factor gradually decreases often to very low values, say 2 to 3. It is obvious that changes of adsorption energy distribution take place simultaneously with these morphological changes.

It is reasonable to assume that a freshly prepared, vertical stack, film is characterized by an almost flat distribution of adsorption energy, because surface atoms of different extra energy are generated irrespective of their energy. A heat treatment at constant temperature leads to a distribution characteristic of that temperature, where however the extension of patches increases with the duration of annealing. This means that the surface recon-struction at constant temperature occurs mainly *via* changes of $E_0 : E_0$ is infinite for the vertical stack film and decreases with film age.

The phenomenology of adsorption can therefore be summarized as in Table 1. By 'isotherm' is meant here the adsorption isotherm in the intermediate range, *i.e.* far from the Henry range and from the full monolayer. Note that the DR behaviour should not be observed for random heterogeneity, while its relevance should increase as the surface becomes more and more patchwise, *i.e.* for prolonged annealing. Otherwise stated, this means that the value of $\chi_0[B]$ should decrease with increasing annealing time: this fact has clearly been observed by Endow and Pasternak[8] for molybdenum films* (Table

---

*These data have however a limited validity in supporting the previous statement since no information is given about T_F.

[50] G. F. Cerofolini, *Z. Phys. Chem. (Leipzig)*, 1978, **259**, 1020.
[51] G. F. Cerofolini, *Thin Solid Films*, 1975, **27**, 297; *ibid.*, 1976, **32**, 177.
[52] K. L. Chopra, Thin Film Phenomena, McGraw Hill, New York, 1969, pp. 185—189.
[53] G. F. Cerofolini, *Thin Solid Films*, 1978, **50**, 69.

Table 1 *Change of structure and properties of films according to the annealing time*

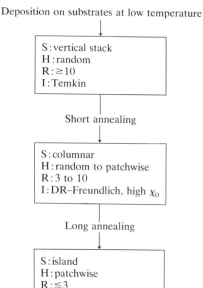

Deposition on substrates at low temperature

S : vertical stack
H : random
R : ≥10
I : Temkin

Short annealing

S : columnar
H : random to patchwise
R : 3 to 10
I : DR–Freundlich, high χ_0

Long annealing

S : island
H : patchwise
R : ≤3
I : DR–Freundlich, low χ_0

S denotes the film structure, H the kind of hetero-geneity, R the roughness, and I the isotherm.

Table 2 *Changes in B and V_m with annealing time*

Annealing time/(days)	$B/(10^{-7} \text{ mol}^2 \text{ cal}^{-2})$	$V_m/(10^{14} \text{ molecules cm}^{-2})$
2	2.2	63
6	2.6	50
12	3.9	29

2). These data are referred to Kr as adsorbed gas and to a constant annealing temperature of 77 K. The geometric area can be assumed to be $1.5 \times 10^{15} \text{ cm}^{-2}$.

Proteins.—Without entering the subject in detail, it may be recalled that proteins are polypetide chains formed by a number of amino-acids ranging from 51 for bovine insulin to about 6000 for myosin. In each protein there are several chemically-different polar sides: the backbone groups of the polypetide chains, $\supset$CO and $\supset$NH, and the sidechain groups of the amino-acid radicals, —COOH, —NH$_2$, —OH, —SH, —S—S—. As each site is subject to different steric and electrostatic constraints, the surface can be reasonably charac-terized by means of a distribution function.[54] This parametrization is however

[54] R. J. D'Arcy and I. C. Watt, *Trans. Faraday Soc.*, 1970, **66**, 1236.

insufficient for the description of adsorption of polar gases. In fact, in this review the heterogeneity has been considered as being fixed *a priori*. However, especially in biological phenomena, induced heterogeneity may be extremely relevant. Induced heterogeneity denotes the following phenomenon: the adsorption energy on a free site of a partially covered, initially homogeneous surface, differs from the adsorption energy of the virgin surface. Induced heterogeneity can be distinguished as due to lateral interactions and to allosteric effects. While lateral interactions can be taken into account even in the expression of the adsorption isotherm (in this case the Fowler–Guggenheim and the Hill–de Boer isotherms replace the Langmuir and the Volmer isotherms, respectively) allostericity needs a particular analysis. Allostericity is that phenomenon where the formation of an adsorption bond modifies the structure of the substrate and, consequently, of the remaining free sites: allostericity is positve or negative according to whether the newly-formed bond increases or decreases the adsorption energy of the other sites.

A detailed analysis of allosteric effects is beyond the aims of this work; however it is possible, in the first approximation, to separate the description of heterogeneity from that of allostericity.[55] In this separate description it should be very interesting to verify whether or not proteins can be described by a distribution such as (28), *i.e.* whether or not proteins can be considered as 'equilibrium surfaces'. If the answer were affirmative, it would be very interesting to look for the biological meaning of the temperature at which the protein was grown under equilibrium conditions.

Topography.—When lateral interactions are excluded, surface topography is irrelevant. Hence the choice of the Langmuir isotherm as kernel in equation (4) allows one to describe patchwise, correlated as well as random heterogeneity.

But, as soon as lateral interactions become relevant, surface topography influences the phenomenon. For instance, the derivation of Section 5, relating the extra energy to the lateral energy, applies only to patchwise heterogeneity. In the subsection on Films it is however explained how, according to the particular preparation method or to the treatment undergone, completely different situations may characterize a given solid surface.

While for patchwise heterogeneous surfaces lateral interactions can again be described simply by equation (4) with a kernel accounting for lateral interactions (for instance, Ross and Olivier[56] considered in detail the case of the Hill–de Boer local isotherm), random heterogeneity requires a new analysis. Much work remains to be done in this field; however, Rudziński and his school have already extended both the condensation approximation[57,58] and the asymptotically correct approximation[59] in order to take random heterogeneity into account. The results obtained by the Polish scientists are encouraging as

[55] G. F. Cerofolini and M. Cerofolini, *J. Colloid Interface Sci.*, 1980, **78**, 65.
[56] S. Ross and J. P. Olivier, 'On Physical Adsorption,' Interscience, New York, 1964.
[57] W. Rudziński, L. Łajtar, and A. Patrykiejew, *Surf. Sci.*, 1977, **67**, 195.
[58] M. Jaroniec and A. Patrykiejew, *Phys. Lett. A*, 1978, **67**, 309.
[59] W. Rudziński, A. Patrykiejew, and L. Łajtar, *Surf. Sci.*, 1978, **77**, L655.

some discrepancies between computations and experimental determinations have been removed.

9 Conclusions

The DR isotherm, initially proposed by Dubinin and Radushkevich for the description of adsorption on porous adsorbents, was found by Hobson to describe adsorption in submonolayer range on non-porous surfaces too. This empirical discovery was a challenge to theoreticians since the DR isotherm was expressed in terms of the Polanyi potential, which is expected not to apply to adsorption in the submonolayer range.

This difficulty can be removed by ascribing the DR behaviour on non-porous surfaces to their intrinsic energy heterogeneity, and modifying the original expression according to the suggestions inferred from this assumption.

The DR isotherm, even when modified to avoid the Polanyi potential, does not describe adsorption at very low pressure because it does not reduce to the Henry behaviour. Soon after the discovery of the applicability of the DR isotherm to submonolayer adsorption, Hobson and Armstrong realized this difficulty and empirically proposed a rule to predict the switching from the DR to the Henry behaviour. Following the line of thought which ascribes the DR behaviour to surface heterogeneity, the Hobson–Armstrong rule is related to a finite upper adsorption energy: the value of this energy can be obtained in a self-consistent way.

The isotherm obtained by modifying the original DR equation describes such a large number of solid-vapour systems to deserve an accurate analysis in order to establish whether or not it is related to a particular surface structure.

A simplified model of equilibrium surface suggests that the DR behaviour is observed in low-pressure adsorption on patchwise, weakly heterogeneous surfaces which were grown in equilibrium conditions and hence were quenched at the adsorption temperature. At higher pressures, these surfaces should exhibit the Freundlich behaviour, while in the case of strong heterogeneity adsorption should be described by the Temkin isotherm. The three classic empirical isotherms, Freundlich, Dubinin–Radushkevich, Temkin, seem therefore to be related to adsorption on equilibrium surfaces, and the explanation of these experimental behaviours can be seen as a new chapter of the theory of adsorption: the theory of physical adsorption on equilibrium surfaces.

Acknowledgments.—Though the responsibility for the opinions herein expressed is only mine, I wish to thank many friends and colleagues for their helpful comments: in particular, Prof. P. Cavallotti (Polytechnic of Milan), Prof. J. G. Dash (University of Washington), Dr. J. P. Hobson (National Research Council, Ottawa), Dr. M. Jaroniec and Dr. W. Rudziński (University Maria Curie Skłodowska, Lublin), Dr. C. Pisani (University of Turin), and Prof. B. W. Wojciechowski (Queen's University, Kingston).

I take the opportunity to thank Ms. Christine Porter for her help in rendering more English my poor English.

3
Adsorption from Solution

BY J. DAVIS AND D. H. EVERETT

1 Introduction

Over the past decade a general consensus has developed concerning the formulation of the thermodynamics of adsorption from solution, and although the various treatments differ in detail there are no major areas of controversy.[1-3] Several papers have, however, appeared recently presenting variations on earlier formulations and these are discussed below.

Two broad approaches may be identified. First, and in many ways preferable, are purely thermodynamic methods in which no appeal is made to physical models of the adsorption process and the derived quantities can be calculated from primary experimental data. However to be meaningful a full thermodynamic analysis requires data of high accuracy covering a range of temperature, preferably supplemented by calorimetric measurements. Furthermore, since adsorption represents an equilibrium between material in the bulk and surface regions, information about the thermodynamic properties of the interface requires knowledge of the properties of the bulk phase. All too often one finds that even when adequate adsorption data are available a proper thermodynamic analysis is severely limited by the absence of reliable information (and in particular activity coefficients) on the bulk equilibrium solution.

The basic equation in this approach, derived from the Gibbs adsorption equation is[4]

$$\sigma - \sigma_2^* = -RT \int_{x_2^l=1}^{x_2^l} \frac{\Gamma_2^{(n)}}{x_1^l x_2^l \gamma_2^l} d(x_2^l \gamma_2^l), \tag{1}$$

where σ and σ_2^* are the interfacial tensions at the {solution of mole fraction (x_1^l, x_2^l)}/solid and pure liquid 2/solid interfaces, respectively. $\Gamma_2^{(n)}$ is the areal surface excess of component 2 (reduced adsorption):

$$\Gamma_2^{(n)} = n^0 \Delta x_2^l / m a_s, \tag{2}$$

where Δx_2^l is the change in mole fraction of 2 in the liquid when a mass m of

[1] D. H. Everett, in 'Colloid Science', ed. D. H. Everett, (Specialist Periodical Reports), The Chemical Society, London, 1973, Vol. 1, Ch. 2.
[2] C. E. Brown and D. H. Everett, in 'Colloid Science', ed. D. H. Everett, (Specialist Periodical Reports), The Chemical Society, London, 1975, Vol. 2, Ch. 2.
[3] D. H. Everett and R. T. Podoll, in 'Colloid Science', ed. D. H. Everett, (Specialist Periodical Reports), The Chemical Society, London, 1979, Vol. 3, Ch. 2.
[4] Ref. 3, p. 63, eqn. (1)

solid of specific surface area a_s is equilibrated with an amount n^0 of solution; and γ_2^l is the activity coefficient of 2 in the bulk solution.

Equation (1) implies that the specific surface area of the solid is known. As pointed out by Schay[5] this is not always known, or even knowable, so that in a more general treatment ε, called by Schay the specific excess Gibbs energy, is used in place of σ, the two being linked formally by the relation $\varepsilon = \sigma a_s$. In equation (1), $\varepsilon - \varepsilon_2^*$ then appears on the left-hand side and the specific surface excess, $n_2^{\sigma(n)}/m$, replaces $\Gamma_2^{(n)}$ on the right-hand side. Despite the formal correctness of this argument it remains conventional to analyse data in terms of σ. One advantage of this procedure is that, even when the surface area of the solid is known only approximately, one can compare the magnitude of σ with those characteristic of liquid/vapour and liquid/liquid interfaces, which can be measured by direct experiment.

If measurements are made over a range of temperature, then the differences $(\Delta_w \hat{h} - \Delta_w \hat{h}_2^*)$ between the enthalpy of immersion of unit area of solid in solution and pure liquid 2 can be calculated:[6]

$$\Delta_w \hat{h} - \Delta_w \hat{h}_2^* = \left[\frac{\partial}{\partial(1/T)} \frac{(\sigma - \sigma_2^*)}{T} \right]_{x_2^l}. \tag{3}$$

The corresponding difference between entropies of immersion is

$$\Delta_w \hat{s} - \Delta_w \hat{s}_2^* = \{(\Delta_w \hat{h} - \Delta_w \hat{h}_2^*) - (\sigma - \sigma_2^*)\}/T. \tag{4}$$

If the analysis is made in terms of ε, then the enthalpies and entropies of immersion refer to unit mass of solid.

An alternative approach which makes less demand on the experimental data, and which is therefore useful in a preliminary study, is that in which simple models of adsorption are postulated and their predictions compared with experimental results. The conclusions to be drawn from an analysis of this kind are of course critically dependent on the assumptions of the model, which may not be amenable to independent verification. The most popular of these models is the 'surface layer' model in which the fluid phase is divided into two regions, supposed homogeneous. In surface layers the mole fractions (in a binary mixture) are x_1^σ and x_2^σ, whereas in the bulk phase they are x_1^l and x_2^l. If the surface layer is pictured as t molecular layers thick then the surface mole fraction can be calculated from the observed areal surface excess, $\Gamma_2^{(n)}$, using the formula[7]

$$x_2^\sigma = \frac{t x_2^l + a_1^* \Gamma_2^{(n)}}{t - (a_2^* - a_1^*)\Gamma_2^{(n)}}. \tag{5}$$

Here a_i^*/N_A is the area subtended at the surface by a molecule of component i. N_A is Avogadro's constant. This calculation calls immediately for knowledge of, or assumptions concerning, a_s, a_1^*, a_2^*, and t. It is usually supposed that a sufficiently reliable value of a_s is available, and a_1^* and a_2^* may be estimated either from molecular models or from the bulk-liquid density assuming the

[5] G. Schay, *J. Colloid Interface Sci.*, 1973, **42**, 478; ref. 2, p. 54.
[6] Ref. 3, p. 65, eqn. (12).
[7] Ref. 1, p. 52, eqn. (9); p. 67, eqn. (63).

molecules to be effectively spherical and close-packed in the liquid. A preliminary assumption often made is that the adsorption is limited to a monolayer. The acceptability of this assumption may be tested by calculating x_2^σ as a function of x_2^l and checking that the following criteria (Rusanov's criteria) are satisfied[8]

$$x_2^\sigma < 1; \qquad (\partial x_2^\sigma / \partial x_2^l) > 0. \tag{6}$$

If these are not satisfied then successive values $t = 2, 3$ are tried until x_2^σ behaves in an acceptable manner. This defines the *minimum* adsorbed layer thickness. Calculation of x_2^σ through equation (5) with $t = 1$ is sometimes called[9] the Schay–Nagy method, although this equation seems first to have been derived by Elton.[10] The simpler calculation applicable to monolayer adsorption of equal sized molecules: $(a_1^* = a_2^* = a^*)$;

$$x_2^\sigma = x_2^l + a^* \Gamma_2^{(n)}, \tag{7}$$

has similarly been called the Everett method.

It is important, as stressed later,[11] to realise that equations (5) and (7) in the case where different molecular areas are attributed to the components must inevitably lead to different results, so that any attempt to 'compare' the two methods is meaningless. When x_2^σ has been calculated it is then possible to derive activity coefficients in the adsorbed layer using the equation[12]

$$\ln \gamma_2^\sigma = \ln \frac{x_2^l \gamma_2^l}{x_2^\sigma} + a_2 (\sigma - \sigma_2^*)/RT, \tag{8}$$

where $(\sigma - \sigma_2^*)$ is calculated from equation (1). If the layer thickness is t, $a_2 = a_2^*/t$. If the molecular areas are equal then an alternative but less convenient equation is[13]

$$\ln \gamma_2^\sigma = x_1^\sigma \ln \frac{x_1^\sigma x_2^l \gamma_2^l}{x_2^\sigma x_1^l \gamma_1^l} - \int_0^{x_1^\sigma} \ln \frac{x_1^\sigma x_2^l \gamma_2^l}{x_2^\sigma x_1^l \gamma_1^l} \, dx_1^\sigma. \tag{9}$$

Again papers which attempt to compare the results of calculations based on equations (8) and (9) are largely irrelevant. If the molecules are of equal size, then identical results should be obtained, if they are of different size then different results are inevitable, and that derived from (9) is wrong anyway.

2 Thermodynamics

General Considerations.—A series of papers by Vernov and Lopatkin[14–17] provide a useful review and restatement of the Gibbs concept of surface excess quantities, particular attention being paid to the variables which must be specified to define unambiguously a suitable reference system. The first

[8] Ref. 1, p. 66.
[9] See p. 132.
[10] G. A. H. Elton, *J. Chem. Soc.*, 1951, 2958.
[11] See p. 132.
[12] Ref. 1, p. 58, eqn. (35); (see errata, ref. 2, p. 314).
[13] Ref. 1, p. 58, eqn. (32).

paper[14] includes a discussion of the application of the phase rule to heterogeneous systems in which interfacial effects are important. In effect while the conventional Gibbs phase rule takes no account of the sizes of the coexisting phases, it is necessary for finely divided systems to make allowance for the finite dimensions of the phases. The second paper[15] is specifically concerned with adsorption where three types of theory are identified: (i) those in which the transition region is treated as a physical phase, (ii) strictly formal theories closely related to Gibbs approach, and (iii) theories directly related to experimental methods of measuring adsorption. These authors adopt an essentially formal algebraic approach and derive standard equations relating this approach to experimental measurements. They stress a point which is becoming increasingly important in the development of the thermodynamics of adsorption, namely that 'the working formulae given by the theory should include only those quantities which can be measured directly or can be calculated from primary experimental data without introducing arbitrary models'. On this basis they reject theories of type (i) because (although logically satisfying) they have the major 'shortcoming that total quantities relating to some artificially defined layer appear in its equations, whereas only excess quantities are measured in adsorption experiments'. In this connection it should be noted that although the apparent physical reality of the models of this kind is obtained by introducing the thickness, τ, of the interfacial layer, in practical applications it is necessary either to adopt some arbitrary figure or, more usually,[18] to allow τ to tend to zero. The resulting equations, relating to an interfacial layer of infinitesimal thickness correspond mathematically to taking a single dividing surface and of course become identical with those resulting from the conventional Gibbs formulation. The third paper[16] deals specifically with adsorption at the gas/solid interface and is not of immediate relevance to the present discussion. The fourth paper[17] returns to the problems of the solid/liquid interface and in particular to various ways in which adsorption from solution can be measured, and to the relationships between them. Three conventional methods of measurement are considered leading to three measures of adsorption:

$$\Gamma_i^{(n)} = \frac{n^0(x_i^0 - x_i^l)}{ma_s}, \tag{10}$$

$$\Gamma_i^{(w)} = \frac{W^0(w_i^0 - w_i^l)}{ma_s M_i}, \tag{11}$$

$$\Gamma_i^{(v)} = \frac{V^0(c_i^0 - c_i^l)}{ma_s}. \tag{12}$$

[14] A. V. Vernov and A. A. Lopatkin, *Zh. Fiz. Khim.*, 1979, **53**, 2333 (*Russ. J. Phys. Chem.*, 1979, **53**, 1327).

[15] A. V. Vernov and A. A. Lopatkin, *Zh. Fiz. Khim.*, 1979, **53**, 3161 (*Russ. J. Phys. Chem.*, 1979, **53**, 1813).

[16] A. V. Vernov and A. A. Lopatkin, *Zh. Fiz. Khim.*, 1980, **54**, 2327 (*Russ. J. Phys. Chem.*, 1980, **54**, 1324).

[17] A. V. Vernov and A. A. Lopatkin, *Zh. Fiz. Khim.*, 1981, **55**, 438 (*Russ. J. Phys. Chem.*, 1981, **55**, 240).

[18] *e.g.* E. A. Guggenheim, 'Thermodynamics', North Holland, Amsterdam, 5th Edn., 1967, pp. 161, 207.

In these equations x_i^0, w_i^0, and c_i^0 are, respectively, the mole fraction, mass fraction, and molar concentration of initial solutions, containing an amount n^0, a mass W^0, or having a volume V^0, which after equilibration with a mass m of solid of specific surface area a_s are characterized by final values of x_i^l, w_i^l, and c_i^l. M_i is the molar mass of component i.

These methods of measurement lead to values for the surface excess of component i defined as the difference between the amount of i actually present in the final equilibrium state, and that which would have been present in the absence of adsorption. Thus the first terms in the numerators $n^0 x_i^0$, $W^0 w_i^0$, $V^0 c_i^0$ respectively establish the total amounts of i actually present, whereas $n^0 x_i^l$, $W^0 w_i^l$, $V^0 c_i^l$ are the amounts of i which would have been present if, respectively, the mole fraction of equilibrium liquid had been the same throughout the n^0 moles of liquid; w_i^l had been constant throughout the mass W^0; and the concentration c_i^l had been constant throughout V^0. Vernov and Lopatkin draw attention to the ambiguity in the third definition if the adsorption process is accompanied by a change in volume. If the final volume of the whole system is V and it is assumed that the volume V^s of the solid is unchanged by adsorption, then the volume V^l attributable to the liquid phase is $(V - V^s)$. The amount of i which would be present if c_i^l were uniform throughout V^l is $c_i^l V^l$ so that the adsorption should be defined by

$$\Gamma_i = \frac{(V^0 c_i^0 - V^l c_i^l)}{ma_s}. \tag{13}$$

Vernov and Lopatkin call Γ_i the Gibbs adsorption, although it is not clear why this term should be assigned specifically to this quantity. If the difference $V^l - V^0$ is called the excess volume V^{ex} then it follows that

$$\Gamma_i = \Gamma_i^{(v)} - \frac{V^{ex} c_i^l}{ma_s}, \tag{14}$$

which can be calculated provided $\Gamma_i^{(v)}$ and V^{ex} are known. Equation (14) is, incidentally, important in the case of a one-component system when $\Gamma_i^{(v)} = 0$ and

$$\Gamma_1 = -\frac{V^{ex}}{ma_s} \cdot \frac{M}{\rho^l}, \tag{15}$$

where ρ^l is the density of the liquid and M its molar mass. The other definitions of adsorption are applicable only to multicomponent systems.

The bulk of this paper is concerned with relationships between these definitions. The more important equations are the following:

$$\Gamma_i^{(n)} = \Gamma_i^{(v)} - \left(\frac{v^l - v^0}{v^0}\right) \frac{c_i^l V^0}{ma_s} = \Gamma_i^{(v)} - x_i^l \sum_{j=1}^{c} \Gamma_j^{(v)}, \tag{16}$$

$$\Gamma_i^{(n)} = \Gamma_i - \frac{(V^0 v^l - v^0 V^l)}{v^0} \frac{c_i^l}{ma_s} = \Gamma_i - x_i^l \sum_{j=1}^{c} \Gamma_j, \tag{17}$$

with similar equations for $\Gamma_i^{(w)}$. Here v^0 and v^l are the mean molar volumes of the liquid at mole fractions x_i^0 and x_i^l. However, the measure of adsorption which plays the key role in the thermodynamics of adsorption is the relative adsorption of i with respect to component 2 defined, in general, by

$$\Gamma_i^{(1)} = \Gamma_i^* - \Gamma_1^* \frac{\Delta c_i}{\Delta c_1}, \tag{18}$$

where Γ_i^* and Γ_1^* are Gibbs adsorptions (in the conventional sense) defined with respect to an arbitrary choice of dividing surface and Δc_i and Δc_1 are the differences between the concentrations of i and 1 between the two phases. For a liquid/(impenetrable solid) interface

$$\Gamma_i^{(1)} = \Gamma_i^* - \Gamma_1^* \frac{c_i^l}{c_1^l} = \Gamma_i^* - \Gamma_1^* \frac{x_i^l}{x_1^l}. \tag{19}$$

It may then be shown that, irrespective of whether a volume change occurs on adsorption,

$$\Gamma_i^{(1)} = \Gamma_i - \Gamma_1 \frac{c_i^l}{c_1^l} = \Gamma_i^{(v)} - \Gamma_1^{(v)} \frac{c_i^l}{c_1^l} = \Gamma_i^{(n)} - \Gamma_1^{(n)} \frac{x_i^l}{x_1^l} = \Gamma_i^{(w)} - \Gamma_1^{(w)} \frac{w_2^l}{w_1^l} \frac{M_1}{M_i}. \tag{20}$$

In the particular case of a binary solution it follows that

$$\Gamma_2^{(1)} = \Gamma_2^{(n)}/x_1^l. \tag{21}$$

It would appear from this analysis of Vernon and Lopatkin's paper that although they are correct in pointing out that the conventional definition (12) of $\Gamma_i^{(v)}$ ignores the change in volume of the system accompanying adsorption, when employing the measured adsorptions in thermodynamic equations (*e.g.* the Gibbs adsorption isotherm) the effect of this discrepancy cancels out in the calculation of the relative adsorption. Any one of the equations (10), (11), (12), or (13) may thus be used in thermodynamic calculations, by substitution in the appropriate equality in (20).

The problems in correlating the various reformulations of adsorption thermodynamics arise partly because the underlying assumptions are not always made explicit, and partly because of the wide range of notations which are employed.

Tolmachev [19] discusses the distinction between what he calls the internal ($\overline{\mu_i}$) and total ($\overline{\phi_i}$) chemical potentials of a component in the surface

$$\overline{\phi}_i = \overline{\mu}_i - s_i\pi = \overline{\mu}_i^0(T, p) + RT \ln \overline{c}_i\overline{\gamma}_i - s_i\pi, \tag{22}$$

where $\overline{\mu}_i^0(T, p)$ is a standard chemical potential, and $\overline{c}_i$, $\overline{\gamma}_i$ refer to the concentration and activity coefficient of component i in the adsorption space.† The interpretation of s_i and π_i depends on the type of system under consideration. If a 'surface layer' model is considered π is the surface tension of the layer and

---

† It is misleadingly stated (at least in the translated version) that $\overline{c}_i$ and $\overline{\gamma}_i$ are the bulk concentration and activity coefficient, respectively, but later it is clear that symbols with a superscript bar refer to the surface whereas c_i and γ_i refer to the bulk.

[19] A. M. Tolmachev, *Zh. Fiz. Khim.*, 1978, **52**, 1050 (*Russ. J. Phys. Chem.*, 1978, **52**, 597).

s_i is the molar area of i; but for micropore filling π is the 'internal pressure' and s_i the molar volume of i.

Tolmachev then considers two alternative ways of describing the system. In the first, the adsorbed components are chosen as the components of the adsorbed phase: the sorbent is taken only as defining the location of the adsorbed phase, and variations in its properties arising from adsorption are included in the $s_i\pi$ terms. In agreement with earlier work, the equilibrium condition is

$$\phi_i = \mu_i \quad \text{(all } i\text{)},\tag{23}$$

or $\mu_i^\sigma = \mu_i^l$ in the notation of refs. 1—3, where

$$\mu_i^l = \mu_i^{0,l} + RT \ln x_i^l \gamma_i^l.\tag{24}$$

Alternatively, the sorbent is included as a component with a chemical potential $\bar{\mu}_R$ per mole of adsorption sites R. Adsorption is then regarded as the formation of adsorption complexes between i and R (denoted by iR): this is the essence of Tolmachev's so-called 'stoichiometric theory' of adsorption. The equilibrium condition now takes the form

$$\frac{(\bar{\mu}_{iR} - \mu_i^l)}{s_i} = \frac{\bar{\mu}_R}{s_R}, \quad \text{(all } i\text{)},\tag{25}$$

where $\bar{\phi}_{iR}$ has been equated with $\bar{\mu}_{iR}$ (the argument leading to this identification is slightly obscure). This corresponds to the process $\beta_i i + R = \beta_i(i R)$, where $\beta_i = s_R/s_i$.

The former method of analysis is said to be not always convenient since, because π is not directly measurable, the adsorption isotherm for a single adsorbate cannot be obtained:

$$\bar{c}_i\bar{\gamma}_i = c_i\gamma_i \exp[(\mu_i^0 - \bar{\mu}_i^0 + s_i\pi)/RT].\tag{26}$$

This argument ignores the fact that by considering the displacement equilibrium at constant surface area, terms in π can be eliminated, and also that in any case use of the Gibbs equation enables π to be calculated. This distinction between alternative formulations is, of course, the essence of the difference between 'adsorption theory' and 'solution theory' of the adsorption of vapours as developed by Hill[20] many years ago.

The model of adsorption complexes is then used by Tolmachev and Trubnikov[21] to discuss the case in which some adsorption sites are unoccupied. Equilibrium constants of the following form, for the adsorption of component A, *i.e.* formation of an adsorption complex AR from A plus a vacant site R are written down:†

$$\frac{(\bar{c}_{AR}\bar{\gamma}_{AR})^{\beta_A}}{(c_A\gamma_A)^{\beta_A}\bar{c}_R\bar{\gamma}_R} = \bar{K}_A = \frac{\bar{\gamma}_{AR}^{\beta_A}}{\gamma_A^{\beta_A}\bar{\gamma}_R} \cdot K_A.\tag{27}$$

---

† The omission in the paper of the index β_A in the denominator of the last term is presumed to be a printing error.

[20] T. L. Hill, *J. Chem. Phys.*, 1950, **18**, 246.
[21] A. M. Tolmachev and I. B. Trubnikov, *Zh. Fiz. Khim.*, 1978, **52**, 746 (*Russ. J. Phys. Chem.*, 1978, **52**, 422).

This introduces the concept of an activity coefficient for free sorption centres which is not further defined.

For the exchange reaction‡

$$A + \beta BR \rightleftharpoons AR + \beta B,$$

where $\beta = \beta_B/\beta_A$, the equilibrium constant is given by

$$\frac{\bar{c}_{AR}\bar{\gamma}_{AR}(c_B\gamma_B)^{\beta}}{c_A\gamma_A(\bar{c}_{BR}\bar{\gamma}_{BR})^{\beta}} = \bar{K}_c = K_c \frac{\bar{\gamma}_{AR}\gamma_B^{\beta}}{\gamma_A\bar{\gamma}_{BR}^{\beta}}. \tag{28}$$

This is essentially the equation widely used in conventional descriptions where the 'adsorption complexes' AR and BR are simply A and B in the adsorbed state.[22] By combination of these equations with a Gibbs–Duhem equation formulated for the sorption phase treated as a homogeneous solution, expressions are derived for $\bar{\gamma}_{AR}$, $\bar{\gamma}_{BR}$, $\ln \bar{K}_A$ and $\ln K_c$ for three cases: (a) adsorption of a single substance (presumably from the vapour), (b) adsorption from a binary mixture in which a certain constant number of adsorption sites remain empty, and (c) adsorption from a binary mixture which fully saturates the surface ($\bar{c}_R = 0$).

The question of alternative definitions of adsorption equilibrium constants is discussed by Avramenko, Glushchenko, and Shesterkin.[23] In very dilute solution an equilibrium constant, which is in effect the limiting Henry's law constant for the process, is defined by

$$K_H^0 = \underset{c_a \to 0}{Lt} (c_a/c_s), \tag{29}$$

where c_a and c_s are the concentrations in the adsorbed and bulk phases, respectively. (Although the authors talk of concentrations and use the symbol c, they later specify that the concentrations should be expressed in terms of mole fractions). Alternatively, by integration of the following expression involving the surface excess isotherm $\Gamma_1^{(n)} = n^0 \Delta x_1^l/ma_s$, we have, as already discussed in previous Reports[24]

$$\ln K_a = a_1 \int_{x_1=0}^{x_1=1} \frac{\Gamma_1^{(n)}}{(1-x_1^l)x_1^l\gamma_1^l} \cdot d(x_1^l\gamma_1^l), \tag{30}$$

where K_a is an equilibrium constant which in terms of an 'adsorbed phase' formulation is

$$K_a = \frac{x_1^{\sigma}\gamma_1^{\sigma}}{x_1^l\gamma_1^l}\left(\frac{x_2^l\gamma_2^l}{x_2^{\sigma}\gamma_2^{\sigma}}\right)^r \tag{31}$$

where r is the ratio (a_1/a_2) of sizes of the two molecules. In dilute solution of

---

‡ The authors' notation $A + \beta BR_{1/\beta_B} = AR_{1/\beta_A} + \beta B$ seems unnecessarily complicated.

[22] See ref. 1, p. 53, eqn. (14').

[23] V. A. Avramenko, V. Yu. Glushchenko, and V. P. Shesterkin, *Zh. Fiz. Khim.*, 1977, **51**, 2304 (*Russ. J. Phys. Chem.*, 1977, **51**, 1348).

[24] Ref. 1, p. 58; ref. 2, p. 68; ref. 3. p. 66.

1, when $x_2^l, \gamma_2^l \to 1$ and also $x_2^\sigma, \gamma_2^\sigma \to 1$

$$K_a = K_H^0 \left(\frac{\gamma_1^\sigma}{\gamma_1^l} \right)_\infty , \tag{32}$$

where $(\gamma_1^\sigma/\gamma_1^l)_\infty$ is the ratio of the activity coefficients at infinite dilution. Thus by comparing K_a calculated from equation (31) with the limiting slope of the adsorption isotherm the ratio of activity coefficients at infinite dilution can be derived, and hence if $(\gamma_1^l)_\infty$ is known, $(\gamma_1^\sigma)_\infty$ is found. This analysis has been applied to the adsorption of propan-1-ol, acetone, and p-dioxan from aqueous solutions by activated charcoal; its application to the adsorption of nitro-phenols is discussed later (p. 112).

In an attempt to elucidate the mechanism of inhibition of the corrosion of metal surfaces by adsorbed substances, Agres and his co-workers have consi-dered the various thermodynamic factors that determine the extent of adsorp-tion.[25,26] In the first of these papers[25] Agres, Altsybeeva, Levin, and Fedorov discuss the competition between various ways in which a given molecule can be adsorbed. If two different modes of attachment (i, j) are possible and can co-exist on the surface, then using standard equations, assuming that the two forms of attachment involve equal areas (a_0) of the surface and have the same activity coefficients in the adsorbed state, it follows that the ratio of the mole fractions of the two forms on the surface is given by

$$\frac{x_{j,ads.}}{x_{i,ads.}} = \exp\left[\frac{(\sigma_i - \sigma_j)a_0}{RT} \right]. \tag{33}$$

Here σ_i and σ_j are the interfacial tension of an adsorption layer consisting, respectively, of the pure i or pure j form. It is supposed that the difference of free energies $(\sigma_i - \sigma_j)$ can be taken as approximately equal to the differences $(U_i - U_j)$ between the energies of interaction of the i and j forms with the adsorbent surface. If several methods of attachment $i, j, k, \ldots$ are possible then

$$x_i : x_j : x_k \ldots = 1 : \exp(U_i - U_j)/RT : \exp(U_i - U_k)/RT : \ldots \tag{34}$$

As an example, the energies of possible modes of attachment of protonated pyridine and several of its methyl derivatives to the [100] face of the body-centered α-iron lattice were calculated by the Hückel molecular orbital method: the form in which the charged N atom lies over an Fe atom in the surface is so much more favourable energetically that other modes of attach-ment are present only to the extent of 1—2%. This argument, however, ignores the possible difference between the entropies of molecules adsorbed in different ways.

The extent of adsorption depends not only on the relative energies of interaction of the solute (2) and solvent (1) with the surface, represented by the difference $(\sigma_2^* - \sigma_1^*)$ between the interfacial tensions, but also by the deviations of the bulk and surface regions from ideality. In a second paper[26] the

[25] E. M. Agres, A. I. Altsybeeva, S. Z. Levin, and V. S. Fedorov, *Zh. Fiz. Khim.*, 1977, **51,** 165 (*Russ. J. Phys. Chem.*, 1977, **51,** 91).
[26] E. M. Agres, A. I. Altsybeeva, V. S. Fedorov, and S. Z. Levin, *Zh. Fiz. Khim.*, 1977, **51,** 168 (*Russ. J. Phys. Chem.*, 1977, **51,** 92).

equilibrium condition for adsorption of a corrosion inhibitor (I) from aqueous solution is expressed in the usual way

$$RT \ln \frac{\gamma_I^\sigma x_I^\sigma}{\gamma_I^l x_I^l} \left(\frac{\gamma_{H_2O}^l x_{H_2O}^l}{\gamma_{H_2O}^\sigma x_{H_2O}^\sigma} \right)^r = (\sigma_{H_2O}^* - \sigma_I^*) a_I^0, \tag{35}$$

and the activity coefficients given a form appropriate to regular behaviour in both bulk and surface regions. Discussion is limited to the case in which $r = 1$ and the interaction parameter of the regular solution theory (denoted here by ε) is assumed to be the same in both regions. In dilute solution the adsorption equation expressed in terms of the fractional coverage θ, by component I becomes

$$\frac{\theta}{1-\theta} \exp\left[\frac{z\varepsilon(2l\theta + m)}{RT} \right] = x_I^l \exp\left[\frac{(\sigma_{H_2O} - \sigma_I) a_I^0}{RT} \right], \tag{36}$$

where z is the co-ordination number of the system and l, m the fractions of neighbouring sites in the same and adjacent lattice planes. This equation is of the same form as the Frumkin equation.[27] High adsorption (high surface activity) of the inhibitor is favoured by low values of σ_I and by increasing values of ε, *i.e.* increased positive deviations of the bulk solution from Raoult's law.[28] These considerations are employed to explain why, on the basis of adsorption factors alone, pyridine (whose solutions in water exhibit in dilute solution large positive deviations from Raoult's law) protects iron from corrosion whereas dioxan (whose activity coefficients are smaller) does not. This interpretation involves the assumption that since the first ionization potentials of the two molecules are roughly equal, their adsorption energies, and hence $(\sigma_I^* - \sigma_{H_2O}^*)$, will be approximately the same. The paper then argues that since the activity coefficients of homologous series in aqueous solution (*e.g.* aliphatic alcohols) increase with increasing chain length, they should be increasingly adsorbed at a metal/aqueous medium interface. This is illustrated by considering the interfacial tensions between solutions of alcohol and an uncharged mercury interface.

It is appreciated that these predictions ignore the effect of size differences which can act in an opposing fashion.[29] Agres[30] distinguishes those cases (*e.g.* straight-chain aliphatic compounds) in which the molecules are adsorbed normal to the surface and the areas they occupy vary little with chain length, from those in which larger solute molecules displace an increasing amount of water. In the former case bulk-phase non-ideality (increasing hydrophobicity) is a major factor affecting adsorption. It is further argued that protonation of an organic base will, in aqueous solution, be accompanied by strong hydration, which will reduce the hydrophobicity and diminish surface activity. The application of these ideas to the mercury/water interface is dealt with in a further paper by Agres.[31]

[27] A. N. Frumkin, *Tr. Inst. Karpova*, 1925, No. 4, 56; 1926, No. 3, 3; *Z. Phys.*, 1926, **35,** 729; *cf.* ref. 3, pp. 70—71.
[28] See *e.g.* D. H. Everett, *Trans. Faraday Soc.*, 1965, **61,** 247, Figure 1.
[29] See ref. 3, p. 70.
[30] E. M. Agres, *Zh. Fiz. Khim.*, 1978, **52,** 985 (*Russ. J. Phys. Chem.*, 1978, **52,** 560).
[31] E. M. Agres, *Zh. Fiz. Khim.*, 1978, **52,** 989 (*Russ. J. Phys. Chem.*, 1978, **52,** 562).

Surface Activity Coefficients.—The problem of the calculation of activity coefficients in the surface region continues to be discussed. The two alternative equations available for this purpose are equations (8) and (9). The latter is applicable only to mixtures of equal size molecules. A more general form of this equation, valid for molecules of different size, has been derived by Kazakevich and Eltekov.[32] The adsorption equilibrium constant is written in the form of equation (31). Taking logarithms, differentiating and combining with the Gibbs–Duhem equation for the surface layer:†

$$x_1^\sigma \, d \ln \gamma_1^\sigma + x_2^\sigma \, d \ln \gamma_2^\sigma = 0 \tag{37}$$

leads to

$$\ln \gamma_1^\sigma = \frac{x_2^\sigma}{rx_1^\sigma + x_2^\sigma} \ln \left[\frac{x_1^l \gamma_1^l}{x_1^\sigma} \left(\frac{x_2^\sigma}{x_2^l \gamma_2^l} \right)^r \right] - r \int_0^{x_2} \frac{\ln \left[\frac{x_1^l \gamma_1^l}{x_1^\sigma} \left(\frac{x_2^\sigma}{x_2^l \gamma_2^l} \right)^r \right]}{(rx_1^\sigma + x_2^\sigma)^2} \, dx_2^\sigma, \tag{38}$$

which reduces to equation (9) when $r = a_1/a_2 = 1$. Although equation (38) is formally correct and thermodynamically equivalent to (8) it is less convenient to use, and equation (8) is generally to be preferred.

In a second paper Kazakevich and Eltekov[33] consider the problem in terms of a layer of finite thickness, a treatment equally applicable to adsorption by microporous bodies where the adsorption volume is constant. They define a new system of activity coefficients (ξ_i) and for both bulk and adsorbed phases write

$$\mu_i^\alpha = \mu_i^{0,\alpha} + RT \ln \phi_i^\alpha \xi_i^\alpha, \tag{39}$$

where ϕ_i^α is the volume fraction of i in the phase α.

They then use as equilibrium condition

$$\mu_i^l = \mu_i^\sigma, \qquad i = 1, 2 \tag{40}$$

and by combining these equations obtain (for preferential adsorption of 2)

$$\frac{\phi_2^\sigma \xi_2^\sigma \phi_1^l \xi_1^l}{\phi_1^\sigma \xi_1^\sigma \phi_2^l \xi_2^l} = K_e. \tag{41}$$

However, the condition that the adsorption space shall remain full, *i.e.* on exchange 1 mole of component 1 of molar volume v_1 is replaced by v_1/v_2 moles of component 2, does not seem to have been taken account of. In effect the ξ_i^σ terms include 'surface tension' terms which do not cancel so that K_e is not a constant. The correct formulation in terms of volume fractions will still

† Since the Gibbs–Duhem equation for a bulk phase reduces to the form $x_1^l \, d \ln \gamma_1^l + x_2 \, d \ln \gamma_2^l = 0$ only when $dp = 0$, it might be expected that for a surface layer equation (37) would be valid only when $d\sigma = 0$. However as a result of defining the activity coefficients by equation (15) of ref. 1, the above equation (37) holds.

[32] Yu. V. Kazakevich and Yu. A. Eltekov, *Zh. Fiz. Khim.*, 1980, **54,** 154 (*Russ. J. Phys. Chem.*, 1980, **54,** 82).

[33] Yu. V. Kazakevich, and Yu. A. Eltekov, *Zh. Fiz. Khim.*, 1980, **54,** 156 (*Russ. J. Phys. Chem.*, 1980, **54,** 83).

be of the form $(r = v_2/v_1)$

$$\left(\frac{\phi_1^l \xi_1^l}{\phi_1^\sigma \xi_1^\sigma}\right)\left(\frac{\phi_2^\sigma \xi_2^\sigma}{\phi_2^l \xi_2^l}\right)^{1/r} = K, \tag{42}$$

with the activity coefficients ξ_i^α related to γ_i^α used previously by

$$\xi_2^l = \gamma_2^l(x_1^l/r + x_2^l) = \gamma_2^l \frac{x_1^l}{r\phi_1^l}. \tag{43}$$

This reformulation of the problem does not appear to have any significant advantage over the more conventional methods.

Ościk and Goworek[34] discuss the calculation of activity coefficients by equations (8) and (9). For systems of molecules of the same size the two methods should be equivalent. The authors have analysed the experimental data of Lu and Lama[35] for the adsorption from benzene + cyclohexane mixtures by silica gel. The graphically presented results indicate that the curves derived from equation (8) and (9) deviate considerably from one another. However, since no details are given of the values assumed for the molar areas it is not possible to identify the origin of these discrepancies, which because of the close similarity of the sizes of the molecules concerned, are unexpected. Madan[36] has used equation (9) to calculate the activity coefficients of benzene and acetic acid adsorbed from their mixtures by tin oxide gel ($a_s = 160 \text{ m}^2 \text{ g}^{-1}$). The results look surprising and consideration of the behaviour of $\ln(\gamma_2^\sigma/\gamma_1^\sigma)$ shows that the two sets of activity coefficients are thermodynamically inconsistent, presumably as a result of a computational error.

A third method of determining activity coefficients suggested by Ościk and Goworek[34] is based on measurements of enthalpies of immersion.

However, this employs a definition of activity coefficient completely unrelated to conventional definitions. Thus for an ideal system the enthalpy of immersion of unit area of solid in a large volume of solution (so that no appreciable change in the bulk solution composition occurs) may be written[37]

$$\Delta_w \hat{h} = x_1^\sigma \Delta_w \hat{h}_1^* + x_2^\sigma \Delta_w \hat{h}_2^*, \tag{44}$$

where $\Delta_w \hat{h}_i^*$ is the enthalpy of immersion of unit area of solid in pure i.

Ościk and Goworek then write for a non-ideal system

$$\Delta_w \hat{h} = f_1^\sigma x_1^\sigma \Delta_w \hat{h}_1^* + f_2^\sigma x_2^\sigma \Delta_w \hat{h}_2^*, \tag{45}$$

where f_i^σ is a quantity that they call an activity coefficient and imply that it is the same as the activity coefficient defined in terms of chemical potentials. They give an approximate method for finding f_1^σ and f_2^σ and present results based on Lu and Lama's experimental data for (benzene + cyclohexane)/silica

[34] J. Ościk and J. Goworek, *Bibl. Lubel. Tow. Nauk.*, (*Folia Societatis Scientiarum Lublinensis*), *Mat-Fiz-Chem.*, 1978, **20**, 17 (*Chem. Abstr.*, 1979, **90**, 61 752).
[35] C. Y. Lu and R. F. Lama, *Trans. Faraday Soc.*, 1967, **63**, 727.
[36] R. L. Madan, *J. Indian Chem. Soc.*, 1978, **55**, 66.
[37] D. H. Everett, *Trans. Faraday Soc.*, 1964, **60**, 1803.

gel: their variation with concentration in the liquid phase bear only a qualitative resemblance to that shown by conventional activity coefficients, and exhibit extreme values which 'are hard to explain'.

Several other papers on the evaluation of surface activity coefficients have appeared, among them that by Levchenko and Kirichenko,[38] who have studied the competitive adsorption of benzene and *p*-toluidine by active carbon from dilute aqueous solutions. The abstract does not give details. Thomas and Eon[39] in a paper discussed more fully later give surface activity coefficients for the three binaries derived from acetone, carbon tetrachloride, and benzene adsorbed by silica gel. They comment that the functions $\gamma_i^\sigma(x_i^\sigma)$ bear no resemblance to the corresponding bulk functions.

It has frequently been postulated that adsorption from solution can be described adequately by assuming that the adsorbed layer behaves ideally, whereas all deviations from ideal adsorption behaviour can be ascribed to the bulk solution. A less restricted condition is that the ratio $\gamma_2^\sigma/\gamma_1^\sigma$ is close to unity.

Experiments on systems which are near-ideal in the bulk show, however, that this approximation is not by any means justified generally. Thus measurements by Everett and Podoll[40] on the adsorption by graphitized carbon black (Graphon) from mixtures of bromobenzene + chlorobenzene show that near-ideal behaviour is shown in both bulk and surface regions. In contrast benzene + 1,2-dichlorobenzene mixtures, which are near-ideal in the bulk, show substantial negative deviations from ideality in the adsorbed state. Li and Gu[41] have similarly made a careful study of five near-ideal systems in which the adsorbent was a silica gel of surface area $420 \, \text{m}^2 \, \text{g}^{-1}$. In all cases the calculated surface activity coefficients assuming ideality in the bulk phase deviated substantially from unity. In Figure 1, $\ln \gamma_1^\sigma/\gamma_2^\sigma$ calculated from their results is shown as a function of x_1^σ. Only in the case of chlorobenzene + bromobenzene does the ratio $\gamma_1^\sigma/\gamma_2^\sigma$ differ from unity by less than 10% over the whole concentration range. It may be commented that if the analysis were completely self-consistent the integral $\int_0^1 \ln(\gamma_1^\sigma/\gamma_2^\sigma) \, dx_1^\sigma$ should be zero. It is clear from Figure 1 that in several cases this is not so, and we suggest that this discrepancy arises from the approximation of setting the bulk activity coefficients equal to unity in the calculations.

Thermodynamic Data.—While ultimately one hopes to understand adsorption phenomena in terms of molecular concepts, it is nevertheless of considerable interest to examine the behaviour of purely thermodynamic quantities as a function of experimental variables such as temperature and composition. Indeed analysis of data in this way is likely to indicate the direction in which theoretical treatments have to be developed.

[38] T. M. Levchenko, and V. A. Kirichenko, *Adsorbtsiya Adsorbenty*, 1977, **5**, 90 (*Chem. Abstr.*, 1978, **89**, 136 300).
[39] G. Thomas, and C. H. Eon, *J. Colloid Interface Sci.*, 1977, **62**, 259.
[40] D. H. Everett and R. T. Podoll, *J. Colloid Interface Sci.*, 1981, **82**, 14.
[41] Pei-Sen Li and Ti-Ren Gu, *Sci. Sin.* (*Engl. Ed.*), 1979, **22**, 1384 (*Chem. Abstr.*, 1980, **92**, 65 253).

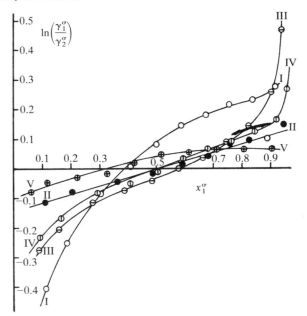

Figure 1 Ln $\gamma_1^\sigma/\gamma_2^\sigma$ as a function of mole fraction x_1^σ for adsorption by silica gel from the systems: I, benzene (1) + toluene (2); II, benzene (1) + chlorobenzene (2); III, toluene (1) + chlorobenzene (2); IV, toluene (1) + bromobenzene (2); V, chlorobenzene (1) + bromobenzene (2)
(Calculated from data in *Sci. Sin. (Engl. Ed.)*, 1979, **22**, 1384).

As stressed by Vernov and Lopatkin[14] it is important that the quantities concerned must be derivable from experimental data without the intervention of a molecular model. Convenient functions satisfying this criterion are (i) the difference $(\sigma - \sigma_2^*)$, between the interfacial tension σ between liquid mixture and the solid and σ_2^*, that between pure component 2 and the solid; (ii) the difference between the areal enthalpy of immersion of the solid in the liquid mixture, $\Delta_w\hat{h}$, and in the pure liquid 2, $\Delta_w\hat{h}_2^*$; (iii) the corresponding difference between the entropies of immersion. Of these (i) is derived from adsorption isotherms using equation (1), but requires a knowledge of the activity coefficients (γ) in the bulk mixture and the specific surface area (a_s) of the solid, whereas (ii) can be derived either from the dependence of $(\sigma_2^* - \sigma)$ on temperature using equation (3) or from direct calorimetric measurement of enthalpies of immersion; (iii) is obtained from (i) and (ii) [equation (4)].

Strictly speaking this procedure involves the use of a theoretical model for the calculation of the surface area of the solid from, for example, gas adsorption data. This objection can be overcome, as advocated by Schay,[5] by working in specific quantities. But it then becomes difficult to intercompare results obtained on different samples of solid.

In the practical application of equation (1) to experimental data, two main problems have to be overcome. In the first place the measurements must be of high accuracy and extend throughout the whole composition range, for it is often found that substantial contributions to the integral come from the extremes of the concentration range so that it is important to be able to extrapolate the integrand reliably to the limits $x_2^l = 0$ and 1. Secondly, the final result depends critically on the existence of reliable values of the activity coefficient, γ_2^l, in the bulk liquid. Unfortunately it often happens that such data are sparse for many systems of interest in adsorption studies: more data are often available in temperature ranges around the normal boiling points of the components, but these have to be extrapolated to the lower temperatures at which adsorption measurements are usually carried out.

The use of equation (1) integrated across the whole composition range to obtain $(\sigma_2^* - \sigma_1^*)$ to check the thermodynamic consistency of measurements on binary systems formed from components 1, 2, and 3 is becoming an accepted procedure. Analysis of data to obtain $(\sigma_2^* - \sigma)$ as a function of x_2^l is less common.

Thomas and Eon[39] have analysed their data on the systems (acetone + carbon tetrachloride), (acetone + benzene), and (benzene + carbon tetrachloride)/silica gel in this way. They find, first, that the data for the three systems are thermodynamically consistent (Table 1). They also investigated the form of the $(\sigma - \sigma_1^*)$ curves in relation to an equation proposed by Semenchenko and Israilov[42] for the liquid/vapour interface:

$$\sigma - \sigma_i^* = J \ln[1 + K(1 - x_i^l)] + L(1 - x_i^l). \tag{46}$$

The experimental data were represented by this equation to a precision of better than 1%. Although Semenchenko[43] gives a derivation of this equation, several approximations are introduced, which mean that it is best regarded as an empirical relationship. It will be of interest to see whether it can be applied to other systems.

Table 1 *Thermodynamic consistency test for the adsorption of acetone* (1), *benzene* (2), *and carbon tetrachloride* (3) *on silica gel* (631 m² g⁻¹)[39]

$i + j$	$(\sigma_i^* - \sigma_j^*)$/mJ m⁻²
2 + 1†	39.2
1† + 3	−56.1
3 + 2†	17.1
$\Sigma(i + j)$	0.2

† Denotes preferentially adsorbed component

[42] V. K. Semenchenko and I. U. Israelov, *Zh. Fiz. Khim.*, 1974, **48**, 3082 (*Russ. J. Phys. Chem.*, 1974, **48**, 1801).
[43] V. K. Semenchenko, *Zh. Fiz. Khim.*, 1973, **47**, 2906 (*Russ. J. Phys. Chem.*, 1973, **47**, 1630).

Table 2 *Thermodynamic consistency tests for the adsorption of ethanol* (1), *n-hexane* (2), *and benzene* (3) *on silica gel* $(a_s = 345\ m^2\ g^{-1})$[44]

$i+j$	$(\sigma_i^* - \sigma_j^*)$/mJ m^{-2}			$(\Delta_w h_i^* - \Delta_w \hat{h}_j^*)$/mJ m^{-2}	
	15 °C	25 °C	35 °C	from $\sigma_i^* - \sigma_j^*$	Calorimetric
$2+1\ddagger$	-58	-53	-48	-199	$(53-222) = -169$
$1\ddagger+3$	36	32	27	170	$(222-72) = 150$
$3\ddagger+2$	23	23	23	0	$(72-53) = 19$
$\Sigma(i+j)$	$+1$	$+2$	$+2$	-10	0

‡ Indicates preferentially adsorbed component

A further example of the use of $\sigma_i^* - \sigma_j^*$ to establish the thermodynamic consistency of adsorption data is given by Ościk and Gorowek[44] for the binary systems derived from benzene, n-hexane, and ethanol adsorbed by silica gel at 15, 25, and 35 °C. The temperature dependence of $\sigma_i^* - \sigma_j^*$ gives the difference between the enthalpies of immersion of the solid in the two pure liquids. Moderate agreement was found between the differences obtained in this way and by direct calorimetry (Table 2).

The results obtained by Li and Gu[41] referred to earlier can be tested in the same way (Table 3).

Data on the systems (benzene + cyclohexane), (cyclohexane + n-heptane), and (benzene + n-heptane)/Graphon, which had previously been analysed in terms of activity coefficients and molecular models,[45] have been re-analysed[46] to give $(\sigma - \sigma_2^*)$, $(\Delta_w \hat{h} - \Delta_w \hat{h}_2^*)$, and $T(\Delta_w \hat{s} - \Delta_w \hat{s}_2^*)$ as functions of x_2^l. The curves shown in Figure 2 illustrate clearly the way in which enthalpy and entropy effects combine to determine the sign of the adsorption. Thus for the (benzene + cyclohexane)/Graphon system the curves of $(\sigma - \sigma_2^*)$ and $(\Delta_w \hat{h} - \Delta_w \hat{h}_2^*)$ virtually coincide: the entropy term makes a negligible contribution. For

Table 3 *Thermodynamic consistency tests for the adsorption of benzene, toulene, chlorobenzene, and bromobenzene by silica gel* $(420\ m^2\ g^{-1})$[41]

$i+j$	$(\sigma_i - \sigma_j)$/mJ m^{-2}		$(\sigma_i - \sigma_j)$ mJ m^{-2}
benzene‡ + toluene	6.05 ± 0.2	toluene†‡ + chlorobenzene	3.78 ± 0.1
chlorobenzene + benzene‡	-10.01 ± 0.2	bromobenzene + toluene†‡	-5.02 ± 0.1
toluene†‡ + chlorobenzene	3.78 ± 0.1	chlorobenzene‡ + bromobenzene	1.54 ± 0.1
$\Sigma(i+j)$	-0.27 ± 0.5	$\Sigma(i+j)$	-0.30 ± 0.3

‡ Indicates preferentially adsorbed component; † over most of the concentration range

[44] J. Ościk and J. Gorowek, *Bibl. Lubel. Tow. Nauk.*, (*Folia Societatis Scientiarum Lublinensis*) *Mat.-Fiz.-Chem.*, 1979, **21**, 17 (*Chem. Abstr.*, 1980, **92**, 83 069).

[45] S. G. Ash, R. Bown, and D. H. Everett, *J. Chem. Soc., Faraday Trans. 1*, 1975, **71**, 123.

[46] D. H. Everett, *J. Phys. Chem.*, 1981, **85**, 3263.

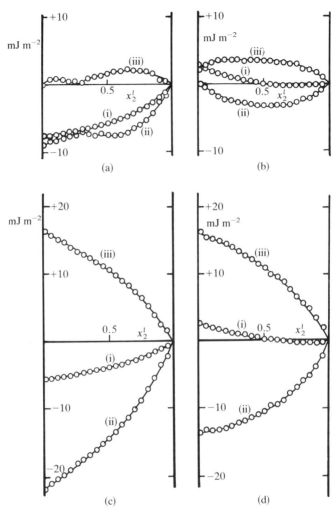

Figure 2 *Thermodynamic functions of wetting at 298 K (i) $(\sigma - \sigma_2^*)$; (ii) $\Delta_w\hat{h} - \Delta_w\hat{h}_2^*$; (iii) $-T(\Delta_w\hat{s} - \Delta_w\hat{s}_2^*)$ as function of mole fraction x_2^l for adsorption by graphitized carbon from the systems: (a) benzene (1) + cyclohexane (2); (b) 1,2-dichloroethane (1) + benzene (2); (c) n-heptane (1) + cyclohexane (2); (d) n-heptane (1) + benzene (2)*
(Redrawn from *J. Colloid Interface Sci.*, 1981, **82**, 14; *J. Phys. Chem.*, 1981, 85, 3263).

the other two systems the entropy term is large and essentially identical in the two systems. However, although in the cyclohexane + n-heptane system the enthalpy term is large enough to ensure that $(\sigma - \sigma_2^*)$ remains of the same sign and has no minimum, for benzene + n-heptane where the enthalpy term is smaller $(\sigma - \sigma_2^*)$ passes through a minimum (and the adsorption changes sign) so that over most of the concentration range benzene is preferentially adsorbed, despite the fact that the enthalpy of immersion of Graphon in benzene is smaller than that in n-heptane: the sign of the adsorption is controlled by the entropy term.

Alternative methods of characterizing adsorption thermodynamically involve the analogue of the 'isosteric enthalpy' used in discussing gas/solid systems. However, as discussed in an earlier Report[47] this method of analysis, which involves calculating the temperature coefficient of the equilibrium value of x_2^l corresponding to a constant surface mole fraction x_2^σ, can be carried out only within the framework of an 'adsorbed phase' model. It is sometimes supposed that the restriction of the partial differentiation to constant x_2^σ can be replaced by specifying constant n_2^σ. However, this implies that both the surface area of the adsorbent, and the thickness of the adsorbed layer are independent of temperature. In many instances this is an unjustified assumption.

Nevertheless this method has been used extensively by Russian authors[48] and recently Ościk and Goworek[49] have analysed their data at 25 °C and 65 °C on the system (benzene + cyclohexane)/silica gel similarly.

3 Dilute Solutions

General.—Adsorption from dilute solutions can, in the limit of sufficiently low concentrations, be described by a linear (Henry's law) isotherm, which may, however, be expressed in various forms. Among the possible definitions are the following:

$$K_H^0 = x_2^\sigma / x_2^l, \text{ Avramenko } \textit{et al.},^{23} \tag{47}\dagger$$

$$K_H' = c_2^\sigma / c_2^l, \text{ Stadnik and Eltekov,}^{50} \tag{48}$$

$$K_H'' = x_2^\sigma / c_2^l, \tag{49}$$

$$K_H''' = \frac{n_2^\sigma}{n_2^{\sigma,0}} \cdot \frac{1}{c_2^l} = \frac{\theta}{c_2^l}, \text{ Stadnik and Eltekov,}^{51} \tag{50}$$

$$K_H'''' = \frac{n_2^\sigma / m}{c_2^l} = \frac{n^0 \Delta x_2^l / m}{c_2^l}. \tag{51}$$

In equation (50), $n_2^{\sigma,0}$ is the adsorption capacity for component 2 and θ the fractional filling of the adsorption space. It is important to distinguish between these alternative forms. In the first three cases some model of the adsorption space must be employed. Thus if (*e.g.* in the case of microporous adsorbents)

---

† Although Avramenko *et al.* write their equation in the form (48) they specify that the concentrations should be expressed in mole fractions.

[47] Ref. 3, p. 67.
[48] Ref. 3, p. 126.
[49] J. Ościk and J. Gorowek, *Bibl. Lubel. Tow. Nauk*, (*Folia Societatis Scientiarum Lublinensis*), *Mat.-Fiz.-Chem.*, 1978, **20**, 11 (*Chem. Abstr.*, 1979, **90**, 77 325).
[50] A. M. Stadnik and Yu. A. Eltekov, *Zh. Fiz. Khim.*, 1975, **49**, 191 (*Russ. J. Phys. Chem.*, 1975, **49**, 105).
[51] A. M. Stadnik and Yu. A. Eltekov, *Zh. Fiz. Khim.*, 1978, **52**, 2100 (*Russ. J. Phys. Chem.*, 1978, **52**, 1211).

the adsorption space is equated to the micropore volume $V^\sigma = v_2 n_2^{\sigma,0}$ where v_2 is the molar volume of component 2. Then

$$c_2^\sigma = n_2^\sigma / v_2 n_2^{\sigma,0} \quad \text{and} \quad K_H' = K_H''' / v_2. \tag{52}$$

Since in dilute solution, $c_2^l = x_2^l / v_1$

$$K_H^0 = \frac{v_2}{v_1} K_H', \tag{53}$$

whereas if the molecules are of the same size ($v_1 = v_2 = v$), $n_2^\sigma / n_2^{\sigma,0} = x_2^\sigma$ and

$$K_H''' = K_H^0 v = K_H' v. \tag{54}$$

The fifth form is the only one that is defined unambiguously in experimental terms. The relationship between the adsorption equilibrium constant K_a [equation (31)] and the Henry's law constant K_H^0 was derived earlier [equation (32)]:

$$K_a = \frac{x_2^\sigma}{x_2^l} \left(\frac{\gamma_2^\sigma}{\gamma_2^l}\right)_\infty = K_H^0 \left(\frac{\gamma_2^\sigma}{\gamma_2^l}\right)_\infty. \tag{55}$$

The various forms of the Henry constant have been used to define various standard free energies of adsorption, corresponding to various possible choices of standard states. Thus Stadnik and Eltekov define the standard free energy as

$$\Delta G^{0.1} = -RT \ln K_H' = -RT \ln (K_H''' / v_2). \tag{56}$$

This is the free energy change when one mole of component 2 is transferred from solution at unit concentration to the surface at unit surface concentration, accompanied by desorption of solvent. The magnitude of $\Delta G^{0.}$ depends on the units in which K_H' is measured. An alternative standard free energy is

$$\Delta G_a^0 = -RT \ln K_a = -RT \ln K_H^0 - RT \ln(\gamma_2^\sigma / \gamma_2^l)_\infty. \tag{57}$$

Two-component Aqueous Systems.—As a direct consequence of the increasing importance of the use of active carbon for water purification, there has been a recent spate of papers on the adsorption of a wide range of aqueous solutions.

Stadnik and Eltekov[51] have considered again their thermodynamic approach to adsorption from dilute aqueous solutions of microporous active carbons.[50] An important problem in the use of the Dubinin–Radushkevich (D.R.) equation for both gas/solid and solution/solid systems is the mode of transition from this equation to the linear (Henry's law) region of the isotherm (see this Vol. p. 65). In a previous papers[50,52] they found an empirical relationship between the quantity B/β^2 in the D.R. equation:

$$\ln(n_2^\sigma / n_2^{\sigma,0}) = -\left(\frac{B}{\beta^2}\right) T^2 [\ln c_2^s / c_2^l]^2, \tag{58}$$

(where c_2^s is the saturation concentration), and the Henry law constant (K_H''') defined in equation (50). For a given solute in a range of active charcoals they

52 A. M. Stadnik and Yu. A. Eltekov, *Zh. Fiz. Khim.*, 1975, **49**, 194 (*Russ. J. Phys. Chem.*, 1975, **49**, 106).

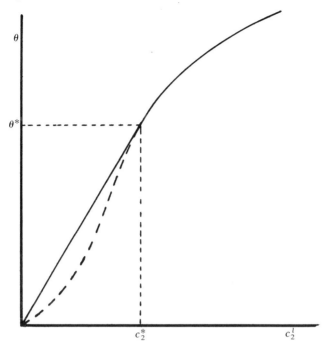

Figure 3 *Transition from Henry's law isotherm in range* $0 \to c_2^*$ *to Dubinin–Radushkevich isotherm* $c_2 > c_2^*$. *Dashed curve, D.R. isotherm in range* $0 \to c_2^*$

found that

$$\left(\frac{B}{\beta^2}\right)^{1/4} \ln K_H''' = \text{constant}. \tag{59}$$

In the present paper[51] they write $n_2^\sigma/n_2^{\sigma,0} = \Gamma/\Gamma_{\max} = \theta$ and consider the transition from the D.R. to the Henry equation as occurring when the tangent to the D.R. equation passes through the origin (Figure 3). Thus from $0 < c_2^l < c_2^*$ the isotherm is linear, whereas for $c_2^* < c_2^l < c_2^s$ it follows the D.R. equation. The concentration c^* is found by seeking the maximum value of θ/c_2 in equation (58); this occurs when

$$c_2^* = c_2^s \exp[-\beta^2/2BT^2]. \tag{60}$$

Now, at c_2^*,

$$\theta^* = \exp[-\beta^2/4BT^2] = K_H''' c_2^*, \tag{61}$$

whence inserting the value for c_2^* leads to

$$\ln K_H''' = \beta^2/4BT^2 - \ln c_2^s. \tag{62}$$

It is stressed that c_2^*, K_H''', and θ^* are unambiguously defined by the constant, B, the affinity coefficient, β, and the temperature, T. however, there appears to be no obvious correlation between the relationships shown in equations (60)—(62) and the previous empirical relation (59).

Combination[53] of equation (62) with the previously derived equation

$$\Delta G_a^0 = -RT \ln K_a, \tag{63}$$

where

$$K_a = K_H'''/v_2, \tag{64}$$

leads to

$$\Delta G^0 = -\frac{R\beta^2}{4BT} + RT \ln(v_2 c_s). \tag{65}$$

In this second paper ΔG^0 is calculated for a series of solutes on KAD (using data obtained by Koganovskii) and AG-3 active carbons (using the authors' own measurements). Values of β for the adsorptives were calculated by an unspecified method described in Stadnik's thesis[54] but the method of estimating B is not stated. The calculated and experimental values of ΔG^0 are compared in Tables: agreement is to within ±6—7% for the KAD charcoal and ±2—4% for the AG-3 charcoal for various organic molecules. It is difficult to judge the validity of this agreement without knowing how β and B were calculated.

Koganovskii and his group have continued their work[55] on the adsorption from aqueous solutions by active carbons. Botsan and Koganovskii[56] have examined the effect of different porosity characteristics of active carbon and carbon black on the adsorption of chloroform, benzene, chlorobenzene, and n-octanol from dilute aqueous solution. The saturation degree of adsorption on the porous carbon, corresponding to filling of the pores, correlated with the size of the adsorbed molecule, whereas no such effects were observed on the non-porous black. It was found that on a non-porous acetylenic carbon black and activated charcoals containing mesopores and 'supermicropores' (*i.e.* having dimensions in the region of overlap of the conventional classification of micro- and meso-pores) the adsorption energies (defined as $\Delta G_{ads.}^0 = -RT \ln c_2^\sigma/c_2^l$) are virtually the same. In the adsorption of the same substances by microporous active charcoal a higher value of $\Delta G_{ads.}^0$ is observed in the region of low adsorption. This is attributed to an increase in the selectivity of adsorption in micropores whose sizes approach that of the organic molecule. Yakimova, Mamchenko, and Koganovskii[57] have compared a carbon black with the almost completely microporous KAD charcoal (iodide activated, showing no hysteresis loop on benzene vapour adsorption) for several aromatic molecules (substituted phenols and p-chloroaniline) and for hexan-1-ol. In analysing these data it is supposed that the charcoal consists of micropores of volume $n_2^\infty(\text{mic}) \cdot v_2$ together with 'supermicropores' of specific surface area $a_s(\text{mic})$: adsorption per unit area in the latter is supposed to be, under given

[53] A. M. Stadnik and Yu. A. Eltekov, *Zh. Fiz. Khim.*, 1978, **52,** 1795 (*Russ. J. Phys. Chem.*, 1978, **52,** 1041); *cf.* ref. 3, pp. 140—141.

[54] A. M. Stadnik, *Thesis*, Moscow, 1976.

[55] See ref. 2, pp. 71—74; ref. 3, 100—103.

[56] V. Ya. Botsan and A. M. Koganovskii, *Dopov. Akad. Nauk Ukr. RSR, Ser. B*, 1978, **8,** 712 (*Chem. Abstr.*, 1978, **89,** 204 706).

[57] T. I. Yakimova, A. V. Mamchenko, and A. M. Koganovskii, *Zh. Fiz. Khim.*, 1980, **54,** 741 (*Russ. J. Phys. Chem.*, 1980, **54,** 422).

conditions, the same as on non-porous carbon. Thus the total adsorption n_2^σ is given by

$$n_2^\sigma = n_2^\sigma(\text{mic}) + \left[\frac{a_s(\text{mic})}{a_s(\text{cb})}\right] n_2^\sigma(\text{cb}), \qquad (66)$$

where $a_s(\text{cb})$ and $n_2^\sigma(\text{cb})$ are the specific surface area and amount adsorbed by carbon black, respectively. Multiplying equation (66) by v_2 and dividing by $\theta_{\text{mic}} = n_2^\sigma(\text{mic})/n_2^\infty(\text{mic})$ gives

$$\frac{n_2^\sigma v_2}{\theta_{\text{mic}}} = n_2^\infty(\text{mic})v_2 + \frac{a_s(\text{mic})}{a_s(\text{cb})} \cdot \frac{n_2^\sigma(\text{cb})v_2}{\theta_{\text{mic}}} \qquad (67)$$

for a given solution concentration. For each system a curve is constructed showing the relation between the volume of solute adsorbed by charcoal $[n_2^\sigma(\text{mic})v_2]$ and that adsorbed by carbon black $[n_2^\sigma(\text{cb})v_2]$ for various bulk solution concentrations. By a method not clearly described θ_{mic} is derived. A graph of the l.h.s. against $n_2(\text{cb})v_2/\theta_{\text{mic}}$ should thus be linear with slope $a_s(\text{mic})/a_s(\text{cb})$ and intercept $n_2^\infty(\text{mic})v_2$. The authors indeed show that data for five systems all fit on a single straight line giving a single value for $a_s(\text{mic})/a_s(\text{cb})$, and enabling the volume of micropores to be estimated. When the total pore volume is known from the limiting adsorption, the volume of supermicropores is obtained.

Once $n_2^\sigma(\text{mic})$ has been obtained as a function of the solution concentration c, the conformity with the Dubinin-Radushkevich equation can be tested. Here the D.R. equation is written in the more general form

$$\ln n_2^\sigma(\text{mic}) \cdot v_2 = \ln V(\text{mic}) - \left(\frac{RT}{E} \ln \frac{c_s}{c}\right)^n, \qquad (68)$$

where n is left as an additional parameter and E is the characteristic adsorption energy. Analysis by Yakimova, Mamchenko, Koganovskii, and Botsan[58] of the data presented in the above paper shows that for the systems under consideration $n = 3$ [in contrast to the more usual value of 2, *cf.* equation (58)]. It is argued that E should increase with the molar refraction $|R|$ of the adsorbed molecule (as found by Manes and Hofer[59]), but since adsorption involves the displacement of water molecules, the energy of adsorption will be decreased in proportion to the volume of water displaced. Thus it is supposed that

$$E = k_R|R| - k_v v_2. \qquad (69)$$

The data obtained in this work conformed to this equation with E/v_2 a linear function of $|R|/v_2$. Although only a small range of values of E and $|R|$ are covered, so that extrapolation is rather uncertain (and not attempted by the authors) it is of interest that a value for k_v of about $130\ \text{J cm}^{-3}$ is indicated, or some $2.3\ \text{kJ mol}^{-1}$, for the energy of displacement of water from the charcoal surface.

[58] T. I. Yakimova, A. V. Mamchenko, A. M. Koganovskii, and V. Ya. Botsan, *Zh. Fiz. Khim.*, 1980, **54**, 476 (*Russ. J. Phys. Chem.*, 1980, **54**, 273).
[59] M. Manes and L. J. E. Hofer, *J. Phys. Chem.*, 1969, **73**, 584; *cf.* ref. 1, p. 83.

The adsorption of m-trifluoromethyl aniline from aqueous solution by a range of active charcoals has been studied by Plekhotkin and Khramchikhina.[60] They have fitted their data to equation (68) with $n = 2$, but have included a constant term $\ln K$ on the right-hand side: K is called a 'correction coefficient', which is introduced to bring the extrapolated intercept into coincidence with $\ln V$(mic). For KAD charcoal a linear relation is followed over the whole range of concentration covered, but the intercept gives a volume less than V(mic). This is in contrast to Koganovskii's observation that $n = 3$ for his carbon. It is suggested either that this means that at saturation the adsorbed phase contains two components, or that some pores are inaccessible to the adsorptive molecule. The graphs for several other charcoals consists of two linear portions. In some cases the linear portion at low concentrations extrapolates to $\sim V$(mic), whereas the section at higher concentrations gives a volume $> V$(mic) suggesting that in these charcoals mesopores begin to fill before saturation is reached.

It is often found that the adsorption of a sparingly soluble substance rises sharply as the saturation concentration is approached. In the case of nonporous adsorbents this may arise from multilayer adsorption, or from some form of structuring at the solid surface. For porous adsorbents it is attributed[61] to 'capillary stratification', which is regarded as analogous to capillary condensation in vapour/porous solid systems. Avramenko and Glushchenko[62] have examined this phenomenon in the case of aqueous solutions of n-propanol, isobutanol, and isopentanol adsorbed by SKT active charcoal. If the capillary stratification is related to the phase separation phenomenon in the bulk liquid phase, then its appearance may correlate with the distance the bulk phase is from its consolute point. This point is identified thermodynamically with the concentration and temperature at which $[\partial^2 g/\partial(x_2^l)^2] \to 0$, where g is the mean molar Gibbs energy.[63] The distance that the system is from this state may be conveniently expressed by the dimensionless quantity

$$f = \frac{x_1^l x_2^l}{RT \, \partial^2 g/\partial(x_2^l)^2},\qquad(70)$$

which tends to infinity at the critical point.

In this paper f is identified with the 'level of concentration fluctuations', and the concentration at which it begins to rise steeply is found to be just that concentration at which the adsorption isotherm has a point of inflexion. Measurements at 293, 308, and 323 K enabled the partial enthalpies and partial entropies of adsorption to be calculated. The former exhibited a minimum in ΔH_{ads} and a maximum in ΔS_{ads} close to the inflexion points. n-Propanol behaved in a similar fashion even though it is miscible with water

[60] V. F. Plekhotkin and G. I. Khramchikhina, Zh. Fiz. Khim., 1979, **53**, 2362 (Russ. J. Phys. Chem., 1979, **53**, 1346).

[61] N. N. Gryazev and A. V. Kiselev, Zh. Fiz. Khim., 1959, **33**, 1581 (Russ. J. Phys. Chem., 1959, **33**, 62), and earlier papers.

[62] V. A. Avramenko and V. Yu. Glushchenko, Izv. Akad. Nauk SSSR, Ser. Khim., 1978, **4**, 759 (Bull. Acad. Sci. USSR, Chem. Sci, 1978, **27**, 654).

[63] I. Prigogine and R. Defay, 'Chemical Thermodynamics', Longmans, London, 1954, pp. 220, 241.

in all proportions. In this case f was correlated with the minimum volume of the adsorbed phase calculated using Rusanov's criteria.[8] It was found that, according to these criteria the minimum phase volume at the maximum in f is much greater than the micropore volume of the charcoal, and tends towards the total pore volume. Studies of the rate of adsorption also showed that when f increases, so does the diffusion coefficient, and that this increase in diffusion rate is accompanied by a decrease in the activation energy for diffusion.

The adsorption of iodine from aqueous solutions by a series of non-activated and zinc chloride-activated charcoals was studied by Elnabarawy, El-Shobaky, Youssef, and Mikhail.[64] The adsorption data are not presented, but the authors estimated the monolayer capacity from the isotherm and, using a value of 0.40 nm^2 for the molecular area of I_2, calculated surface areas of the charcoals which agreed moderately well with values obtained by the BET-N_2 method. However, for less active carbons ($a_s < 600 \text{ m}^2 \text{ g}^{-1}$) the I_2-area tended to be substantially (15% or more) higher than the BET-N_2 areas, whereas for zinc chloride-activated charcoals of surface area $> 600 \text{ m}^2 \text{ g}^{-1}$ they were consistently 3—4% high. The adsorption of methylene blue from aqueous solution was also measured and was directly proportional to the BET-N_2 surface area: the molecular area of methylene blue was estimated to be 2.15 nm^2 in agreement with the value (2.08 nm^2) suggested by Gleysteen and Scheffler.[65] Results for the adsorption of methylene blue from aqueous solution (and quinizarin red from xylene) by active carbon have been reported by Trkulja.[66]

Several papers by Swiatkowski and co-workers,[67—75] all in Polish, have dealt with various aspects of iodine adsorption by active carbon. Measurements of the adsorption of I_2 from 0.5 M-NaI solutions in aqueous methanol[68] showed that the shift of the inflection point on the isotherm towards lower equivalent iodine concentrations with increase in water concentration in the solvent, was caused partly by the decrease in the solubility of iodine and partly by the effect of water on the polyiodide complex equilibria. For adsorption from solutions of NaI in ethanol, isopropanol, and acetonitrile,[70] the isotherms were of the

[64] T. Elnabarawy, G. A. El-Shebaky, A. M. Youssef, and R. S. Mikhail, *Indian J. Chem., Sect. A*, 1979, **17**, 509.

[65] L. F. Gleysteen and G. H. Scheffler, 'Proc. 4th Conf. on Carbon', Pergamon Press, London, 1960, 47.

[66] S. Trkulja, *Zb. Rad. Poljopr. Fak., Univ. Beogradu*, 1978, **23—26**, 111 (*Chem. Abstr.*, 1980, **92**, 153 589).

[67] A. Swiatkowski, *Przem. Chem.*, 1977, **56**, 599 (*Chem. Abstr.*, 1978, **88**, 55 377).

[68] A. Swiatkowski, *Buil. Wojsk. Akad. Tech.*, 1977, **26**, 77 (*Chem. Abst.* 1978, **88**, 126 796.

[69] H. Jankowska, A. Swiatkowski, and S. Zietek, *Buil. Wojsk. Akad. Tech.*, 1978, **27**, 23 (*Chem. Abstr.*, 1978, **89**, 95 466).

[70] A. Swiatkowski, S. Gruchalski, and E. Wisniewska, *Przem. Chem.*, 1978, **57**, 310 (*Chem. Abstr.*, 1978, **89**, 136 289).

[71] H. Jankowska and A. Swiatkowski, *Przem. Chem.*, 1978, **57**, 624 (*Chem. Abstr.*, 1979, **90**, 110 409).

[72] A. Swiatkowski and M. Kwasny, *Buil. Wojsk. Akad. Tech.*, 1978, **27**, 55 (*Chem. Abstr.*, 1979, **90**, 110 503).

[73] A. Swiatkowski, S. Gruchalski, and E. Wisniewska, *Przem. Chem.*, 1979, **58**, 46 (*Chem. Abstr.*, 1979, **90**, 157 478).

[74] A. Swiatkowski and S. Zietek, *Buil. Wojsk. Akad. Tech.*, 1979, **28**, 111 (*Chem. Abstr.*, 1979, **90**, 210 632).

[75] M. Mioduska. S. Pietrzyk, A. Swiatkowski, and T. Zmijewski, *Buil. Wojsk. Akad. Tech.*, 1979, **28**, 109 (*Chem. Abstr.*, 1980, **92**, 65 194).

L-4 type in Giles' classification,[76] but the precise shape of the isotherm was affected by the degree of iodide complex formation, and in the case of adsorption from solutions of NaCl, NaBr, and NaI in methanol[67] deviations from the Langmuir isotherm increased with increasing stability of the complexes $ClI_2^- < BrI_2^- < I_3^-$, which were formed in solution. The isotherm shape was not, however, significantly affected by the porous structure of the carbon adsorbent.[72] Adsorption and desorption isotherms of iodine on activated carbon from solutions of KI in water,[69] ethanol, and acetonitrile[74] showed hysteresis, and some adsorption studies were made using mixed solvents.[73] The data[75] for the adsorption of iodine from aqueous KI solution by a series of carbon adsorbents were fitted to the Dubinin–Raduschkevich equation and correlated with the benzene vapour adsorption by the same carbons. The use of iodine adsorption for evaluating the properties of carbon adsorbents is also reviewed.[71]

Akhrimenko[77] has studied the kinetics of adsorption of iodine from aqueous 0.1 M-KI solutions on single carbon grains, and found that the adsorption rate was greatly affected by the pore dimensions of the charcoal. Kinetic adsorption isosteres and isosteric enthalpies of adsorption were determined.

For several other adsorption systems the work is only readily available in abstract form. Activated carbon was the adsorbent used with the following systems: p-(C_7–C_9)alkylphenols from aqueous solution,[78] 2,4-dichlorophenoxyacetic acid from aqueous solution,[79] isoamyl alcohol from ethanol + water mixtures,[80] and the herbicide atrazine from aqueous solution.[81]

Ammons, Dougharty, and Smith[82] have studied the adsorption of methylmercuric chloride from aqueous solution by activated carbon by both batch and flow techniques. The data were analysed to investigate the factors that control the breakthrough curves. Axial dispersion was found to contribute no more than 15% to the second moment of the breakthrough curve, while liquid-to-particle mass transfer contributed about 60%. On a similar topic Benediktov, Vlasov, and Yurkevich[83] in a paper of which only the title is abstracted, discuss the determination of the degree of exhaustion of active carbon with respect to organic substances during adsorption from aqueous solutions.

Adsorption of Weak Electrolytes.—The effect of pH on the adsorption of weak electrolytes by activated carbon has been studied by several workers. The early

[76] See ref. 3, p. 93.
[77] V. E. Akhrimenko, *Zh. Fiz. Khim.*, 1978, **52**, 2122 (*Russ. J. Phys. Chem.*, 1978, **52**, 1228, VINITI No. 623—78).
[78] A. N. Turanov, I. N. Kremenskaya, A. M. Reznik, and V. I. Bukin, *Zh. Prikl. Khim. (Leningrad)*, 1978, **51**, 2364 (*Chem. Abstr.*, 1979, **90**, 29 518).
[79] J. de D. Lopez Gonzalez, C. Valenzuela Calahorro, and A. Jimenez Lopez, *An. Edefol. Agrobiol.*, 1977, **36**, 835 (*Chem. Abstr.*, 1978, **88**, 1498).
[80] P. Ya. Bachurin, S. P. Kalinkina, A. A. Kniga, and V. M. Perelygin, *Fermentn. Spirt. Prom-st.*, 1979, **5**, 30 (*Chem. Abstr.*, 1979, **91**, 138 807).
[81] P. Fusi, M. Franci, and A. Malquori, *Ann. Chim. (Rome)*, 1977, **67**, 241 (*Chem. Abstr.*, 1978, **88**, 126 806).
[82] R. D. Ammons, N. A. Dougharty, and J. M. Smith, *Ind. Eng. Chem., Fundam.*, 1977, **16**, 253.
[83] A. P. Benediktov, V. A. Vlasov, and A. A. Yurkevich, *Poluch., Strukt. Svoistva Sorbentov*, 1977, 59; from *Ref. Zh. Khim.*, 1978, Abstr. No. 101453 (*Chem. Abstr.*, 1978, **89**, 80 718).

work of Phelps and Peters[84,85] and Kipling[86] demonstrated the importance of pH, and indicated that the effect was related to the differing strengths of adsorption of the neutral and ionized forms of the molecule. It was first believed that only the neutral form was adsorbed, but Getzen and Ward[87,88] recognised that the charge on the carbon surface was an important factor which varied with pH and which could enhance or prevent the adsorption of ionized species. These workers combined separate Langmuir adsorption isotherms for the neutral and ionized forms with the dissociation constant determining the relative amounts of each form at equilibrium. For carboxylic acid solutions a decrease in pH led to H^+ adsorption, which in turn enhanced the adsorption of electrolyte anions although the effect was limited since the bulk anion concentration decreased with decreasing pH. More recently Baldauf, Frick, and Sontheimer[89] used separate Freundlich isotherms for neutral and ionic forms to describe the effect of pH on the adsorption of acids and bases.

Rosene and Manes[90] studied the effect of pH on the total adsorption from aqueous solutions of sodium benzoate + benzoic acid by activated charcoal. They interpreted their data in terms of the Polanyi potential theory applied to bisolute adsorption (see later p. 117), in which the concentrations of neutral benzoic acid and benzoate anions depend on the pH of the solution (activity coefficient corrections were ignored). They confirmed that, at constant total equilibrium concentration, the adsorption dropped from a relatively high plateau for pH < 2 down to a small adsorption at pH ≥ 10. The analysis of results is somewhat more complex than with essentially non-electrolyte adsorption, and in this case there were additional effects involving chemisorption of benzoate ion by residual ash in the carbon which had, therefore, to be eliminated. Even with ash-extracted carbon there was evidence of some residual chemisorption. The theoretical analysis correlated satisfactorily with the experimental data on the basis that at pH > 10 sodium benzoate is not physically adsorbed and that the effect of pH is completely accounted for by its effect on the concentration of free acid. In addition the theory explains successfully the increase in pH (called by the authors 'hydrolytic adsorption') when solutions of sodium benzoate are treated with neutral carbon. However, no account is taken in this paper of the effect of pH on the surface charge of the carbon.

A study of the effect of pH on the adsorption of the lower fatty acids (C_1 to C_8) and poly(ethylene glycol)s (PEG) by active carbon is reported by Chudoba, Hrncir, and Remmelzwaal.[91] The experimental isotherms were analysed in terms of the Freundlich isotherm. The constant K increased with increasing

[84] H. J. Phelps and R. A. Peters, *Proc. R. Soc. London, Ser. A.*, 1929, **124**, 554.
[85] H. J. Phelps, *Proc. R. Soc. London, Ser. A.*, 1931, **133**, 155.
[86] J. J. Kipling, *J. Chem. Soc.*, 1948, 1483.
[87] F. W. Getzen and T. M. Ward, *J. Colloid Interface Sci.*, 1969, **31**, 441.
[88] T. M. Ward and F. W. Getzen, *Environ. Sci. Technol.*, 1970, **4**, 64.
[89] G. Baldauf, B. Frick, and H. Sondheimer, *Vom Wasser*, 1977, **49**, 315.
[90] M. R. Rosene and M. Manes, *J. Phys. Chem.*, 1977, **81**, 1651.
[91] J. Chudoba, B. Hrncir, and E. J. Remmelzwaal, *Acta Hydrochim. Hydrobiol.*, 1978, **6**, 153 (*Chem. Abstr.*, 1978, **89**, 31 333).

molecular weight of both fatty acids and PEG's. The exponent n remained constant for the acids but decreased with increasing molecular weight of the polymers. The pH affected strongly the adsorption of fatty acids, little adsorption being found at pH > 8.2, but it had little effect on the polymer adsorption.

An important contribution to the problem is made in a paper by Müller, Radke, and Prausnitz[92] who present a new theoretical model for the adsorption of weak organic electrolytes on activated carbon. Unlike previous models the theory takes into account surface heterogeneity and the effect of pH on surface charge. The solid surface is assumed to consist of three types of adsorption site, neutral, basic, and acidic, the relative proportions of which vary with pH and are characterized by q, the surface charge density per unit area. This is related to the surface potential ψ_0 by simple diffuse double-layer theory, assuming that the surface charge is balanced only by the counter charge of the double layer.

$$q = \sqrt{8\varepsilon RTI} \sinh \frac{zF\psi_0}{2RT}. \tag{71}$$

Here ε is the bulk permittivity of water, F is Faraday's constant, I is the ionic strength, and z the ionic charge. For adsorption on a homogeneous surface the neutral and ionized forms of the electrolyte are competing for the same number of sites, and Langmuir isotherms giving the fractional coverage θ of each species may be written:

$$\theta_m = (c_m/K)/(1 + c_{m\pm}/K' + c_m/K_m), \tag{72}$$

$$\theta_{m\pm} = (c_{m\pm}/K')/(1 + c_{m\pm}/K' + c_m/K_m). \tag{73}$$

The concentrations of each species $c_m, c_{m\pm}$ are roughly determined by the dissociation equilibrium constant K_m and the equilibrium constant of the adsorption process K' can be written:

$$K' = \exp(-\Delta G^0/RT + z_{m\pm}F\psi_0/RT), \tag{74}$$

where ΔG^0 is the standard Gibbs energy of desorption which is independent of the electrical state. The constant K' can also be separated into K_m, the adsorption equilibrium constant for the neutral species, and K_q, the portion of K' due only to charge.

$$K' = [K_m][K_q] = [A_0 \exp(-U/RT)][\exp(z_{m\pm}F\psi_0/RT)]. \tag{75}$$

Here U is the adsorption potential and A_0 a pre-exponential factor. One implication of equation (75) is that the affinity of a charged species for a neutral surface and of neutral molecules for a charged surface are assumed to be the same.

The heterogeneity of the surface is introduced by assuming a large number of homogeneous patches.[93] Each patch i is characterized by an adsorption potential U_i for which the Langmuir isotherm is locally valid. An exponential

[92] G. Mueller, C. J. Radke, and J. M. Prausnitz, *J. Phys. Chem.*, 1980, **84**, 369, (*Chem. Abstr.*, 1978, **89**, 31 333).
[93] See Chapter 1.

distribution for the reduced adsorption potential is chosen,

$$f(U_i/RT) = \sin n\pi/\{\pi[\exp(U_i - U_0)/RT - 1]^n\}, \tag{76}$$

such that when $n = 1$ the surface is homogeneous with an adsorption potential of U_0. The overall adsorption isotherm is then found by applying Langmuir isotherms to each patch i and integrating over all adsorption potential energies. The total coverage is given by

$$\theta(c, \text{pH}) = \theta_m + \alpha\theta_{m\pm}. \tag{77}$$

The fraction α is the fraction of the total number of adsorption sites available to co-ion species and takes account of the fact that co-ions cannot adsorb on sites where the adsorption potential U_i is less than the potential due to electrostatic repulsion, *i.e.*

$$\alpha = n^a_{max, m\pm}/n^a_{max} = \int_{U_i/RT}^{\infty} f(U_i/RT)\, \mathrm{d}(U_i/RT). \tag{78}$$

Then,

$$\theta(c, \text{pH}) = \int_{U_0/RT}^{\infty} \theta_{m,i} f(U_i/RT)\, \mathrm{d}(U_i/RT) + \alpha \int_{z_{m\pm}F\psi_0/RT}^{\infty} \theta_{m\pm} f(U_i/RT)\, \mathrm{d}(U_i/RT). \tag{79}$$

With the approximation that $z_{m\pm}F\psi_0/RT \leqslant U_0/RT$, this reduces to

$$\theta(c, \text{pH}) = \frac{c_\alpha}{c_1}\left(\frac{c_1}{c_1 + K_0}\right)^n. \tag{80}$$

Here $c = c_m + \alpha c_{m\pm}/K_q$ and is equal to c_1 when $\alpha = 1$. K_q is calculated from surface titration experiments and $K_0 = A_0 \exp(-U_0/RT)$. Once n^a_{max}, K_0, n, and U_0 are specified and the variation of ψ_0 with pH known, then a plot of adsorption against pH can be constructed. Equation (80) reduces to Freundlich or Langmuir isotherms in given limiting cases, and when ψ_0 varies with pH in a manner parallel to the electrolyte dissociation curve the equation predicts adsorption curves that agree with the experimental observations of early workers.

The dependence of the form of the adsorption curve on pH was investigated for a range of weak acids and bases at typical concentrations and ionic strengths. The shape depends on the pH at which the surface charge is zero, pH (pzc), and the dissociation equilibrium constant. The surface is positively charged in the lower pH range, pH < pH (pzc), and the adsorption of weak acids which dissociate in this range increases with increasing pH because of enhanced anion adsorption. This will continue until the decrease in surface charge becomes larger than the corresponding increase in anion concentration and the maximum will be at pH = pK_m. Above pH (pzc) the adsorption decreases because of repulsion and desorption of anions from the increasingly negatively charged surface. For weak acids that dissociate at pH > pH (pzc) there is no pH effect in the low pH range and the decrease in adsorption is only apparent when dissociation begins. The same arguments are used to predict adsorption against pH curves for weak organic bases. The theory accounted satisfactorily for new experimental measurements on benzoic acid and *p*-nitrophenol but the authors say that more data are needed for a comprehensive test of the theory.

Glushchenko and co-workers have continued their work on the adsorption of nitro-compounds by graphite and carbon blacks from aqueous solution, often phosphate-buffered at pH 6.4.[94] Adsorption studies of nitrobenzene, nitrobenzaldehydes, nitroanilines, nitrophenols, and nitrobenzoic acids on DG-100 carbon black and graphite were combined with a polarographic study using a graphite rod as the working electrode against a reference calomel electrode.[95,96] The adsorption equilibrium constant was calculated from the initial segment of the isotherm using $K = \lim[c^\sigma/c^l]_{c^l \to 0}$ and hence $\Delta G_{ads.}$ was obtained (see p. 101). The values for the free energy of adsorption were found to vary linearly with the ionization potential of the adsorbate and with the half-wave potential measured polarographically. Noting the correlation between the ability of a molecule to capture electrons from a graphite surface and its 'adsorbability' the authors concluded that adsorption involved the carbon acting as an electron donor. The anomalous behaviour of *p*-nitrophenol, previously explained[97] by the suggestion that stable complexes could be formed between the molecule and surface oxygen complexes, was reconsidered by Glushchenko and Khabalov,[98] who postulated a kind of proton-exchange mechanism involving *p*-nitrophenol, surface carboxyl, or carbonyl groups and a surface hydroxyl group. The different adsorption behaviour of *o*-nitrobenzoic acid from the *meta*- and *para*-isomers was explained by conjugation in the adsorbate molecule.[99]

The thermodynamic parameters of adsorption of nitrophenols on highly purified graphite from buffered aqueous solution have been tabulated by Kozlov and Glushchenko.[100] The adsorption isotherms were determined at 293, 308, and 323 K by a chromatographic technique. $\Delta G_{ads.}^0$ is the quantity defined in equation (56), and the differential enthalpy was calculated from $[\partial \ln c_2^l/\partial(1/T)]_\Gamma = q/R$ (see ref. 3 p. 68). It is not stated how c_2^σ was calculated from the observed adsorption. They argue that because nitrophenols are essentially dissociated when they adsorb at pH 6.5 the entropy of adsorption includes contributions arising from an entropy of ionization and entropy of hydration. In another paper[101] Avramenko, Glushchenko, and Kozlov have studied adsorption at two pH's (1.60 and 6.50) of a series of nitrophenols from

[94] See ref. 3, p. 102.
[95] V. Yu. Glushchenko and V. V. Khabalov, *Izv. Akad. Nauk SSSR, Ser. Khim.*, 1976, 25, 2169 (*Bull. Acad. Sci., USSR, Chem. Ser.*, 1976, **25**, 2026).
[96] V. Yu. Glushchenko and V. V. Khabalov, *Kolloid. Zh.*, 1978, **40**, 765 [*Colloid J. (USSR)*, 1978, **40**, 635].
[97] See ref. 3, p. 103, ref. 80.
[98] V. Yu Glushchenko and V. V. Khabalov, *Zh. Fiz. Khim.*, 1978, **52**, 1022 (*Russ. J. Phys. Chem.*, 1978, **52**, 580).
[99] V. V. Khabalov and V. Yu Glushchenko, Deposited Doc., 1974, VINITI 442-74 (*Chem. Abstr.* 1977, **86**, 146 422).
[100] S. G. Kozlov and V. Yu. Glushchenko, *Kolloid. Zh.*, 1976, **38**, 577 [*Colloid J. (USSR)*, 1976, **38**, 523].
[101] V. A. Avramenko, V. Yu. Glushchenko, and S. G. Kozlov, *Izv. Vyssh. Uchebn. Zaved., Khim., Khim. Tekhnol.*, 1979, **22**, 1246 (*Chem. Abstr.*, 1980, **92**, 83 097).

aqueous solution. They have used equations (30) and (47) to calculate two forms of the adsorption equilibrium constant and hence the corresponding free energies of adsorption ΔG_{12} and ΔG_1 and, using equation (32) the activity coefficient of the adsorbed component at infinite dilution, γ_∞^α. The differences between the enthalpies and entropies of immersion of graphite in water and in pure nitrophenol (calculated from K_a) and of adsorption from the standard state of unit concentration in the liquid to unit concentration in the adsorbed state [calculated from K_H', equation (48)] are tabulated. The effect of pH on the free energies can be understood qualitatively. In those cases in which pK_a lies between 1.6 and 6.5 the predominant form present in solution, and hence it must be assumed in the adsorbed state, is the uncharged species at pH 1.6 and the anion at pH 6.5. In the case of m-dinitrobenzene ($pK = 16.8$) the neutral molecule will be involved at both pH's, whereas 2,4-dinitroaniline ($pK = 9.5$) will be in the cationic form. It appears from the dependence of ΔG_{12} on $(pH - pK)$ (see Figure 4) that when the form in solution is the same at both pH's, then pH has a relatively small effect on adsorption, but a change from neutral phenol molecule to the anion (*i.e.* pH-pK changes sign) involves a shift of ΔG to less negative values, *i.e.* to weaker adsorption. This is consistent with the evidence that a graphite surface is negatively charged in

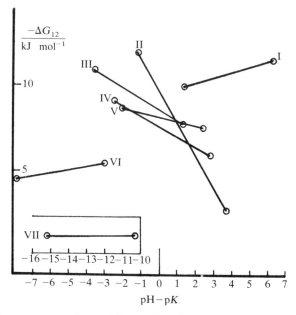

Figure 4 *Free energy of exchange ΔG_{12} at pH 1.60 and 6.5 as function of pH-pK for adsorption from aqueous solutions on to graphite of: I, trinitrophenol; II, 2,4-dinitrobenzoic acid; III, 2,5-dinitrophenol; IV, 2,4-dinitrophenol; V, 2,6-dinitrophenol; VI, 2,4-dinitroaniline; VII, m-dinitrobenzene*

(Calculated from data in *Izv. Vyssh. Uchebn, Zaved., Khim., Khim. Tekhnol.*, 1979, **22**, 1246).

water, although no consideration appears to have been given to the effect of pH on the surface charge on the graphite. The activity coefficient data are also broadly in agreement with this interpretation: at pH 1.6 the neutral adsorbed molecules exhibit small positive deviations from ideality; at pH 6.5 the ionic form deviates in a strong negative direction.† The enthalpy and entropy terms are less easy to understand. The anomalous behaviour of 2.5-dinitrophenol, for which ΔG passes through a minimum as the temperature is raised, is analogous to that of p-nitrophenol. The surface activity coefficients of 2,4-dinitrophenol are also presented graphically as a function of x^σ: they exhibit a very curious behaviour, $\ln \gamma^\sigma$ passing through a minimum around $x \approx 0.2$—0.3 at pH 1.6. At pH 6.5 the large negative value at infinite dilution changes sign at $x \approx 0.005$, passes through a maximum at $x \approx 0.15$, and then declines. The interesting behaviour suggested by this work should be confirmed since the adsorption isotherms determined by the classical ampoule technique and by chromatography differed by up to 10%. Greater precision is needed to enable reliable thermodynamic data to be obtained.

Glushchenko and Radaev[102] have extended their work on aromatic nitro-compounds to the adsorption from aqueous solutions of the nitroalkanes, from nitromethane to nitrobutane, on oxidized and unoxidized graphite. Polarographic work again revealed a correlation between the half-wave potential and $\Delta G_{ads.}$ obtained from adsorption isotherms. Electro-osmotic transfer through graphite diaphragms was measured[103] to gain information about the nature of the adsorption layer and the authors suggested that the nitroalkane molecules were oriented normally to the oxidized and tangentially to the unoxidized graphite surface. The adsorption of amines and polyamines on carbon black has been studied by Verlinskaya and Shuter[104] who noted that adsorption was largely irreversible and increased with increasing molecular weight of the amine. Salt formation with anionic oxygen groups on the carbon surface was said to be the adsorption mechanism.

The adsorption of phenols from aqueous electrolyte solution has been studied from a different point of view by Peschel, Belouschek, Kress, and Reinhard.[105] Instead of measuring the adsorption as a function of pH or of solute concentration they measured the adsorption of fixed concentrations (2×10^{-4} M) of phenol, p-cresol, and 4-ethylphenol by active carbon at four temperatures between 20 and 38 °C as a function of supporting electrolyte (NaCl) concentration. The objective of this work was to seek evidence for multimolecular hydrate layers at the carbon surface. Earlier work[106] on the

---

† It is, however, not clear from the paper what values were taken for the bulk activity coefficients at the two pH's. If these were activity coefficients derived from vapour pressure measurements of unbuffered solution then the calculated values of γ^σ_∞ are meaningless.

[102] V. Yu. Glushchenko and E. F. Radaev, *Izv. Akad. Nauk, SSSR, Ser. Khim.*, 1978, **27**, 538 (*Bull. Acad. Sci., USSR, Chem. Ser.*, 1978, **27**, 463).
[103] V. Yu. Glushchenko and E. F. Radaev, *Izv. Akad. Nauk, SSSR, Ser. Khim.*, 1978, **27**, 535 (*Bull. Acad. Sci., USSR, Chem. Ser.*, 1978, **27**, 461).
[104] R. M. Verlinskaya and L. M. Shuter, *Visn. L'viv. Politekh. Inst.*, 1977, **111**, 32 (*Chem. Abstr.*, 1978, **88**, 12 385).
[105] G. Peschel, P. Belouschek, B. Kress, and R. Reinhard, *Prog. Colloid Polym. Sci.*, 1978, **65**, 83.
[106] G. Peschel and P. Belouschek, *Prog. Colloid Polym. Sci.*, 1976, **60**, 108.

forces between quartz surfaces had indicated that the thickness of the structured hydration layer reached a maximum at a NaCl concentration of about 10^{-2} M. This thickness also varied with temperature, reaching a maximum at 32 °C. It was argued that if a similar zone existed at the carbon/water interface then the adsorption of organic molecules should pass through a minimum when the thickness of the zone reaches a maximum. The raw experimental are not easy to interpret unambiguously since any specific effects of temperature have to be seen against a background of a decrease in adsorption with increasing temperature. In broad terms the adsorption increases in the series phenol, p-cresol, 4-ethylphenol, and the effect of salt concentration and temperature decrease in this sequence. In the case of phenol the adsorption shows two peaks in the regions $10^{-3.5}$ to 10^{-3} M and $10^{-1.5}$ to $10^{-0.5}$ M with a minimum between. The location of the minimum varies between 10^{-1} to $10^{-2.5}$ M, however, somewhat erratically with temperature, while the adsorption at the minimum tends for phenol and p-cresol to show a maximum at 32 °C and only for 4-ethylphenol to show a minimum. The authors also present curves of $\Delta H_{ads.}$ and $\Delta S_{ads.}$, which appear to show a dependence on c_{NaCl} sharply differentiating the data at 35 °C (where $\Delta H_{ads.}$ shows a very deep minimum at $10^{-1.3}$ M, suggesting strong adsorption, offset by a deep minimum in $\Delta S_{ads.}$) from those at other temperatures. The broad conclusion from this work would appear to be that adsorption does indeed pass through a minimum in the concentration range 10^{-1} to $10^{-2.5}$ M-NaCl, but that the effects of temperature on these phenomena are far from clear. The authors suggest that in some instances a second minimum may occur at higher salt concentrations.

The interaction between the adsorption of strong electrolytes (HCl, KI, or AgNO₃) and phenols and aromatic carboxylic acids from dilute aqueous solutions by carbon has been examined by Skripnik and Strazhesko,[107] who find that phenols affect the adsorption to a lesser degree than carboxylic acids, which can cause a marked decrease (96% in the case of picric acid) in electrolyte adsorption. The results are interpreted in terms of the orientation of adsorbed organic molecules. The adsorption of strong acids (HNO₃, HCl, HClO₄) by coconut charcoal was studied by Kraus and Nelson;[108] they found that this charcoal behaved in a similar manner to anion exchangers.

Multisolute Adsorption from Dilute Aqueous Solutions.—Earlier work on adsorption from multicomponent systems has been reviewed in references 2 (p. 103) and 3 (p. 75). The work of Radke and Prausnitz,[109] Fritz and Schlünder,[110] and of Myers and Minka[111] has been developed in a paper on the thermodynamics of multisolute adsorption from dilute aqueous solutions by Jossens, Prausnitz, Fritz, Schlünder, and Myers.[112] The experimental work

[107] Z. D. Skripnik and D. N. Strazhesko, *Adsorbtsiya Adsorbenty*, 1977, **5**, 14 (*Chem. Abstr.*, 1978, **89**, 136 298).

[108] K. A. Kraus and F. Nelson, *Ext. Abstr. Program-Bienn. Conf. Carbon*, 1975, **12**, 201 (*Chem. Abstr.*, 1978, **88**, 66 263).

[109] C. J. Radke and J. M. Prausnitz, *A.I.Ch.E. J.*, 1972, **18**, 761.

[110] W. Fritz and E. U. Schlünder, *Chem. Eng. Sci.*, 1974, **29**, 1279.

[111] C. Minka and A. L. Myers, *A.I.Ch.E. J*, 1973, **19**, 453.

[112] L. Jossens, J. M. Prausnitz, W. Fritz, E. U. Schluender, and A. L. Myers, *Chem. Eng. Sci.*, 1978, **33**, 1097.

included single solute isotherms and six bisolute systems drawn from phenol, p-nitrophenol, p-chlorophenol, benzoic acid, phenylacetic acid, 2,4-dichlorophenol, o-phenylphenol, and dodecylbenzene sulphonic acid. The adsorbent was a commercial activated carbon of surface area 1400—1500 $m^2 g^{-1}$ and a pore volume of $1.5 cm^3 g^{-1}$ consisting of roughly equal volumes of pores less than and greater than 10 nm in size. Various concentration ranges were covered for the different systems between the limits 5×10^{-8} and 10^{-2} mol dm^{-3}. The method of Radke and Prausnitz for the calculation of multisolute adsorption requires a knowledge in analytical terms of the single solute isotherms. In this work three previously proposed empirical adsorption isotherms and one new isotherm were tested. The two most successful were the Tóth equation,[113] originally devised for gas–solid systems:

$$\frac{n_i}{n_i^\infty} = c_i/[b + c_i^M]^{1/M}, \tag{81}$$

where b and M are constants, and a new equation:

$$n_i \exp(kn_i^p) = Hc_i, \tag{82}$$

where p is a constant and H and k are functions of temperature only. Here n_i refers to the surface excess, which for dilute systems is equal to the total amount adsorbed, and c_i is the bulk liquid concentration.

Tóth's equation is related to the behaviour of the function

$$\Psi = \frac{d \ln c_i}{d \ln n_i} - 1, \tag{83}$$

which is given, empirically, the form

$$\Psi = \frac{1}{b} c_i^M. \tag{84}$$

Introduction of (84) into (83) and integration gives (81). The experimental data for single solute isotherms are compared in figures with both (81) and (84).

The new isotherm is related thermodynamically[114] to a variation of the isosteric enthalpy, q_{st}, of the form

$$q_{st} - q_{st}^0 = -Cn_i^p \tag{85}$$

with $$C = -RT^2(dk/dT); \quad q_{st}^0 = -RT^2(d \ln H/dT). \tag{86}$$

Here q_{st}^0 is the limit of q_{st} at zero surface coverage. This equation has the advantage that the spreading pressure, π, needed in the calculations, can be expressed in an analytical form:

$$\frac{\pi A}{RT} = n_i + \left(\frac{kp}{p+1}\right)n_i^{p+1}, \tag{87}$$

whereas for the Tóth equation analytical integration is tedious and numerical integration is used.

[113] J. Toth, *Acta Chim. Acad. Sci. Hung.*, 1971, **69**, 311.
[114] A. L. Myers, *A.I.Ch.E. J.*, in press.

Application of the Radke and Prausnitz procedure to the individual isotherms led to calculated total and individual isotherms for the bisolute systems, which agreed with experiment to within 2—20%. The discrepancies showed a tendency to increase with increase in the difference between the acidities of the two compounds, suggesting that ionic equilibria played a part in determining the adsorption.

An appendix to this paper discusses thermodynamic consistency tests for binary solute adsorption. Provided that the data are within the Henry's law region of the isotherm, it is shown that application of the Gibbs adsorption equation leads, for any closed path, to

$$\oint (n_1 \, d \ln c_1 + n_2 \, d \ln c_2) = 0. \tag{88}$$

Alternatively, considering a locus of total bulk phase composition, starting from $c_1 = 0$ and ending at $c_2 = 0$,

$$\frac{A(\pi_1 - \pi_2)}{RT} = \int_{c_1=0}^{c_2=0} \left[\frac{n_1}{c_1} + \frac{n_2}{c_2} \right] dc_1, \quad [c_1 + c_2 = \text{const.}] \tag{89}$$

The left-hand side is obtained by integration of the single solute data up to a concentration c, and the right-hand side from bisolute isotherms at constant $(c_1 + c_2)$. The first of these criteria, using the data in the form of the empirical equation for bisolute adsorption previously proposed by Fritz and Schlünder,[110] indicated that the data reported in this paper are thermodynamically consistent within the accuracy of experiment. No mention is made of the application of the second criterion.

An alternative approach to multisolute adsorption is by application of the Polanyi potential theory. Its use for binary solute systems was outlined in reference 3 (pp. 111–114); Rosene and Manes[115] have now extended consideration to ternary solute systems. The same principles apply as in the binary solute case: the driving force for adsorption per unit volume is

$$-\frac{1}{v_i} \frac{\partial G}{\partial n_i^\sigma} = \frac{|\varepsilon_i^\ominus|}{v_i} - \frac{RT}{v_i} \ln \frac{c_i^{\text{sat}}}{c_i}. \tag{90}$$

This leads to the expectation that the component that will be adsorbed in a given element of adsorption space will be the component with the highest adsorption driving force. Thus in a multicomponent system the kth component will displace the jth if

$$\frac{RT}{v_k} \ln \left(\frac{c^{\text{sat}}}{c} \right)_k \leqslant \frac{|\varepsilon_k|}{v_k} - \frac{|\varepsilon_j|}{v_j} + \frac{RT}{v_j} \ln \left(\frac{c^{\text{sat}}}{c} \right)_j. \tag{91}$$

As before the correlation curves $\varepsilon_i(\phi)$, where ϕ is the volume of adsorption space occupied, are calculated from the single solute isotherms and, for given bulk phase compositions the free-energy curves, as functions of ϕ, calculated for each component. The fraction of adsorption space occupied by a given component is that in which its free-energy curve is the lowest. In this way,

[115] M. R. Rosene and M. Manes, *J. Phys. Chem.*, 1977, **81**, 1646.

varying the composition of the ternary along a given locus, the variation in the volume of adsorption space occupied by a given component adsorbed can be calculated; knowing the density of the adsorbed component (derived from the saturation value in single solute adsorption) the amount adsorbed is obtained.

This paper reports isotherms for the aqueous binary solute system *p*-nitrophenol (PNP) + benzamide and the ternaries PNP + benzamide + glucose; PNP + benzamide + methionine;   PNP + thiourea + acrylamide.   The   other binaries needed for the calculation had been determined previously.[116] Typical graphs show a comparison of the calculated and observed isotherms on active carbon, in which the concentration of two components is kept constant, while that of the third is varied, and in general good agreement is found except where the two characteristic curves are close together.

A basic feature of this model is that the solutes are adsorbed as pure solid-like material in specified regions of the pore space: it will not be valid if solid solutions are formed. Other limitations will arise if chemisorption or molecular sieving effects intervene.

A further complication arises from shifts in chemical equilibria between the adsorbing components, and this is particularly marked in the case of organic acids and their salts, where the effect of pH must be taken into account (see earlier p. 109). Other accounts of the application of the Polanyi potential theory are given by Rosene[117] and Ozcan.[118]

Gryazev *et al.* have extended their earlier work on the adsorption of carboxylic acids from ternary systems.[119]

Adsorption from the four-component mixture acetic acid + propionic acid + butyric acid + cyclohexane by ASK silica gel is reported on by Gryazev *et al.*[120] G.l.c. was used for the analysis of the equilibrium solutions and the experiments were based on a 'simplex-grid' design.

4 Experimental Techniques

A great deal of the work reported recently is based on the traditional batch method, although a widening range of techniques is being employed for analysis of the equilibrium solution. This method is tedious and difficult to apply over a range of temperatures. Improved techniques (*cf.* ref. 2, p. 81) have been designed to enable more precise data to be obtained in the absence of atmospheric contamination. A new apparatus which is said to be easier to operate is described by Larionov, Chmutov, and Shayusupova.[121] In this the outgassed solid is sealed into a bulb the tube from which is closed with a glass membrane. This tube is connected to a small reservoir closed by a Teflon

[116] See refs. 95 and 96 in ref. 3, and ref. 53 in ref. 2.

[117] M. R. Rosene, *Diss. Abstr. Int. B*, 1977, **38**, 2203.

[118] M. Ozcan, *Tech. J. Ankara Nucl. Res. Cent.*, 1976, **3**, 67 (*Chem. Abstr.*, 1976, **85**, 131 124); M. Ozcan, *Tech. J. Turk. A.E.C., Ankara Nucl. Res. Train. Cent.*, 1976, **6**, 97 (*Chem. Abstr.*, 1980, **92**, 65 212).

[119] See ref. 1, p. 102; ref. 2, p. 78; ref. 3, p. 107.

[120] N. N. Gryazev, T. N. Rzyanina, M. N. Rakhlevskaya and E. N. Moleva, *Zh. Fiz. Khim.*, 1976, **50**, 2960 (*Russ. J. Phys. Chim.*, 1976, **50**, 1762).

[121] O. G. Larionov, K. V. Chmutov and M. Sh. Shayusupova, *Zh. Fiz. Khim.*, 1978, **52**, 2118 (*Russ. J. Phys. Chem.*, 1978, **52**, 1225).

stopper through which liquid can be injected with a micro-syringe, and to one cell of a differential interferometer. Solvent is introduced and the membrane broken with an internal glass hammer; increments of the second component are added. The liquid can be decanted first into contact with the solid, and the whole unit shaken to achieve equilibrium, and then, after the solid has settled, transferred by tipping into the interferometer cell. In this way both a range of temperature and range of concentrations can be covered. The technique has been used to study the adsorption from p-xylene + cyclohexane and *m*-xylene + cyclohexane mixtures by NaX zeolite in the temperature range 303—363 K. The technique seems to be particularly applicable to relatively dilute solutions ($x_2 < 0.1$), and clearly cannot be used if the adsorbent does not settle readily. Another novel technique, the slurry technique, is described by Nunn, Schecter, and Wade.[122] In this, after equilibration of an excess of solution with a mass m of solid, most of the liquid phase is removed by decantation or centrifugation and a mass m^s of slurry recovered. The concentration of liquid phase in the slurry is the same as that in the equilibrium solution so that if the whole of the liquid in the slurry of mass $(m^s - m)$ were at this concentration (weight fraction w_i) the system would contain a mass $(m^s - m)w_i$ of component i. The total amount of i actually present is obtained by diluting the slurry and analysing for i by a convenient technique. If the mass of i actually present is m_i, then the surface excess amount is

$$n_i^\sigma = \frac{m_i - (m^s - m)w_i}{M_i}, \qquad (92)$$

where M_i is the molar mass of i. The method is clearly analogous to the McBain microtome method for measuring adsorption at the liquid/vapour interface.

In this work the adsorptive was radioactive and m_i was obtained using scintillation counting. By using an excess of liquid in the initial equilibration the solution concentration changed infinitesimally so that w_i could be taken as the initial concentration. This technique was compared with the conventional batch method in experiments using sodium laurate and hexanol in solution in water and n-decane; close agreement was found. This method is seen as having particular advantages when applied to multicomponent systems where the initial composition of the liquid can be established accurately, and since it does not change appreciably the planning and analysis of experiments is easier.

Liquid chromatographic techniques are increasingly used for the determination of adsorption isotherms (*cf.* ref. 3, pp. 115—118). Wang, Duda, and Radke[123] have given a very thorough assessment of the use of frontal analysis techniques in which, instead of changing from pure solvent to solution, as in Sharma and Fort's method,[124] the solution concentration is changed stepwise and the chromatogram recorded. Each step then gives a value for the slope of the chord of the isotherm between the two concentrations, and enables the whole isotherm to be built up in one series of experiments. Special care is

[122] C. Nunn, R. S. Schecter, and W. H. Wade, *J. Colloid Interface Sci.*, 1981, **80**, 598.
[123] H. L. Wang, J. L. Duda and C. J. Radke, *J. Colloid Interface Sci.*, 1978, **66**, 153.
[124] S. C. Sharma and T. Fort, *J. Colloid Interface Sci.*, 1973, **43**, 36.

devoted in the analysis to the need to take account of volume changes which result in the inlet and outlet volume rates of flow being different. This method can be used with conventional h.p.l.c. equipment with only minor modification. The technique has been evaluated by comparing the results obtained for the (n-hexane + n-hexanol)/wide pore silica gel systems with those measured by the conventional batch technique. The root mean square deviation between the two methods is of the order of 8%, the static adsorption data tending to be somewhat higher.

The calculation of vapour adsorption isotherms from gas-chromatographic experiments has been discussed by Huber and Gerritse.[125] The validity of this method of analysis when adapted for solution adsorption using liquid chromatography has been tested by Chuduk and Eltekov[126] using the systems (anisole + n-heptane)/macroporous silica (Silochrome S-80) and (anisole + n-heptane)/silica coated with monolayer of poly(ethylene glycol). The method originally developed by Glückauf[127] gives good results provided that the leading and rear edges of the peaks coincide when superimposed; that of Huber provides a fuller description of the adsorption isotherm and eliminates the effect of diffusion spreading. With the systems studied, static and dynamic methods gave results in close agreement.

Other examples of the use of liquid chromatography to study adsorption isotherms and surface properties are the work of Hammers, Kos, Brederode, and De Ligny[128] on the adsorption of a large number of organic compounds from n-hexane and dichloromethane by N-2-cyanoethyl-N-methylamino-silica and Ościk and Chojacka[129] on the adsorption of six aromatic hydrocarbons from some mixed-solvent mobile phases by silica gel.

A laboratory technique for measuring adsorption from solution at high pressures is described by Ozawa, Kawahara, and Ogino[130] and applied to studies of the (methylcyclopentane + ethanol)/activated carbon system. Jankowska, Swiatkowski, and Zietek[131] have developed a method for determining the isotherms of iodine adsorption from solution using a single sorbent sample, but no details are available. A paper on an optical method of studying adsorption from benzene + n-hexane solutions by Egorev and Fedorov is also abstracted.[132]

5 Surface Area Determination

The use of adsorption from solution for the determination of surface areas has been discussed in earlier reports (*cf.* ref. 1, pp. 76—81; ref. 3, pp. 73—78).

[125] J. F. K. Huber and R. G. Gerritse, *J. Chromatogr.*, 1971, **58**, 134.
[126] N. A. Chuduk and Yu. A. Eltekov, *Zh. Fiz. Khim.*, 1979, **53**, 1032 (*Russ. J. Phys. Chem.*, 1979, **53**, 586).
[127] E. Glueckauf, *Nature (London)* 1945, **156**, 748.
[128] W. E. Hammers, C. H. Kos, W. K. Brederode, and C. L. De Ligny, *J. Chromatogr.*, 1979, **168**, 9.
[129] J. Ościk and G. Chojnacka, *Chromatographia*, 1978, **11**, 731.
[130] S. Ozawa, K. Kawahara, and Y. Ogino, *High Pressure Sci. Technol.*, AIRAPT Conf. 6th, 1977 (Publ. 1979), `1`, 593 (*Chem. Abstr.*, 1979, **90**, 153 760).
[131] H. Jankowska, A. Swiatkowski, and S. Zietek, *Buil. Wojsk. Akad. Tech.*, 1978, **27**, 75 (*Chem. Abstr.*, 1978, **88**, 19 830).
[132] M. E. Egorov and A. F. Fedorov, *Fiz.-Khim. Izuch. Neorg. Soedin.* (*Cheboksary*) 1978, (6), 82; from *Ref. Zh. Khim.*, 1979, Abstr. No. 10B 1435 (*Chem. Abstr.*, 1979, **91**, 63 198).

Schay, Nagy, and Foti[133] have reconsidered the problem of estimating adsorption capacities from measurements on dilute solutions. The linear equation [*cf.* ref. 1, p. 55, equation (24)]:

$$\frac{x_1^l x_2^l}{n_2^{\sigma(n)}} = \frac{1}{n^\sigma}\left[x_2^l + \frac{1}{K-1}\right] \tag{93}$$

which applies to ideal adsorption of equal size molecules can be written more generally [*cf.* ref. 1, p. 79, equation (86)]:

$$\frac{x_1^l x_2^l}{n_2^{\sigma(n)}} = \frac{x_1^l}{(S-1)n_1^{\sigma,0}} + \frac{Sx_2^l}{(S-1)n_2^{\sigma,0}}, \tag{94}$$

where $n_1^{\sigma,0}$, $n_2^{\sigma,0}$ are the saturation capacities of the surface for components 1 and 2; S the separation factor has the same form as K, the ideal adsorption constant, but is not constrained to a constant value. If $S \gg 1$ and $x_1^l \to 1$, then*

$$\frac{x_1^l x_2^l}{n_2^{\sigma(n)}} = \frac{1}{Sn_1^{\sigma,0}} + \frac{x_2^l}{n_2^{\sigma,0}}. \tag{95}$$

It is then argued that since in most experimental situations the first term is much smaller than the second, the fact that S is not a constant will have little influence on the linearity of the graph of the left-hand side against x_2^l. They then give examples of the use of the slope of this graph to obtain $n_2^{\sigma,0}$. Rather than take an assumed value for the surface area occupied by an adsorbed molecule to calculate the specific surface area of the solid, they invert the calculation and use a BET-N_2 surface area to calculate an apparent molecular area. In the case of benzoic acid adsorption from aqueous solution by activated carbon, the molecular area calculated from measurements on a highly activated carbon (0.47 nm² molecule⁻¹) is substantially higher than that found with a lower surface-area carbon (0.36 nm² molecule⁻¹). It is suggested as a possible explanation that the more active carbon contains micropores into which water, but not benzoic acid, can penetrate. Measurements are also reported for the adsorption from dilute butanol + water and anthracene + cyclohexane by activated carbons. The authors conclude that for dilute solutions of strongly adsorbed solutes, the use of equation (95) provides a means of obtaining a reliable estimate of the adsorption capacity. 'On the other hand any mechanical application without due criticism might lead to erroneous conclusions.' It must be pointed out, however, that although equation (95) is derived from a more general model, the final result (except for x_1^l on the l.h.s.) is identical with the linearized form of the Langmuir equation: in that respect the method is not novel.

An alternative method of determining the surface phase capacity of heterogeneous surfaces has been proposed by Dabrowski and Jaroniec.[134]

---

* It is not clear why the authors retain x_1^l on the left-hand side of this equation, having put it equal to 1 on the right-hand side.

133 G. Schay, L. G. Nagy, and G. Foti, *Acta Chim. Acad. Sci. Hung.*, 1979, **100**, 289.
134 A. Dabrowski and M. Jaroniec, *Z. Phys. Chem. (Leipzig)*, 1980, **261**, 359.

According to equation (5), for molecules of equal size

$$x_1^\sigma = x_1^l + \frac{n_1^{\sigma(n)}}{n^\sigma} \tag{96}$$

Theoretical treatments of adsorption on heterogeneous surfaces lead to the following approximate expressions for the surface mole fraction:

$$\ln x_1^\sigma = -B\left[\ln \frac{a_1^l}{\alpha a_2^l}\right]^2 \tag{97}$$

and

$$x_1^\sigma = A\left(\frac{a_1^l}{a_2^l}\right)^c \bigg/ \left[1 + A\left(\frac{a_1^l}{a_2^l}\right)^c\right]; \qquad 0 < c \leqslant 1, \tag{98}$$

where a_1^l, a_2^l are the activities of the two components in the liquid phase, A and α are related to the difference between the adsorption energies of the two components and B and c are parameters characterizing the heterogeneity. These equations can be linearized to give

$$[-\ln x_1^\sigma]^{1/2} = B^{1/2} \ln \alpha - B^{1/2} \ln(a_1/a_2), \tag{99}$$

and

$$\ln\left(\frac{x_1^\sigma}{1 - x_1^\sigma}\right) = \ln A + c \ln\left(\frac{a_1}{a_2}\right). \tag{100}$$

These equations are used in the following way: values of x_1^σ are calculated from experimental measurements of $n_1^{\sigma(n)}$ choosing a succession of values of n^σ. These are substituted into equations (99) and (100) and the value of n^σ that gives the smallest standard deviation from linearity is taken as the true value of n^σ. This method was applied to typical data from the literature. It was found that the standard deviation S plotted against the chosen value of n^σ showed in all cases a sharp minimum and gave values of n^σ, which in most cases of type II isotherms agree moderately well with those calculated by the so-called Everett plot [equation (24) of ref. 1]; for type IV isotherms the Schay–Nagy method (*cf.* ref. 1, p. 56) tended to give lower values than those derived using equations (99) and (100). Two comments may be made. First the method (like others) is valid only for molecules of similar size. Secondly it is not clear how the authors have decided which of equations (99) and (100) to employ for a given system since in no case is a comparison made between the two equations applied to the same system. This new method is clearly one whose applicability should be more widely examined since the equations employed appear to be able to reproduce experimental data with high accuracy.

The use of methylene blue dye adsorption from aqueous solution has been used by Lirige and Tyler[135] for measuring the surface area of chalcopyrite. At a dye concentration of 10^{-5} M the plateau of the isotherm is apparently reached enabling a one-point method to be used. Comparison with BET-N_2 areas led to an effective area of 2.4 ± 0.1 μ mol m^{-2} for methylene blue (*cf.* p.

[135] H. G. Lirige, and R. J. Tyler, *Proc. Australas. Inst. Min. Metall.*, 1979, **271**, 27.

107). Brand, Koenig, and Lehmann[136] use phenol adsorption from heptane to measure the surface area of alumina.

An unusual application is described by Fedoseeva, Nechaev, and Strel'tsova,[137] who measured the adsorption of thirty-five organic substances on snow surfaces. Toluene was used as solvent since it does not dissolve ice to any appreciable extent, and has a lower density than ice. Of the 34 substances studied only formic, monochloracetic, and trichloracetic acids, and methyl and ethyl alcohols were adsorbed. The type II isotherm for trichloracetic acid seemed to exhibit a 'B-point', use of which led to a surface area of the snow sample of $2.8 \pm 0.3 \, m^2 \, g^{-1}$, which is similar to the value obtained from the low-temperature N_2 adsorption.[138] Further work on the use of calorimetry for the determination of surface areas is described by Rahman,[139] who compares flow and static measurements of the heats of adsorption of stearic acid from n-heptane by Fe_2O_3 and FeS. It is reported that heats of adsorption determined by these two methods differ by a constant factor that depends to some extent on the operating conditions used in the flow calorimetry. Under standardized conditions and using an Fe_2O_3 sample as a reference material, it is considered that flow calorimetry using the adsorption of stearic acid from n-heptane is a reliable method of finding the surface area of oxides.

6 Spectroscopic Studies

Recent years have seen a continued growth in the use of i.r. spectroscopic methods for the study of the effect of adsorption both on the spectral characteristics of adsorbed molecules and the surface chemical groups of the solid. The report published in Vol. 3 of this series covered work up to 1975. A review by Rochester[140] covers the period up to 1979. In this, recent developments in the experimental technique are described, followed by discussions of the perturbation of surface hydroxyl groups of silica and other oxides by immersion in liquids, and the adsorption by silica of molecules containing specific functional groups. The use of i.r. spectroscopy for determining adsorption isotherms developed mainly by Rochester is described, together with an account of polymer adsorption on silica.

I.r. methods have also been combined with gas chromatography and adsorption methods by Uvarov, Chuduk, and Eltekov[141] to study the chemistry of silica and rutile surfaces modified by poly(ethylene glycol).

Gubina, Kiselev, and Lygin[142,143] have continued the work described in Vol.

[136] P. Brand, A. Koenig, and R. Lehmann, *Freiburg. Forsehungsh. A*, 1979, **616,** 125 (*Chem. Abstr.* 1980, **92,** 153 616).
[137] V. I. Fedoseeva, E. A. Nechaev, and O. A. Streltsova, *Kolloid. Zh.*, 1977, **39,** 1009 [*Colloid J. (USSR)*, 1977, **39,** 894].
[138] S. W. Adamson, L. M. Dormant, and M. Oram, *J. Colloid Interface Sci.*, 1967, **25,** 206.
[139] M. A. Rahman, *Bangladesh J. Sci. Ind. Res.*, 1977, **12,** 151.
[140] C. H. Rochester, *Adv. Colloid Interface Sci.*, 1980, **12,** 43.
[141] A. V. Uvarov, N. A. Chuduk, and Yu. A. Eltekov, *Kolloid. Zh.*, 1978, **40,** 386 [*Colloid J. (USSR)*, 1978, **40,** 325].
[142] L. N. Gubina, A. V. Kiselev, and V. I. Lygin, *Zh. Fiz. Khim.*, 1978, **52,** 161 (*Russ. J. Phys. Chem.*, 1978, **52,** 84).
[143] L. N. Gubina, A. V. Kiselev, and V. I. Lygin, *Sorbtsiya Khromatogr.*, 1979, 106 (*Chem. Abstr.*, 1980, **92,** 83 089).

3, p. 124 on the use of u.v. spectroscopy to study adsorption from solution by silica. They found that adsorption of *o*- and *p*-nitroanilines from cyclohexane onto Silochrome shifted the u.v. maximum to longer wavelengths, but there was no effect on the spectrum of the *m*-isomer. These shifts were correlated with the basicity of the series of isomers. The second paper deals with the adsorption of aniline and amino-phenols where evidence was found for hydrogen-bond formation and interaction of the unshared electron pair of nitrogen with the surface.

7 Studies of Specific Systems

Carbon.—Much of the recent work on adsorption from solution on carbon has been designed to correlate the surface chemistry of carbons with their adsorption properties. Thus Puri and co-workers[144] have studied the adsorption of bromine by carbon blacks and sugar charcoal, and the adsorption of chlorine by carbon blacks, charcoals, and activated carbons from solution in carbon tetrachloride.[145] In all cases both reversible and irreversible adsorption was observed. The authors identified the latter as arising partly from chemisorption involving substitution of surface hydrogen, leading to the production of HCl or HBr, and addition to unsaturated sites, and partly from essentially irreversible adsorption by entrapment in pores or permeation into the internal structure of the carbon particles. The amount reversibly adsorbed from a 0.35 M solution was defined as that which could be removed by standing for 30 min with an excess of sodium thiosulphate and potassium iodide. In the case of bromine this was directly proportional to the surface area (BET-N_2) of the carbon blacks employed, although the proportionality constant led to a molecular area for Br_2 of 0.489 nm^2, which is twice the value calculated from bond lengths and van der Waals radii. With chlorine the amount adsorbed reversibly could not be related to the surface area: in this case the carbons studied varied widely in porosity and pore-size distribution.

The influence of high-temperature outgassing was studied: the amount of HBr formed on Br_2 adsorption decreased when the pretreatment temperature was raised to 600 and 1000 °C, while the surface unsaturation, leading to irreversible adsorption, increased. Most of the irreversibly adsorbed bromine could be recovered by heating the carbon in a stream of nitrogen. In the case of chlorine, the amount of HCl formed was a maximum for pretreatment at 600 °C, but decreased at 1000 °C, while the surface unsaturation again increased on heat treatment. However, as much as half the chemisorbed chlorine remained even after outgassing at 1000 °C, indicating the presence of carbon–chlorine complexes of high thermal stability.

The relationship between phenol adsorption and the chemical state of carbon surfaces is unclear. Some workers[146,147] have claimed that the presence of surface oxides decreases the adsorption of phenol, others that it enhanced

[144] B. R. Puri, D. L. Gandhi, and O. P. Mahajan, *Carbon*, 1977, **15**, 173.
[145] B. R. Puri, D. D. Singh, and V. M. Avora, *J. Indian Chem. Soc.*, 1978, **55**, 488.
[146] D. Graham, *J. Phys. Chem.*, 1955, **59**, 896.
[147] R. W. Coughlin, F. S. Ezra, and R. N. Tan, *J. Colloid Interface Sci.*, 1968, **28**, 386.

adsorption,[148] and some that it has no effect.[149] According to Puri, Bhardwaj, Kumar, and Mahajan[150] the differences are important only at low phenol concentrations (up to 5×10^{-3} M), when the presence of acidic surface oxides (giving CO_2 on heating) decreases phenol adsorption whereas the CO-complex enhances adsorption. Studies at higher concentrations (up to 0.4 M) have been made by Puri, Bhardwaj, and Gupta[151] on a range of carbon blacks and activated carbon. Only a small amount of irreversible adsorption was observed, and plots of the reversible adsorption against concentration showed in all cases a well defined plateau around 0.16 M followed in many instances by a further rise. The data were well represented by a modified BET equation in which p/p^0, appropriate to gas adsorption, is replaced by c/c_s, where c_s is the saturation concentration of solute. The surface areas calculated in this way, taking the molecular area of phenol as $0.522 \, nm^2$, were in good agreement with the conventional BET-N_2 areas. This agreement was unaffected by degassing the carbon blacks at 600 °C, provided that account was taken of the increased amount of irreversible adsorption found with heat-treated samples. It may be noted however that according to Singh[152] phenol is adsorbed in an 'edge-on' orientation on carbon black with an effective molecular area of $0.30 \, nm^2$.

To eliminate the competition between water and phenol for surface sites, Puri, Singh, and Gupta[153] have studied the adsorption of phenol from benzene solution by three carbon blacks before and after degassing at 600 °C and 1000 °C. They found adsorption of phenol preferred over the whole concentration range, the isotherms rising sharply and then falling linearly in the range of weight fraction (w) of phenol of 0.1 to 0.6; no experiments were made at higher concentrations. In all cases outgassing decreased the phenol adsorption even though the BET-N_2 surface areas were not greatly affected. To estimate the saturation uptake of phenol, use was made of the Schay–Nagy procedure of extrapolating the linear section of the isotherm to $w = 0$. Assuming that the BET-N_2 areas represent the true physical area, the uptakes per unit area were calculated. For Mogul carbon black, for example, the values were 0.60, 0.49, and 0.31 mg m^{-2}. Since monolayers of phenol molecules oriented perpendicular and parallel to the surface would predict uptakes of 0.62 and 0.30 mg m^{-2}, Puri *et al.* have deduced that phenol is adsorbed perpendicular to oxygen-covered surfaces and parallel to oxygen-free surfaces. Similar results were found for the other two carbon blacks. However, the surface excess isotherms are unusual in that extrapolation of the linear section to $w = 1$ gives *positive* intercepts whereas the Schay–Nagy procedure is only applicable when this intercept is negative or zero. The above conclusions, although demonstrating an influence of surface oxides, may not be quantitatively justified. The change in uptake with oxygen content could, incidentally, be equally well accounted for by a change in the thickness of the adsorbed layer rather than by a change in

[148] B. D. Epstein, E. D. Malle, and J. S. Mattson, *Carbon*, 1971, **9**, 609.
[149] A. Claus, H. P. Boehm, and U. Hofmann, *Z. Anorg. Chem.*, 1957, **290**, 35.
[150] B. R. Puri, S. S. Bhardwaj, V. Kumar, and O. P. Mahajan, *J. Indian Chem. Soc.*, 1975, **52**, 26.
[151] B. R. Puri, S. S. Bhardwaj and U. Gupta, *J. Indian. Chem. Soc.*, 1976, **53**, 1095.
[152] Ref. 2, ref. 158.
[153] B. R. Puri, D. D. Singh, and U. Gupta, *J. Indian Chem. Soc.*, 1979, **56**, 1193.

orientation. This type of argument has been used by Bansal and Dhami[154] who have studied the adsorption of methanol + benzene mixtures by a range of charcoals of varying oxygen content. They assumed fixed molecular areas (not given) and calculated adsorbed layer thicknesses, which for all but the oxygen-free sample indicated the formation of multilayers of methanol, which they attributed to specific interactions between methanol and surface oxygen complexes. The amounts of physisorbed and chemisorbed methanol were found by pretreating the carbons with methanol, evacuating at 120 °C to remove physisorbed material, and redetermining the adsorption isotherms. Application of the Schay–Nagy method to the resulting S-shape surface excess isotherms gave surface areas (apparently assuming monolayer adsorption) in good agreement with BET-N_2 values.

The adsorption of hydroquinone by carbon may be related to the probable existence of quinone–hydroquinone systems on carbon surfaces.[155] To examine this question in more detail, Puri *et al.*[156] have studied the interaction of a series of carbon blacks with hydroquinone in aqueous solution. Both physisorption and chemisorption were observed and particular attention was paid to the effect of previously degassing the carbon on the extent of chemisorption. It is suggested that hydroquinone is chemisorbed on quinonic surface groups, an effect that is enhanced when other less-stable oxygen groups are eliminated by degassing. For carbons degassed at temperatures up to 700 °C, treatment with boiling sodium sulphite solution under reflux recovered about 70% of the chemisorbed hydroquinone, reducing surface quinonic groups in the process. For 1000 °C degassed carbon it is suggested that the hydroquinone is adsorbed at sites vacated by quinonic groups, and that some aerobic oxidation to quinonic structures occurs subsequently in the experiment.

The adsorption of binary organic mixtures by a porous carbon was studied by Takeuchi and Furaya,[157] the components being chosen such that some were accepted by and some excluded from 0.5 nm micropores. The experimental results could be represented by a combination of Freundlich adsorption isotherms, the parameters of which could be related to the physical properties of the adsorptives, and it was found possible to predict satisfactorily the adsorption isotherms of new systems. The adsorption by and desorption from active carbon has been reported by Andreikova, Kondratov, and Kogan,[158] who found that adsorption from (unspecified) organic solvents decreased in the series   phenol > quinoline > phenanthrene > acenaphthene > naphthalene.   In desorption, acidic compounds are best desorbed with a mixture of methanol and dichloroethane but for basic compounds benzene is most effective.

A series of papers by Abe and his co-workers[159—161] deal with the mechan-

[154] R. C. Bansal and T. L. Dhami, *Carbon*, 1977, **15**, 153.
[155] V. A. Garten and D. E. Weiss, *Aust. J. Chem.*, 1955, **8**, 68.
[156] B. R. Puri, S. K. Sharma, I. S. Dosanjh, and D. L. Gandhi, *J. Indian Chem. Soc.*, 1976, **53**, 486.
[157] Y. Takeuchi and E. Furaya, *J. Chem. Eng. Jpn.*, 1977, **10**, 268.
[158] G. L. Andreikova, V. K. Kondratov, and L. A. Kogan, *Zh. Fiz. Khim.*, 1976, **50**, 1059 (*Russ. J. Phys. Chem.*, 1976, **50**, 643; VINITI No. 3839–75).
[159] I. Abe, K. Hayashi, and M. Kitagawa, *Kagaku To Kogyo (Osaka)*, 1979, **53**, 274 (*Chem. Abstr.*, 1979, **91**, 199 408).
[160] I. Abe, K. Hayashi, M. Kitagawa, and T. Urahata, *Bull. Chem. Soc. Jpn.*, 1979, **52**, 1899.
[161] I. Abe, K. Hayashi, M. Kitagawa, and T. Urahata, *Nippon Kagaku Kaishi*, 1979, **7**, 830 (*Chem. Abstr.*, 1979, **91**, 97 119).

ism of adsorption on activated carbon. The results are discussed in relation to the free energy of adsorption and comparison of the characteristic curves for adsorption from the gas and liquid phases.

Competitive adsorption on active carbon of two organic substances dissolved in water is discussed in two papers by Fritz, Merk, and Schlünder.[162,163] Experiments were made on aqueous solutions of phenol and *p*-nitrophenol, and interpreted using the ideal adsorbed solution theory of Radke and Prausnitz.[109,164] A review in Japanese of the adsorption of multiple component aqueous solutions by activated carbon is presented by Tàkeuchi.[165] The stability of the sorption properties of active carbons under cyclic operating conditions is discussed by Chubarova *et al.*[166]

Oxides.—The adsorbent properties of silica gel are known to be highly dependent on the degree of hydroxylation of the surface, and this dependency has been widely investigated in relation to gas adsorption.[167] Khopina and Eltekov[168] have continued earlier work of Kiselev and his collaborators[169] on the influence of the surface chemistry of silica on adsorption from solution. In the present paper a comparison is made of the adsorption of a series of aromatic hydrocarbons (benzene, naphthalene, biphenyl, phenanthrene, *o*- and *m*-terphenyl) from n-heptane solution by hydroxylated silica, dehydroxylated silica, and graphitized carbon black, using data obtained in the present work together with earlier published data.[170,171] The results are analysed in terms of the separation factor[172] (partition coefficient)

$$S = \frac{x_1^\sigma x_2^l}{x_1^l x_2^\sigma}, \tag{101}$$

where the surface mole fractions are calculated according to the usual formula,[173] assuming monolayer adsorption and molecular surface areas corresponding to planar orientation on the solid surface.† The separation factor is not a true equilibrium constant and is found to vary with the equilibrium concentration in the bulk.‡ Nevertheless it is a useful parameter for comparing qualitatively the variation of the adsorbent properties of the three adsorbents.

---

† Equation (4) of this paper appears to contain a printing error. In the denominator α_{m_1} should read α_{m_2}.

‡ It would be interesting to know whether the ideal equilibrium constant $K = \frac{x_1^\sigma}{x_1^l} \cdot \left(\frac{x_2^l}{x_2^\sigma}\right)^{1/\beta}$ is indeed constant for these systems.

[162] W. Fritz, W. Merk, and E. U. Schluender, *Chem.-Ing.-Tech.*, 1978, **50**, 119.
[163] W. Fritz, W. Merk, and E. U. Schluender, *VDI-Ber.*, 1977 (Publ. 1978), **315**, 183.
[164] Ref. 2, pp. 68, 69.
[165] Y. Takeuchi, *Kogyo Yosui*, 1978, **233**, 4 (*Chem. Abstr.*, 1978, **89**, 95 293).
[166] T. F. Chubarova, G. M. Belotserkovskii, I. D. Dashkovskii, V. I. Yakovlev, N. A. Emelyanova, and V. A. Proskuryakov, *Zh. Prikl. Khim. (Leningrad)*, 1978, **51**, 939 (*Chem. Abstr.*, 1978, **89**, 31 321).
[167] Ref. 2, p. 34 *et. seq.*
[168] V. V. Khopina and Yu. A. Eltekov, *Zh. Fiz. Khim.*, 1979, **53**, 1806 (*Russ. J. Phys. Chem.*, 1979, **53**, 1023).
[169] Ref. 1, p. 92; ref. 2, p. 87.
[170] A. V. Kiselev and I. V. Shikalova, *Kolloid Zh.*, 1962, **24**, 687.
[171] A. V. Kiselev and I. V. Shikalova, *Kolloid Zh.*, 1970, **32**, 702 [*Colloid J. (USSR)*, 1970, **32**, 588].
[172] Ref. 1, p. 59 eqn. (42).
[173] Ref. 1, eqn. (63) with $t = 1$.

The authors comment that whereas the values of S for the series of hydrocarbons on hydroxylated silica gel vary from 13—200, the values on dehydroxylated silica gel are 'altogether much lower'. It seems worthwhile to make a more quantitative analysis of the data by comparing the values of ln S for each compound on the two silica surfaces. The mean difference $\Delta(\ln S)$ when S is calculated at a fractional coverage of 0.1 is 1.86, and at a fractional coverage of 0.2 it is 1.80, with a mean deviation of about ±0.2. Thus on average dehydroxylation of silica reduces the separation coefficient of these substances by a factor of about 6. The authors also compare the static adsorption data with those obtained chromatographically and find general agreement, although some discrepancy is found for the o- and m-terphenyls.

Ościk and Goworek[174] have measured the adsorption of aliphatic ketones (acetone, methylethylketone, diethylketone, methylpropylketone, and dipropylketone) from n-heptane and from benzene on silica gel ($a_s =$ 345 m^2 g^{-1}; outgassed at 180 °C for 24 h). They used their data to estimate the adsorbed layer thickness using equation (63) of ref. 1, together with Rusanov's criteria.[8] For this purpose it is necessary to estimate the molecular surface areas of the various components. The usual formula[175,176] assumes the close-packing of spherical molecules. Since the assumption of a spherical shape for the higher ketones seems unrealistic the authors used McClellan and Harnsberger's[177] empirical equation

$$a_i^0 = (a_i^* - 6.16)/0.596, \tag{102}$$

where a_i^* is the area calculated assuming spherical molecules. On this basis it was concluded that whereas the adsorption of these ketones from benzene solution is consistent with a monolayer model, for solutions in heptane a double layer is indicated. Although it seems plausible to suggest that the first layer of ketone molecules is bound by hydrogen bonds and that these form the substrate for the second layer, it is not at all clear why adsorption from benzene solution should differ so markedly from the behaviour in the presence of heptane. In a second paper Ościk and Goworek[178] report work on the adsorption of aliphatic alcohols (C_1 to C_5) from toluene solution on the same silica gel. Again estimates of the adsorbed layer thickness were made, in this case assuming that the alcohol molecules were adsorbed in a perpendicular orientation occupying a molecular area of 0.20 nm^2. According to this analysis methanol forms a double layer, whereas propanol and butanol form monolayers. In the case of ethanol a thickness of 1.5 layers is quoted. It may be noted that if it were assumed that the alcohols adopted a parallel orientation, then it would be concluded that the adsorbed layer was even thicker in terms of molecular layers: in effect from this point of view one monolayer of perpendicularly oriented cylindrical molecules is equivalent to h/d layers of

[174] J. Ościk and J. Goworek, *Pol. J. Chem.*, 1978, **52**, 1781.
[175] S. Brunauer, in 'Physical Adsorption of Gases and Vapours', Oxford University Press, Oxford, 1943, p. 287.
[176] S. J. Gregg and K. S. W. Sing, in 'Adsorption, Surface Area and Porosity', Academic Press, London and New York, 1967, p. 67.
[177] A. L. McClellan and H. F. Harnsberger, *J. Colloid Interface Sci.*, 1967, **23**, 577.
[178] J. Ościk and J. Goworek, *Pol. J. Chem.*, 1978, **52**, 775.

molecules lying parallel to the surface where h is the length of a molecule and d its diameter. It is interesting and possibly significant that the present data would be consistent with the idea that the adsorbed layer has roughly the same absolute thickness in each system, *i.e.* approximately twice the longest dimension of a methanol molecule. The author's use of the so-called Everett equation[179] to assess monolayer capacities cannot be justified since this equation is applicable only to mixtures of molecules of equal size. The angles of orientation calculated for butyl and amyl alcohols can have little true validity.

A series of papers by Nasuto[180—183] have the objective of presenting a thermodynamic analysis of several systems by silica gel. The experimental work involved both the determination of surface excess isotherms and enthalpies of immersion of silica gel in both pure liquids and liquid mixtures. The data were analysed in the following way. First the results were plotted using the Everett equation to obtain the total amount of adsorbed material n^{σ} and the adsorption equilibrium constant K (although no values of the latter are quoted). Since no contrary statement is made it is presumed that linear graphs were obtained and hence n^{σ} was independent of solution composition (this also implies that the components occupy essentially the same areas in the adsorbed state). The surface mole fraction x_1^{σ} is calculated, presumably assuming monolayer adsorption using equation (5). In all cases x_1^{σ} rose asymptotically to unity, thus showing that the data are consistent with a monolayer model. The 'individual isotherms', n_1^{σ} as a function of x_1^{l} are calculated and shown graphically. Unfortunately, there appears to be some unexplained inconsistency in the analysis since in several instances these curves exhibit maxima, implying that n^{σ} does not remain constant. In at least one case a smooth curve of x_1^{σ} against x_1^{l} is accompanied by an individual isotherm showing a point of inflexion; in another a curve of x_1^{σ} rising asymptotically to unity is shown with an almost linear individual isotherm. In view of these discrepancies it is difficult to accept the interpretations given for the various shapes of individual isotherms given in these papers. Similar problems arise in the interpretation of the enthalpies of immersion. These were analysed by supposing that the enthalpy of immersion could be written in the form

$$\Delta_w \hat{h} = \Delta_w \hat{h}_1 + \Delta_w \hat{h}_2 + \Delta h(\text{mix}), \qquad (103)$$

where $\Delta h(\text{mix})$ takes account of the enthalpy change accompanying the change in solution composition in the wetting process. The enthalpy change, corrected for this term, may be represented by $\Delta_w \hat{h}'$.[184] For ideal systems it has been postulated[185] that

$$\Delta_w \hat{h}' = x_1^{\sigma} \Delta_w \hat{h}_1^* + x_2^{\sigma} \Delta_w \hat{h}_2^*, \qquad (104)$$

where $\Delta_w \hat{h}_1^*$ is the enthalpy of immersion in pure liquid i. Nasuto assumes that

[179] Ref. 1, p. 55, eqn. (24).
[180] R. Nasuto, *Rocz. Chem.*, 1977, **51**, 319.
[181] R. Nasuto, *Rocz. Chem.*, 1977, **51**, 515.
[182] R. Nasuto, *Rocz. Chem.*, 1977, **51**, 525.
[183] R. Nasuto, *Rocz. Chem.*, 1977, **51**, 761.
[184] Ref. 2, p. 55.
[185] D. H. Everett, *Trans. Faraday Soc.*, 1964, **60**, 1803.

the term for the solvent (2 in his notation) is valid, but that $\Delta_w \hat{h}_1$ is not necessarily given by $x_1^\sigma \Delta_w \hat{h}_1^*$. Thus he writes

$$\Delta_w \hat{h}_1 = \Delta_w \hat{h}' - x_2^\sigma \Delta_w \hat{h}_2^* \tag{105}$$

and examines the variation of $\Delta_w \hat{h}_1$ with x_1^σ. If this is linear then ideality of the surface layer is demonstrated.

Deviations from ideality are attributed entirely to the behaviour of $\Delta_w \hat{h}_1$, although logically it could equally well reflect the breakdown of the assumption that $\Delta_w \hat{h}_2$ is proportional to x_2^σ. The differential molar enthalpy is defined by

$$\overline{\Delta_w \hat{h}_1} = [\partial(\Delta_w \hat{h}_1)/\partial n_1^\sigma]_{T,p,n_2^\sigma}, \tag{106}$$

and a differential free energy of adsorption by

$$\Delta G_1^\sigma = RT \ln(x_1^\sigma/x_1^l), \tag{107}$$

which is said to 'characterize the work of transition of adsorbate molecules from bulk solution into the adsorption layer'. Reference to the definition of the chemical potential of adsorbed species in an ideal system:[186]

$$\mu_i = \mu_i^{\ominus,l} + RT \ln x_1^\sigma - (\sigma - \sigma_i^*)a, \tag{108}$$

shows that

$$\Delta G_1^\sigma = (\sigma - \sigma_1^*)a. \tag{109}$$

No more general free energy can be defined without the introduction of activity coefficients.

In some instances the enthalpy curves (*e.g.* for toluene + hexane mixtures) were linear indicating a constant value of $\Delta_w \hat{h}_1^*$; in most, however, a break was detected and interpreted as a jump in $\Delta_w \hat{h}_1^*$ at a certain surface coverage. What is not explained is how $\Delta_w \hat{h}_2$ can exhibit a break when the analysis is based on the assumption that $\Delta_w \hat{h}_2$ is directly proportional to x_2^σ. Nor is it clear in the case, for example, of the acetone + nitrobenzene system how a smooth curve of x_1^σ against x_1^l can give rise to a sharp break in ΔG_1^σ. Attempts to relate various breaks and discontinuities in the thermodynamic curves to the presence of two types of adsorption site cannot have any validity. Without access to the original data it is difficult to assess this work, which contains some potentially useful experimental results. In any case it is important to remember, as stressed by the author 'that the thermodynamic quantities of adsorption are calculated from individual isotherms, which depend strongly on the method of their determination'.

Davydov, Kiselev, and Sapozhnikov[187] have presented adsorption isotherms for acetone + water, dioxan + water, and dioxan + acetone mixtures on KSK silica. All the isotherms exhibited azeotropic points at weight fractions of ≈ 0.45 acetone, ≈ 0.6 dioxan, and ≈ 0.9 dioxan, respectively, the first-named component of each pair being preferentially adsorbed up to this point. No check on the thermodynamic consistency of the three sets of results is

[186] Ref. 1, p. 53, eqn. (15) with $\gamma_i^\sigma = 1$.
[187] V. Ya. Davydov, A. V. Kiselev, and Yu. M. Sapozhnikov, *Kolloid. Zh.*, 1979, **41**, 333 [*Colloid J. (USSR)*, 1979, **41**, 271].

reported. One objective of this work was to show that whereas the adsorption isotherms for the acetone + dioxan system are determined mainly by the strengths of the hydrogen bonds between each component and the surface, in the case of acetone + water and dioxan + water the adsorption is also dependent on the strengths of the hydrogen bonds between the components. Some measurements were also made on adsorption from the ternary system. It was found that addition of acetone reduces the adsorption of dioxan and *vice versa*, although the effect is relatively small.

In work related to flotational studies Chibowski[188] has measured the effect on the surface charge of n-alcohol and n-alkane adsorption on quartz particles from aqueous KCl solution. Maximum effect was found with C_6—C_8 n-alcohols, where the pH at the point of zero charge shifted from 5.7 to 8—9; and for C_8—C_{13} n-alkanes, with a corresponding shift to pH 7.

Furuyama *et al.*[189] have reported the asymmetric adsorption of optically active alanine and alanine hydrochloride onto quartz from ethanol solution, the L-isomer being preferentially adsorbed by *l*-quartz and the D-isomer by *d*-quartz. An asymmetric adsorption factor is defined $A_s = n_1^{\sigma} n_2^l / n_1^l n_2^{\sigma}$, where component 1 is the preferentially adsorbed isomer and calculated values range from 1.2—2.2. It is suggested that this phenomenon is related to the genesis of the first optically active molecules in nature.

Madan and Sandle[190] have used a tin oxide gel as the adsorbent with a series of binary mixtures where one component was an aliphatic alcohol. They found that methanol was preferentially adsorbed from benzene, toluene, and *p*-xylene at 30 °C over the complete mole fraction range. Ethanol and propanol were also adsorbed from benzene, the rather sparse data for these two systems suggesting the existence of two maxima in the surface excess isotherm. A superficial examination of the results by Schay and Nagy's method and Everett's equation for an ideal system provided tables of the amounts of each component in the adsorbed phase. Indeed the latter method produced figures indicating preferential benzene adsorption from alcohols at low alcohol mole fraction. Clearly assuming ideality for these systems is of limited use and a more thorough analysis of the results is to be preferred. A table indicates that the thickness of the adsorbed layers is in the range of 2—4 monolayers for the various systems.

The same authors[191] have also studied adsorption of stearic acid from cyclohexane solutions on tin and antimony oxide gels and Sn–Sb mixed oxides. Adsorption isotherms at 30 °C showed a limiting plateau at stearic acid mole fractions of $\approx 10^{-2}$. These limiting values were compared with predicted monolayer values assuming different orientations of the adsorbate. The conclusions were that stearic acid is adsorbed in parallel fashion in SnO_2 and mixed oxides with a high Sn–Sb ratio and perpendicular or perhaps tilted to the surface on oxides with a high antimony content. It is suggested that the more polar the surface, the greater is the tendency for perpendicular orientation.

[188] E. Chibowski, *Przem. Chem.*, 1978, **57**, 647 (*Chem. Abstr.*, 1979, **90**, 110 502).
[189] S. Furuyama, H. Kimura, M. Sawada, and T. Morimoto, *Chem. Lett.*, 1978, **4**, 381.
[190] R. L. Madan and N. K. Sandle, *Indian J. Technol.*, 1976, **14**, 549.
[191] R. L. Madan and N. K. Sandle, *J. Indian Chem. Soc.*, 1977, **54**, 1113.

Suri, Brar, and Ahuja[192] have studied the adsorption from benzene + cyclohexane mixtures by cobalt oxide and cobalt sulphide; they found that benzene was preferentially adsorbed by the oxide, and cyclohexane by the sulphide. A comparision is made between surface mole fractions (x_1^σ, x_2^σ) and total amount adsorbed (n^σ) calculated by the so-called Schay–Nagy and Everett methods. It should be emphasized that the method for finding x_1^σ attributed to Schay and Nagy is that which employs equation (5) with $t = 1$: the authors erroneously suppose that the assumption of monolayer adsorption is not involved. That attributed to Everett uses the form (7) to which this equation reduces when the two molecules occupy equal areas. Clearly if, as in the present case, different molecular areas are assumed for the two components when applying the Schay–Nagy method, then the two methods cannot give the same results.

Work on the following systems has also been abstracted: the adsorption of cyclohexanol from heptane on alumina and silica,[193] of methyl orange from aqueous solution on various alumina powders,[194] and of lauryldimethylbenzyl-ammonium chloride from aqueous solution on silica.[195]

Chemisorption phenomena are not considered in detail in this Report, but some interesting observations of Nechaev and his group[196—204] on the adsorption of a range of organic compounds by oxides are worthy of comment. They have studied up to thirty-nine organic substances adsorbed from aqueous solution by some seventeen oxides, with the objective of explaining the selectivity of these surfaces for particular adsorptives. The authors note that in many cases adsorption of organic compounds is negligible (with the exceptions of dissociated species attracted by surface charge, and adsorptives of high molecular weight) and argue that to displace water from the oxide surface a 'donor–acceptor' bond must be formed between the adsorbate and adsorbent surface. The selectivity is explained by assuming that for charge transfer, and

[192] S. K. Suri, A. S. Brar, and L. D. Ahuja, *J. Colloid Interface Sci.*, 1979, **69**, 347.
[193] I. V. Smirnova, G. N. Filatova, V. V. Zadymov, and K. V. Topchieva, *Vestn. Mosk. Univ., Ser. 2: Khim.* 1980, **21**, 3 (*Chem. Abstr.*, 1980, **92**, 221 290).
[194] K. Nishibe and Y. Ohtsuka, *Aruminyumu Kenkyu Kaishi*, 1976, **112**, 16 (*Chem. Abstr.*, 1977, **87**, 29 491).
[195] V. I. Afanasev, Yu. P. Gladkikh, K. F. Pans, and A. A. Pavlov, *Sb. Tr. Belgorod. Tekhnol. In-t Stroit. Materialov*, 1975, 16; from *Ref. Zh. Khim.* 1977, Abstr. No. 23B1698 (*Chem. Abstr.*, 1978, **88**, 111 029).
[196] E. A. Nechaev, O. A. Strel'tsova and N. F. Fedoseev, *Zh. Fiz. Khim.*, 1977, **51**, 2307 (*Russ. J. Phys. Chem.*, 1977, **51**, 1350).
[197] E. A. Nechaev and V. A. Volgina, *Electrokhimiya*, 1978, **14**, 974 (*Sov. Electrochem.*, 1978, **14**, 844; VINITI, No. 349—78).
[198] E. A. Nechaev, *Zh. Fiz. Khim.*, 1978, **52**, 1494, (*Russ. J. Phys. Chem.*, 1978, **52**, 858).
[199] E. A. Nechaev, V. A. Volgina, and O. A. Strel'tsova, *Sorbtsiya Khromatogr.*, 1979, 83 (*Chem. Abstr.*, 1980, **92**, 83 085).
[200] E. A. Nechaev and N. F. Fedoseev, *Zh. Fiz. Khim.*, 1978, **52**, 1250 (*Russ. J. Phys. Chem.*, 1978, **52**, 712).
[201] E. A. Nechaev and O. A. Strel'tsova, *Kolloid. Zh.*, 1978, **40**, 148 [*Colloid J. (USSR)*, 1978, **40**, 123].
[202] E. A. Nechaev, R. Ya Shaidullin, and V. A. Volgina, *Sorbtsiya Khromatogr.*, 1979, **75** (*Chem. Abstr.*, 1980, **92**, 99 969).
[203] E. A. Nechaev, *Kolloid. Zh.*, 1980, **42**, 371 [*Colloid J. (USSR)*, 1980, **42**, 311].
[204] E. A. Nechaev, V. A. Volgina, and N. G. Bakhchisaraits'yan, *Electrokhimiya*, 1973, **9**, 1825 (*Sov. Electrochem.*, 1973, **9**, 1717).

hence adsorption, to occur the energies of the highest occupied molecular orbital of the solute, $E_{h,0}$, estimated from the ionization potential, and the acceptor level in the solid must be equal. In general, on a given solid surface, physisorption does not occur, but a few organic substances having appropriate values of $E_{h,0}$ may be chemisorbed. Figure 5(a) shows a typical plot for alumina of the degree of adsorption from 0.025 M solutions as a function of $E_{h,0}$.[196] Similar results have been presented for lead dioxide,[197] silica gel,[198] and zinc oxide.[199] In the case of silica, adsorption from iso-octane also showed a similar behaviour,[205] Figure 5(b), although the selectivity is here less sharp.

It is suggested that the peaks in these figures define an energy level, E_{res}, which is characteristic of the adsorbent and independent of surface charge. In Figure 5(a) the naphthols and cresols did not adsorb over a wide pH range, but chemisorption of catechol and pyrogallol decreased towards both high and low pH, and was a maximum at the point of zero charge. Table 4 lists typical examples of the selectivity reported in these papers. Adsorption of nitrophenols and benzoic acids by the dioxides of titanium, zirconium, and manganese was limited to the pH range in which the solute remained in molecular form.[200] However, the effect of pH is complicated by the simultaneous effect on the surface charge on the oxide: the view is put forward, however, that in the adsorption of organic solutes from aqueous solutions the charge on the surface is of secondary importance (see however p. 110).

Other papers have dealt with the adsorption of organic sulphur compounds on a range of oxides,[20] a comparison between adsorption on the oxides and hydroxides of cadmium, tin, nickel, and lead,[202] and the degree of adsorption of methylene blue on a series of 29 metal oxides.[203] Table 4 shows that only certain oxides adsorb methylene blue. Consequently its use for the determination of surface areas is limited; in particular it cannot be used for alumina.

A general conclusion is that organic compounds are not adsorbed from aqueous solution by oxides when $E_{h,0}$ differs from a characteristic resonance frequency, E_{res}, of the oxide by more than about 0.1 eV. When adsorption occurs it is probably accompanied by the formation of a chemical bond as a result of the coincidence of donor levels of the adsorbate and acceptor levels of the adsorbent. This behaviour in adsorption from solution contrasts with adsorption from the gas phase where no correlation has been found between the enthalpy of adsorption, or the shift of vibration frequency of the OH groups accompanying adsorption, with the electronic properties of oxides. No explanation of this difference is offered.

Della Gatta, Stradella, and Venturello[206] have used a calorimetric technique to study iodine adsorption from n-pentane solution at 20 and 27 °C, and from cyclohexane at 35 and 50 °C on η-Al_2O_3 and bayerite. Adsorption isotherms, plotted as a function of equilibrium relative concentration c/c_0, showed a steep rise to a plateau region with a further increase at higher c/c_0. On porous η-Al_2O_3 there was a trend towards increased adsorption at higher temperatures, whereas on non-porous bayerite the plateau region was less well defined,

[205] R. K. Iler, in 'Colloid Chemistry of Silica and Silicates', Cornell Univ. Press, 1955.
[206] G. Della Gatta, L. Stradella, and G. Venturello, *Z. Phys. Chem.* (*Frankfurt am Main*), 1977, **106**, 95.

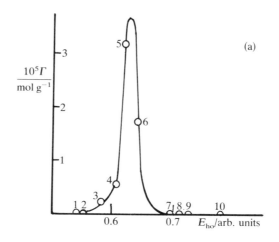

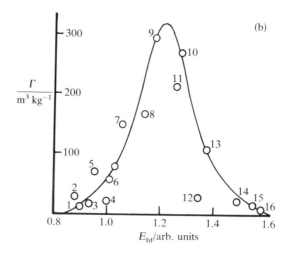

Figure 5 (a) *Adsorption (Γ) on alumina from 0.025 M solution in 0.1 M-KCl at pH 5 of the following substances as a function of the energy of the highest-occupied orbital (E_{ho}): 1, β-naphthol; 2, α-naphthol; 3, phloroglucinol; 4, hydroquinone; 5, pyrogallol; 6, catechol; 7, p-cresol; 8, resorcinol; 9, o-cresol; 10, phenol*
(Redrawn from *Zh. Fiz. Khim.*, 1977, 51, 2307).
    (b) *Adsorption (Γ) on silica from a 0.2% solution in iso-octane as a function of the energy of the highest-filled molecular orbiral (E_{hf}) for 1,1,6-dichlorobenzene; 2, toluene; 3, chlorobenzene; 4, benzene; 5, chlorobenzene; 6, 2,4,6-trichlorobenzene; 7, dioxan; 8, nitrobenzene; 9, butylamine; 10, ethanol; 11, pyridine; 12, chlorobutane; 13, nitropropane; 14, tetrachloroethane; 15, methylene chloride; 16, chloroform*
(Redrawn from *Zh. Fiz. Khim.*, 1978, **52**, 1494; based on data from 'Colloid Chemistry of Silica and Silicates', Cornell Univ. Press, 1955).

Table 4 *Typical selectivities exhibited in adsorption by oxides from aqueous solution* [196—205]

(a) Oxide	Adsorbed	Not appreciably adsorbed
SiO_2	benzophenone, furfural, 9,10-anthraquinone	a total of 37 aromatic amines, naphthols, phenols, nitrophenols, benzamide, benzaldehyde
MnO_2	benzaldehyde, butynediol	benzoic acid, p-nitrophenol
TiO_2	2,4-dinitrophenol, benzoic acid	allyl alcohol
ZrO_2	benzoic acid	butynediol, p-nitrophenol
$Th(OH)_4$	m-nitrophenol, benzoic acid	butynediol
Al_2O_3	pyrogallol, catechol, hydroquinone	other hydroxy aromatics (naphthols, cresols, phenol)
PbO_2	amines, ketones	—

(b) Compound	Adsorbed by	Not adsorbed by
Allyl thiourea	Bi_2O_3	PbO, CdO ... Al_2O_3, SiO_2, etc.
Thiourea	CdO, NiO, Bi_2O_3, Co_2O_2	PbO, ZnO, CuO, Al_2O_3, SiO_2, etc.
Thioacetamide	CdO, NiO, Bi_2O_3, Co_2O_2	PbO, ZnO, CuO, Al_2O_3, SiO_2, etc.
Methylene blue	SiO_2, MnO_2, TiO_2, SnO_2, ZrO_2, PbO_2, In_2O_3, Bi_2O_3, CdO, V_2O_5, Nb_2O_5, WO_3	PrO_2, ThO_2, Al_2O_3, Fe_2O_3, Co_2O_3, Cr_2O_3, La_2O_3, Ga_2O_3, Yb_2O_3, ZnO, PbO, SnO, NiO, CuO

and for $c/c_0 > 0.4$ the adsorption decreased with increasing temperature. Plots of the integral molar heats of adsorption, q_{int}, were interpreted on the assumption that both surfaces, previously degassed at 25 °C, were essentially covered with hydroxyl groups. There appeared to be some chemisorption at very low coverages followed by a decrease in q_{int} as iodine occupied the less active sites. A small maximum in q_{int} at coverages roughly corresponding to the second rise on the adsorption isotherm was thought to be due to iodine molecules clustering around more strongly bound ones. The effect of increasing temperature on the q_{int} curves for iodine adsorption on η-Al_2O_3 was to increase the measured heats over the complete coverage range and the authors propose that this indicates enhanced availability of the internal surfaces of the porous sample to the adsorbate.

More recently[207] the same group has studied adsorption of pyridine and 4-methylpyridine from n-heptane solutions, the adsorbent again being η-Al_2O_3, this time outgassed at either 400 or 650 °C. High-temperature outgassing leads to progressive dehydroxylation of the surface leaving acidic sites where basic adsorbates can chemisorb. The integral heats of adsorption curves consisted of linear sections, each corresponding to a particular type of adsorption. For the 400 °C-treated sample the two sections were ascribed to chemisorption of pyridine on strong acidic sites and hydrogen bonding with remaining hydroxyl groups. The higher outgassing temperature seemed to produce two types of

[207] L. Stradella, G. Della Gatta, and G. Venturello, *Z. Phys. Chem.* (*Frankfurt am Main*), 1979, **115**, 25.

acidic centre characterized by differential heats of adsorption ≈ 425 and ≈ 170 kJ mol^{-1} for pyridine (and about 10 kJ mol^{-1} higher for 4-methyl-pyridine because of its greater basicity). Some physisorption also occurred.

Clays.—The adsorption properties of clays have been extensively studied in the past and the situation up to 1976 is dealt with in van Olphen's book.[208] A major feature of recent work has been the study of the effect on their adsorption characteristics of modification of clays by exchangeable organic cations. The earlier work of Dekany and his collaborators, outlined in a previous Report (Vol. 3, p. 147) and a publication in Hungarian,[209] has been continued. They have been concerned mainly with the adsorption of benzene + alkane and benzene + alcohol mixtures on montmorillonites and kaolinites treated to varying extents with hexadecylpyridinium chloride (HDPCl) to form HDP-complexes of known HDP-content and having a partially organophilic surface.

The surface excess isotherms for methanol adsorption from benzene on untreated montmorillonites and kaolinite[210] were both of type II in the Schay classification.[211] Pretreatment of montmorillonite with water + isopropanol or with methanol (50 h contact with boiling liquid) lowered the initial part of the isotherm indicating that pretreatment had covered some of the more active sites, but the final slope as x (alcohol) $\rightarrow 1$ was unchanged showing that the total surface area available to methanol was unaffected. Estimation of the specific surface area from the limiting slope or from the linear equation (93) led to a surface area of 325 m^2 g^{-1}, which is nearly half the basal plane area (752 m^2 g^{-1}) obtained from crystallographic data. This area is contrasted with a value of 54 m^2 g^{-1} available for adsorption from benzene + n-heptane mixtures, see Table 5. This alcohol-pretreated material was used as the reference with which to compare the HDP-modified samples. Treatment of kaolinite with methanol, however, had a more drastic effect changing the isotherm from type II to type IV with an azeotropic point. Application of the Schay–Nagy method of finding the surface area indicated an increase from 100 m^2 g^{-1} to 162 m^2 g^{-1} on methanol treatment.

Modified montmorillonites with increasing HDP-content led to a sequence of isotherms for methanol + benzene adsorption in which the azeotropic point moved successively to lower methanol mole fractions, demonstrating the increasingly organophilic nature of the clay surface. The surface areas obtained from the linear sections of these isotherms are included in Table 5. The increase in surface area with HDP-content is interpreted as meaning that the organo-complex is more prone to swelling and disaggregation the larger the degree of coverage of the silicate surface; when the HDP-content exceeds 0.8 mequiv. g^{-1} (corresponding to a fractional coverage of 76% by HDP) the

[208] H. van Olphen, in 'An Introduction to Clay Colloid Chemistry', Wiley-Interscience, New York, 2nd Edn., 1977.

[209] I. Dekany, F. Szanto, and L. G. Nagy, *Magy. Kem. Foly.*, 1976, **82**, 491 (*Chem. Abstr.*, 1977, **86**, 34 630).

[210] I. Dekany, F. Szanto, and L. G. Nagy, *Prog. Colloid Polym. Sci.*, 1978, **65**, 125.

[211] Ref. 3, p. 74.

Table 5 *Specific surface areas for montmorillonite and HDP organocomplexes* (θ = *coverage of HDP*$^+$ *ions*)

Liquid mixture	$a_s/m^2 g^{-1}$
Adsorbent: montmorillonite	
benzene + n-heptane	54
methanol + benzene	327
ethanol + benzene	234
n-propanol + benzene	169
n-butanol + benzene	107
Adsorbent: HDP–montmorillonite I	
($\theta = 0.38$)	
methanol + benzene	541
Adsorbent: HDP–montmorillonite II	
($\theta = 0.53$)	
benzene + n-heptane	422
methanol + benzene	636
Adsorbent: HDP–montmorillonite III	
($\theta = 0.66$)	
methanol + benzene	733
ethanol + benzene	700
n-propanol + benzene	659
n-butanol + benzene	653
Adsorbent: HDP–montmorillonite IV	
($\theta = 0.79$)	
methanol + benzene	764
Adsorbent: HDP–montmorillonite V	
($\theta = 0.87$)	
benzene + n-heptane	581
methanol + benzene	778

area is close to the basal plane area. Dekany, Szanto, and Nagy[212] interpret these isotherms as the summation of adsorption of alcohol on the free silicate surface and benzene on the organophilic surface. This view is supported by experiments in which a clay of intermediate HDP-content is compared with a mixture of untreated clay and one of higher HDP-content in a ratio such that the overall HDP-content is the same. The adsorption isotherms of the two were found to be of similar shape although there were quantitative differences arising, it was supposed, from the difference between the swelling properties of the untreated sample and the silicate portions of the modified clay.

When the adsorption from ethanol, n-propanol, and n-butanol + benzene mixtures is compared the adsorption of alcohol decreases with increase in molecular weight both on untreated and HDP-content clay: this is attributed to steric factors.

Similar work but with montmorillonite modified with dimethyldihexadecyl

[212] I. Dekany, F. Szanto, and L. G. Nagy, *Colloid Polym. Sci.*, 1978, **256**, 150.

ammonium chloride (DMDH) is described by Dekany, Szanto, and Nagy.[213] Dekany and Nagy[214] have also reported work on the adsorption of isopropyl alcohol from water by HDP-modified montmorillonite.

A theoretical analysis of these experiments is presented by Dekany, Nagy, and Schay[215,216] and by Dekany and Nagy.[217] In their presentations the assumptions involved are not all stated explicitly, and some of these seem to be mutually incompatible. The following somewhat more complete argument is suggested, which brings out these points more clearly. First it is assumed, with the authors, that the molecules are the same size and that the ratio of the surface activity coefficients can be taken as unity. If adsorption occurs independently on the two types of site A, B, with adsorption equilibrium constants for alcohol adsorption K_A and K_B, respectively, then the surface mole fractions of alcohol on the two types of site are

$$x_1^\sigma(A) = \frac{a_1^l}{K_A a_2^l + a_1^l} = \frac{x_1^l}{K_A \left(\dfrac{\gamma_2^l}{\gamma_1^l}\right)(1 - x_1^l) + x_1^l} \; ; \tag{110}$$

$$x_1^\sigma(B) = \frac{a_1^l}{K_B a_2^l + a_1^l} = \frac{x_1^l}{K_B \left(\dfrac{\gamma_2^l}{\gamma_1^l}\right)(1 - x_1^l) + x_1^l} \; , \tag{111}$$

where a_1^l, a_2^l are the activities of components 1 and 2 in the bulk liquid. If the sites of types A and B can accommodate n_A^σ and n_B^σ moles, respectively, then the total amount of component 1 adsorbed is

$$n_1^\sigma = x_1^\sigma(A)n_A^\sigma + x_1^\sigma(B)n_B^\sigma, \tag{112}$$

or the overall mole fraction of component 1 on the surface is

$$x_1^\sigma = \phi_A x_1^\sigma(A) + \phi_B x_1^\sigma(B) \tag{113}$$

where ϕ_A, ϕ_B are respectively $n_A^\sigma/(n_A^\sigma + n_B^\sigma)$ and $n_B^\sigma/(n_A^\sigma + n_B^\sigma)$, *i.e.* the fractions of the surface sites of types A and B. The authors now consider the bulk liquid composition when $a_1^l = a_2^l$, at which the surface mole fractions are

$$x_1^\sigma(A) = \frac{1}{(K_A + 1)} \; ; \qquad x_1^\sigma(B) = \frac{1}{(K_B + 1)} \; , \qquad (a_1^l = a_2^l), \tag{114}$$

which on substitution in (113) and rearranging gives

$$\left(\frac{x_1^\sigma}{1 - x_1^\sigma}\right)_{a_1^l = a_2^l} = \frac{\phi_A K_A(1 + K_B) + \phi_B K_B(1 + K_A)}{(1 + K_A)(1 + K_B) - \phi_A K_A(1 + K_A) - \phi_B K_B(1 + K_A)} \; . \tag{115}$$

[213] I. Dekany, F. Szanto, and L. G. Nagy, *Tr-Mezhdunar Kongr. Poverkhn-Akt. Veshehestvam, 7th*, 1976, (Publ. 1978), **2**(I), 630 (*Chem. Abstr.*, 1979, **91**, 63 169).

[214] I. Dekany and L. G. Nagy, *Kolo. Ert.*, 1977, **19**, 336 (*Chem. Abstr.*, 1978, **89**, 95 421).

[215] I. Dekany, L. G. Nagy, and G. Schay, *Magy. Kem. Foly.*, 1978, **84**, 333 (*Chem. Abstr.*, 1978, **89**, 136 301).

[216] I. Dekany, L. G. Nagy, and G. Schay, *J. Colloid Interface Sci.*, 1978, **66**, 197.

[217] L. Dekany and L. G. Nagy, *Acta. Phys. Chem., Acad. Sci. Hung.*, 1977, **23**, 485.

If it is now assumed that the overall isotherm can be expressed in terms of an *effective* equilibrium constant K^*:

$$x_1^\sigma = \frac{a_1^l}{K^* a_2^l + a_1^l} = \frac{x_1^l}{K^* \left(\dfrac{\gamma_2^l}{\gamma_1^l}\right) x_2^l + x_1^l} \qquad (116)$$

then when $a_1^l = a_2^l$

$$K^* = \left(\frac{x_1^\sigma}{1 - x_1^\sigma}\right)_{a_1^l = a_2^l} \qquad (117)$$

and is given by the right-hand side of equation (115).

To derive the equation used by Dekany and Nagy it is necessary to assume that $K_A \gg 1$ and $K_B \ll 1$, *i.e.* the two types of site are strongly specific for compounds 1 and 2, respectively. Then we obtain

$$\left(\frac{x_1^\sigma}{1 - x_1^\sigma}\right)_{a_1^l = a_2^l} = K^* = \frac{\phi_A}{1 - \phi_A}. \qquad (118)$$

If the molecules are of the same size then ϕ_A, ϕ_B will be the fractional areas (θ_A, θ_B) occupied by the two sites. However, if the molecules are of different sizes then

$$\frac{\phi_A}{1 - \phi_A} = \frac{\theta_A}{1 - \theta_A} \cdot r = K^*, \qquad (119)$$

where r is the ratio (a_2/a_1) of the areas occupied by the two kinds of molecule. Equation (119) is that obtained by Dekany and Nagy: it involves the assumptions (i) that, even though the molecules are of different size, equations (110) and (111) are valid; (ii) the ratio of surface activity coefficients is unity; (iii) that the overall isotherms can be expressed in the form (116); and (iv) that $K_A \gg 1$, $K_B \ll 1$.‡ The appearance of the size ratio r in equation (119) seems particularly illogical since K^* refers to an equation relating to equal size molecules. According to equation (116) the azeotropic point is observed when $x_1^\sigma = x_1^l$ and hence

$$K^* \left(\frac{\gamma_2^l}{\gamma_1^l}\right)_{x_1(\text{az})} = 1. \qquad (120)$$

More generally,[218]

$$K^* \frac{\gamma_2^l \gamma_1^\sigma}{\gamma_1^l \gamma_2^\sigma} = 1. \qquad (121)$$

Dekany and Nagy give a different formula:

$$x_1(\text{az}) = \frac{K^*}{1 + K^*} = \frac{\theta_1 r}{\theta_2 + \theta_1 r} \qquad (122)$$

‡ This is implied in equation (119) since when $\theta_A \to 1$, $K^* \to K_A \to \infty$, and when $\theta_A \to 0$, $K^* \to K_A \to 0$.

[218] See ref. 1, p. 54, eqn. (20).

which seems, however, to have been derived by supposing that $x_1^\sigma = x_1^l$ when the ratio of the activities of the components in the bulk is unity. In their formula the azeotropic point is independent of the departure of the bulk from ideality and depends only on the mosaic structure of the surface. This is incompatible with equation (116) according to which the azeotropic point is determined by the relative magnitudes of K^* and the deviation of the bulk phase from ideality. It must be stressed that equation (116) cannot lead to an azeotropic point if the bulk system is ideal. On the other hand, Siskova and Erdös[219] showed long ago that the presence of two types of surface of widely differing properties can, even in the case of ideal adsorption from ideal solutions, lead to azeotropic behaviour and a more detailed theory has been developed by Dabrowski, Ościk, Rudzinski, and Jaroniec.[220] It follows that equation (116) must be regarded as an approximate expression. Despite this, using equation (119) to calculate K^* knowing θ_A, derived from analytical data on the modified clay, and knowing the activity coefficients of the bulk liquid,† x_1^σ and hence the excess isotherms can be calculated. They are in reasonable agreement with experiment (Figure 6). It is possible that even better agreement might be achieved using other activity coefficients. If those of Scatchard, Wood, and Mochel[221] are used, together with K^* from equation (118), we find that the observed azeotropic mole fractions are reproduced, using equation (120), with an average deviation of ± 0.03. The reason why equation (121) also gives $x_1(az)$ with similar precision is not easy to understand. The general validity of equation (118) for K^* is confirmed by calculating K^* from the integration of equation (30): agreement to within $\pm 6\%$ is found except for the highest substituted samples where the discrepancies are 33% and 40%. The authors also calculated the variation of the excess energy changes $\varepsilon - \varepsilon_1^*$ as a function of solution mole fraction,‡ although no detailed discussion of these curves is presented. The variation of $\ln K^*$ as θ_A goes from $0 \to 1$ might have been expected to be linear in θ_A; but equation (119) shows that it is of the form $\ln[\theta_A/(1-\theta_A)]$.

We have dealt in some detail with this group of papers because they illustrate well the way in which quantitative analysis of data can be developed. They also stress that although useful relationships can be derived from oversimplified models the fundamental interpretation of the resulting parameters must be approached with caution.

Work of a similar kind has been carried out by Mysak, Nikulichev, and Tarasevich.[222,223] Adsorption of benzene from heptane was studied on natural

† The authors do not state the origin of the activity coefficients they employed.

‡ ε used here, and by Schay (ref. 2, p. 54) is equivalent to σa_s used previously where a_s is the specific surface area of the solid.

[219] See ref. 1, p. 69.
[220] A. Dabrowski, J. Ościk. W. Rudzinski, and M. Jaroniec, J. Colloid Interface Sci., 1979, **69**, 287; ibid., 1976, **56**, 403.
[221] G. Scatchard, S. E. Wood, and J. M. Mochel, J. Am. Chem. Soc., 1946, **68**, 1957.
[222] A. E. Mysak, Yu. G. Nikulichnev, and Yu. I. Tarasevich, Ukr. Khim. Zh. (Russ. Ed.), 1977, **43**, 1066 (Chem. Abstr., 1978, **88**, 95 295).
[223] A. E. Mysak, Yu. G. Nikulichev, and Yu. I. Tarasevich, Ukr. Khim. Zh. (Russ. Ed.), 1978, **44**, 820 (Chem. Abstr., 1978, **89**, 204 708).

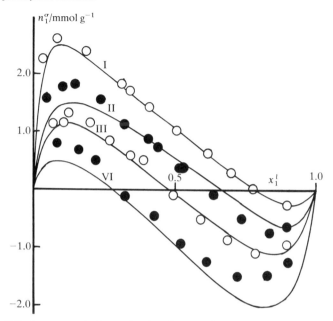

Figure 6 *Surface excess isotherms for the adsorption from methanol* (1) + *benzene* (2) *by HDP-modified montmorillonites. Points, experimental; lines calculated from equations* (116) *and* (119). *Fraction of ion exchange by hexadecylpyridinium ions: I, 0.326; II, 0.536; III, 0.635; VI, 0.842*
(Redrawn from *Acta Phys. Chem., Acta Sci. Hung.*, 1977, **23**, 485).

and organosubstituted montmorillonites. The capacity and selectivity for benzene increases in the organo-modified materials whereas the adsorption of nitrogen decreases. A study of adsorption from methanol + benzene mixture by montmorillonites modified with HDP led to conclusions qualitatively similar to those described above, although no detailed results are available.

Gerstl and Mingelgrin[224] as a result of work on the adsorption of pesticides on modified attapulgite have concluded that the general terms hydrophilic and organophilic used to describe clay surfaces treated with different cations may be misleading and are not always sufficient to predict the relative adsorption of organic molecules from aqueous or organic solution. Steric effects and, more often, specific interactions with the surface may also be controlling factors. They also discuss the usefulness of representing adsorption isotherms of sparingly soluble components in terms of the 'reduced concentration', c/c(sat) (see p. 117).

Bykov and co-workers[225—227] have investigated the adsorption of some

[224] Z. Gerstl and U. Mingelgrin, *Clays Clay Miner.*, 1979, **27**, 285.
[225] V. T. Bykov, T. I. Gavrishchik, V. G. Gerasimova, G. A. Krizhanenko, N. N. Petrenko, and V. P. Teleshova, Deposited Doc., 1974, VINITI 923—74 (*Chem. Abstr.*, 1977, **86**, 146 433).
[226] V. T. Bykov, V. G. Gerasimova, G. N. Vasilenko, G. A. Krizhenenko, and V. P. Teleshova, Deposited Doc., 1975, VINITI 3660—75, 9 pp. (*Chem. Abstr.*, 1978, **88**, 55 387).
[227] V. T. Bykov, V. G. Gerasimova, G. N. Vasilenko, G. A. Krizhanenko, and V. P. Teleshova, Deposited Doc., 1974, VINITI 3034—74 (*Chem. Abstr.*, 1977, **86**, 96 496).

ternary mixtures on palygorskite and kaolinite, the systems studied being phenol +, anisole +, phenetole +, and phenyl sulphide + (pyridine + hexadecane). Adsorption of pyridine was always decreased by addition of the other organic compound, the effect being greater on kaolinite because, the authors claim, of bonding to specific sites. The temperature dependence of adsorption of phenyl sulphide, methylphenyl sulphide, and pyridine from hexadecane solutions on kaolinite and tripoli was also measured. Adsorption of the former two compounds decreased whereas that of pyridine increased with increase in temperature. Adsorption from the ternary system toluene + ethanol + hexadecane on montmorillonite and polygorskite was reported by Gor'kovskaya et al.,[228] who noted that ethanol was more strongly bound than toluene to the polar surface of the clays.

The treatment of bentonite with hexadecylpyridinium chloride and trimethylhexadecylammonium bromide has been considered by Pal and Ghoshal,[229] who noted that above the cation-exchange capacity adsorption of the associated form of the surfactant occurs, mainly on the external surface of the clay.

In a series of papers by Aragon and Gomez[230—232] particular attention is given to the orientation of adsorbed molecules between the silicate layers of clay minerals. The liquid mixtures used were butylamine with octanol, nonanol, decanol, and undecanol; hexylamine with octanoic, nonanoic, decanoic, undecanoic, and dodecanoic acids; and octylamine with decanol and undecanol. The adsorbents were montmorillonite and vermiculite. Knowledge of the adsorbed amount and basal plane spacing (from X-ray diffraction) suggested that the interlamellar spaces contained a double layer of adsorbed molecules with aliphatic chains at an angle of 50—55° to the surface.

For the adsorption of pyridine and phenol from hexane solution on an unspecified clay Maidanovskaya and Korobkova[233] claim that pyridine is oriented perpendicular and phenol parallel to the surface.

Methods of estimating the orientation of adsorbed molecules based solely on adsorption measurements are subject to considerable uncertainty. Where the adsorbing surface can be oriented the method of polarized i.r. attentuated total reflectance (ATR) can be applied.[234,235] Raupach and Janik[236] have used this method to elucidate the orientation of the amino-acids ornithine and 6-aminohexanoic acid adsorbed on vermiculite. The measurements were made on oriented flakes of clay, and a comparison made of the spectra of deuteriated and undeuteriated complexes. This yielded the transition moment directions

[228] V. T. Gor'kovskaya, E. K. Borisova, and B. A. Frolov, Deposited Doc., 1977, VINITI 4407—77 (*Chem. Abstr.*, 1979, **91**, 97 206),

[229] B. K. Pal and D. N. Ghoshal, *Indian J. Chem.*, Sect. A, 1977, **15**, 680.

[230] F. Aragon de la Cruz and A. M. De Andres Gomez de Barreda, *An. Quim.*, 1978, **74**, 1207.

[231] F. Aragon de la Cruz and A. M. De Andres Gomez de Barreda, *An. Quim.*, 1978, **74**, 744.

[232] A. M. De Andres Gomez de Barreda and F. Aragon de la Cruz, *An. Quim.*, 1979, **75**, 476.

[233] L. G. Maidanovskaya and O. I. Korobkova, *Tr. Tomsk. Gros. Univ.*, 1975, **264**, 23 (*Chem. Abstr.*, 1977, **86**, 127 849).

[234] L. E. Wolfram and J. G. Grasseli, *Appl. Spectrosc.*, 1970, **24**, 263.

[235] T. Takenaka, K. Nogami, H. Gotoh, and R. Gotoh, *J. Colloid Interface Sci.*, 1971, **33**, 395.

[236] M. Raupach and L. J. Janik, *Clays Clay Miner.*, 1976, **24**, 127.

for the i.r. band assignments. Ornithine was found to lie almost flat on the surface and form two layers in the interlamellar space, while 6-aminohexanoic acid formed one layer with the molecule sloping at an angle of 36° to the surface.

Several papers have dealt with adsorption from aqueous solutions. Stul *et al.*[237,238] have looked at the adsorption of dilute butanol, hexanol, and octanol solutions on RNH_3-montmorillonite, where R is a C_8—C_{18} alkyl group. The isotherms show that adsorption increased with increasing chain length of both the alkylammonium ion and the alcohol. The results were discussed by treating the interlammellar space as an organic solvent and comparing the distribution of an alcohol between water and an alkane and between water and the adsorbed phase. Standard free energies for the transfer of an alcohol molecule between the two phases at infinite dilution are tabulated and indicate the interlamellar phase as the more effective solvent for alcohol. This was thought to be due to specific interactions between the hydroxyl group of the alcohol and the $-NH_3^+$ group of the cation. In the second paper the swelling of the clay at different bulk alcohol concentrations was measured. The adsorbed alcohol molecules appeared to be oriented perpendicular to the surface and a model describing the association of the alcohols in the interlamellar space was outlined.

The adsorption of benzene, phenol, and two chlorobenzenes from dilute aqueous solution on montmorillonites treated with tetra-alkylammonium ions was studied by McBride, Pinnavaia, and Mortland[239] in work related to pollutant control. Maximum adsorption of benzene and phenol occurred on the tetramethylammonium complex and was three times larger than on a natural soil. However the soil (which contained organic matter) was a better adsorbent for the chlorobenzenes presumably because these molecules were too large to penetrate the interlamellar spaces of the montmorillonite.

Aripov and co-workers[240,241] have studied the adsorption of carboxylic acids from aqueous solution onto untreated bentonite and bentonite treated with alkylbenzyl-, alkylnaphthyl-, and alkylanthracyl-pyridinium ions. For natural bentonite the adsorption decreased in the order acetic, butanoic, and hexanoic acid, and for the modified surface in the order hexanoic, acetic, propionic, and pentanoic acid.

Other work abstracted has included the adsorption of the dyes crystal violet and malachite green on vermiculite;[242] the adsorption of aniline and *p*-chloroaniline on montmorillonite;[243,244] the adsorption of collagen hydrolysate

[237] M. S. Stul, A. Maes and B. Uytterhoeven, *Clays Clay Miner.*, 1978, **26,** 309
[238] M. S. Stul, J. B. Uytterhoeven, J. B. de Bock, and P. L. Huyskens, *Clays Clay Miner.*, 1979, **27,** 377.
[239] M. M. McBride, T. J. Pinnavaia, and M. M. Mortland, *Adv. Environ. Sci. Technol.*, 1975, **8,** 145.
[240] E. Sh. Iminova, N. Yu. Nasyrova, M. A. Rakhmanova, and E. A. Aripov, Deposited Doc., 1974, VINITI 2833—74 (*Chem. Abstr.*, 1977, **86,** 96 494).
[241] E. A. Aripov and Tadzhieva, *Uzb. Khim. Zh.*, 1979, (6), 32 (*Chem. Abstr.*, 1980, **92,** 221 426).
[242] D. K. De, J. L. Das Kanungo, and S. K. Chakravarti, *J. Indian. Chem. Soc.*, 1979, **56,** 608.
[243] A. Moreale and R. Van Bladel, *Proc.-Eur. Clay. Conf. 3rd*, 1977, 131.
[244] P. Cloos, C. Broers, and A. Moreale, *Proc. Eur. Clay. Conf. 3rd*, 1977, 42.

on montmorillonite[245] and the effect of pre-heating polygorskite on naphthenic acid adsorption from cyclohexane solution.[246]

Zeolites.—Zeolites are widely used for their molecular sieving and ion-exchange properties. In this review only the former, which manifests itself as an adsorption phenomenon will be considered.

Although extensive data are available on vapour adsorption by zeolites, relatively little information is available on adsorption from solution. Among earlier papers dealing with this problem are those by Peterson and Redlich[247] who report somewhat fragmentary data on heptane + decane mixtures, with 5 Å molecular sieve, and Sundstrom and Krautz[248] who studied adsorption by 5 Å molecular sieve of mixtures of lower alkanes ($C_7, C_{10}, C_{12}, C_{14}$), the alkane of lower mol. wt. being preferentially adsorbed.

Larionov and his co-workers have extended their earlier work on silica, graphite, and charcoals[249] to a study of the adsorption properties of NaX-zeolite. Gulazhenko, Larionov, Chmutov, and Shayusupova[250] have measured the adsorption isotherms from benzene + iso-octane solutions at 303, 338, and 363 K. At all three temperatures the type II isotherms rise very sharply at low benzene mole fractions and reach maxima at less than 0.1 mole fraction. The data are analysed in two ways. Using equations (1), (3), and (4), free energy, enthalpy, and entropy of wetting are calculated. In systems of this kind, where the concept of surface area of the zeolite has little significance, the free energy quantity obtained is ε (see p. 85) denoted in this paper by ϕ. As indicated by the shape of the isotherms most of the changes in these quantities occur in dilute solution [x(benzene) < 0.1]. The alternative is to adopt an adsorbed-phase model where the extent of the adsorbed phase is defined by the pore volume of the zeolite. In this way activity coefficients in the adsorbed phase can be calculated, and from their temperature coefficients the excess thermodynamic functions of mixing in the adsorbed state are obtained: in the adsorbed state the mixture shows negative deviations from ideality. In a similar study Larionov, Chmutov, and Shayusupova[251] have studied the benzene + cyclohexane system on the same zeolite at the same three temperatures. Again the isotherms show a sharp maximum below a mole fraction of 0.1 of benzene. The data are analysed by the same method as previously employed: again negative deviations from ideality are found. The excess enthalpy and entropy curves exhibit several maxima and minima, which suggests that their precision

[245] L. M. Shuter and R. M. Verlinskaya, *Ukr. Khim. Zh.* (*Russ. Ed.*), 1977, **43**, 1281 (*Chem. Abstr.*, 1978, **88**, 79 659).

[246] A. A. Abdullaev, A. A. Agzamkhodzhaev, and E. A. Aripov, Deposited Publ., 1973, VINITI, 6217—73 (*Chem. Abstr.*, 1976, **85**, 68 744).

[247] D. L. Peterson and O. Redlich, *J. Chem. Eng. Data*, 1962, **7**, 571.

[248] D. W. Sundstrom and F. G. Krautz, *J. Chem. Eng. Data*, 1968, **13**, 223.

[249] See ref. 3, pp. 82—93, 138—140.

[250] V. P. Gulazhenko, O. G. Larionov, K. V. Chmutov, and M. Sh. Shayusupova, *Zh. Fiz. Khim.*, 1978, **52**, 467 (*Russ. J. Phys. Chem.*, 1978, **52**, 262).

[251] O. G. Larionov, K. V. Chmutov, and M. Sh. Shayusupova, *Zh. Fiz. Khim.*, 1978, **52**, 1527 (*Russ. J. Phys. Chem.*, 1978, **52**, 881).

is not high. Two other papers by Larionov and his co-workers[252,253] deal with the adsorption of naphthalene, α-methylnaphthalene, and anthracene (from what solvent is not stated in the abstract), and of benzene from solution in hexane, iso-octane, and heptane, and of toluene and 1-hexane from heptane. The molar free energy of wetting increases with increasing number of aromatic rings. It was found that the presence of residual water in the zeolite (2.1 mol kg^{-1}) caused a significant decrease in selective adsorption of benzene.

The selective adsorption by NaX of benzene from n-heptane and cyclohexane has also been demonstrated by Tryapina, Seidova, and Agapova.[254] The use of a fluidized bed of NaX for the adsorption of binary mixtures is described by Kvasha *et al.*[255] and an extensive review of the adsorption of hydrocarbons from a liquid phase by zeolites had been published by Samoilov and Forminykh.[256] The adsorption properties of Linde 5 Å molecular sieve seem to differ markedly from those of NaX. According to Gupta, Kunzru, and Saraf[257] n-pentane, n-hexane, n-heptane, and n-octane are all preferentially adsorbed from mixtures with benzene. The effect of temperature and concentration on the adsorption was studied.

Wolf, Pilchowski, and Karch[258] have studied the adsorption of n-hexanol from toluene by NaA zeolite partially ion exchanged with Ca^{2+} and Mg^{2+}. The isotherms rise to a sharp plateau, which is reached at a concentration of about 15 mg hexanol/g toluene ($x = 0.013$). Pure NaA exhibits no adsorptive capacity for hexanol but, as divalent ions are introduced, this rises, initially slowly, then more sharply to reach, at about 30% degree of ion exchange, a maximum value that remains unchanged from 40% upwards. The paper also deals with the kinetics of the adsorption process, and presents calculations of effective diffusion coefficients and break-through curves under dynamic adsorption conditions. Some experiments with butan-1-ol and ethanol are also reported.

Atwood and Kempton[259] have studied the ternary system water + methanol + n-butanol with Linde 5 Å molecular sieve. In this system butanol is excluded from the zeolite cavities, so that one may study the competitive adsorption of water and methanol. Experiments were made at mole fractions of butanol of 0.05, 0.10, and 0.15 for a wide range of mole fractions of water

[252] V. P. Gulazhenko, O. G. Larionov, and K. V. Chmutov, *Sorbtsiya Khromatogr.*, 1979, 68 (*Chem. Abstr.*, 1980, **92**, 47 768).
[253] Yu. F. Berezkina, V. P. Gulazhenko, P. S. Deineko, O. G. Larionov, L. F. Fominykh, K. V. Chmutov, and M. Sh. Shayusupova, *Sorbtsiya Khromatogr.*, 1979, 46 (*Chem. Abstr.*, 1980, **92**, 83 082).
[254] L. I. Tryapina, R. M. Seidova, and G. A. Agapova, *Azerb. Khim. Zh.*, 1976, **1**, 94 (*Chem. Abstr.*, 1977, **86**, 79 120).
[255] V. I. Kvasha, A. A. Seballo, Yu. S. Lezin, T. G. Plachenov, and E. I. Baranov, *Sb. Tr. Leningrad Tekhnol. In-t Im Lensoveta*, 1975, **3**, 88—94; from *Ref. Zh. Khim.*, 1976, Abstr. No. 13I 136 (*Chem. Abstr.*, 1976, **85**, 113 010).
[256] N. A. Samoilov and L. F. Fominykh, *Nauchao-Temat. Sb.-Ufim. Neft. Inst.*, 1975 (Publ. 1976), **26**, 156 (*Chem. Abstr.*, 1979, **91**, 44 847).
[257] R. K. Gupta, D. Kunzru, and D. N. Saraf, *J. Chem. Eng. Data*, 1980, **25**, 14.
[258] F. Wolf, K. Pilchowski, and J. Karch, *Chem. Tech.* (Leipzig), 1977, **29**, 554.
[259] G. A. Atwood and W. E. Kempton, *A.I.Ch.E. Symp. Ser.*, 1975, **71**, 40.

and methanol. It was found that n-butanol did not interfere with the adsorption equilibria, and that the separation factor

$$\alpha = \frac{x_i^s}{x_i^l} \cdot \frac{x_k^l}{x_k^s}, \qquad (123)$$

where i, k are water and methanol, respectively, was approximately constant (7.83 ± 0.23 at $30\,°C$). Ternary systems of acetic acid + myristic or palmitic acid + tetradecane adsorbed by NaX and CaX zeolites have been examined by Gryazev, Kucherova, and Rakhlevskaya.[260] It was found that the apparent density of the adsorbed acids fell below that of bulk liquid, the discrepancy increasing with molecular weight. Increase in the concentrations of myristic and palmitic acid led to an insignificant displacement of acetic acid from the zeolite, whereas increase in the acetic acid concentration in solution led to displacement of the other two acids. This indicates that acetic acid is appreciably more strongly adsorbed than the other two. Reference is made to equations characterizing the adsorption equilibria, but they are not given in the paper.

Miscellaneous Adsorbents.—Ivanishchenko and Gladkikh[261] have measured the adsorption of monocarboxylic acids (C_{10}—C_{18}) on chalk from isopropyl alcohol solutions with a view to understanding the hydrophobic properties of powders treated in this way. After adsorption both the electrical resistance across powder diaphragms in 0.1 N-KCl solution and the moisture uptake of a powder sample in a humid atmosphere were measured. Both were found to vary with uptake in a periodic zig-zag fashion, the first maximum in resistance corresponding with a minimum in moisture uptake at a surface coverage equivalent to a monolayer of adsorbate (Figure 7). Subsequent maxima and minima corresponded with further layers of adsorbed molecules which, the authors concluded, must be strictly oriented. Thus the first layer is adsorbed with the alkyl chain away from the surface, presenting a hydrophobic barrier, while the second layer is adsorbed in opposite fashion, providing an attractive surface for moisture uptake. The structuring appears to extend as far as the sixth or seventh layer, although the water repellency of the powder is not significantly improved once the first layer has been completed. In another paper[262] the desorption of fatty acids from hydrophobic chalk was studied.

The adsorption of aliphatic acids (valeric, caproic, enanthic, caprylic, pelargonic) from 1% solution from cyclohexanone on an alumino-silicate catalyst, silica gel (KSK-2), and Al_2O_3 has been studied under dynamic conditions by Narmetova, Khashimova, Narimova, and Ryabova.[263] The lower molecular weight acids were preferentially adsorbed. Both physical and chemisorption

[260] N. N. Gryazev, N. P. Kucherova, and M. N. Rakhlevskaya, *Zh. Fiz. Khim.*, 1976, **50**, 1276 (*Russ. J. Phys.*, 1976, **50**, 765).

[261] O. I. Ivanishchenko and Yu. P. Gladkikh, *Kolloid. Zh.*, 1979, **41**, 774 [*Colloid J. (USSR)*, 1979, **41**, 660].

[262] Yu. P. Gladkikh, O. I. Ivanishchenko, and K. F. Paus, *Sb. Tr. Belgorod Tekhnol. In-t, Stroit Materialov* 1975, **15**, 39 (Russ.); from *Ref. Zh. Khim.* 1977, Abstr. No. 23B1699 (*Chem. Abstr.*, 1978, **88**, 111 030).

[263] G. R. Narmetova, M. A. Khashimova, N. D. Narimova, and N. D. Ryabova, Deposited Publ. 1973, VINITI 5784—73 (*Chem. Abstr.*, 1976, **85**, 149 435).

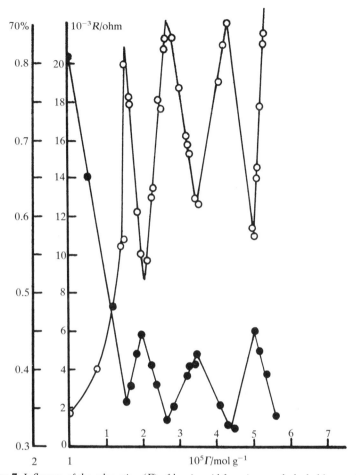

Figure 7 *Influence of the adsorption (Γ) of lauric acid from isopropyl alcohol by powdered chalk on moisture uptake (wt. %) of dried powder, filled points; and electrical resistance (R) of powder diaphragm in 0.1 N-KCl, open circles*
(Redrawn from *Kolloid. Zh.*, 1979, **41**, 774).

were observed. The degree of chemisorption was about 5% of the total on the catalyst, but some 60% on Al_2O_3. House and Nehmer[264] have measured the adsorption of acetic and propionic acid from carbon tetrachloride solutions by solid metal complexes. Here again the adsorption varied with mol. wt., acetic acid being the most strongly adsorbed whereas butyric acid was adsorbed to a negligible extent. On potassium ferricyanide and ferrocyanide acetic acid adsorption followed the Langmuir isotherm.

[264] J. E. House, jun. and W. L. Nehmer, *Trans. Ill. State Acad. Sci.*, 1978, **71**, 95 (*Chem. Abstr.*, 1979, **90**, 142 531).

Aristov, Feizulova, and Frolov,[265] using β-phthalocyanine as adsorbent, studied the adsorption of pentyl, hexyl, and octylamines from aqueous solution. The apparent monolayer capacity was largest for pentylamine, although for the other two amines the isotherms subsequently showed a sharp rise possibly indicative of multilayer adsorption. It was argued that these molecules are oriented normal to the surface. Octylammonium acetate and dodecylammonium acetate were also studied. Adsorption of alkylamines from aqueous solution by iron leads to changes in the metal/solution/air contact angle: according to the work of Sangiergi, Passerone, and Lorenzelli[266] the effect is strongly dependent on the pH of the solution, the contact angle being zero except for an intermediate range of several pH units in which the contact angle reaches a peak.

Martynova and Gatilova[267—269] have studied the effect of adsorption of ethanolamines on the dispersability of synthetic diamond powder in various solvents. They found that triethanolamine was adsorbed from acetone solutions but negatively adsorbed from water or ethanol, though in one paper[268] they reported that triethanolamine was positively adsorbed from ethanol at concentrations above 0.4 M. Monoethanolamine was found to be the most effective in dispersing diamond powder in acetone. The adsorption of poly(ethylene polyamine) on diamond was also studied, but the solvent used is not mentioned in the abstract.

Adsorption by polymers (polypropylene, polycarbonate, and epoxy resins) of barium dialkyldithiophosphate from hexane has been measured by Zharinova *et al.*[270] Adsorption was strongest on oxygen-containing surfaces. Amino-acids are adsorbed by monocarboxycellulose. According to Belaya, Kaputski, and Yurkshtovich[271] the adsorption occurs through the zwitterionic form of the amino-acid which reacts with surface carboxyl groups. The adsorption, which is almost independent of temperature, depends on the structure and electrical properties of the acid.

Extraction of 2-naphthol from dilute aqueous solution by a macroporous ion-exchange resin proceeds, according to Korenman, Alymova, and Polumestnaya,[272] by both molecular and ion-exchange mechanisms. The naphthol is desorbed with a mixture of sodium hydroxide and chloride: this forms the basis of concentrating the naphthol for analysis. Amberlite XAD-4 adsorbs gibberellic acid from aqueous solution but the work of

[265] B. G. Aristov, R. K. G. Feizulova, and Yu. G. Frolov, *Zh. Fiz. Khim.*, 1979, **73**, 1859 (*Russ. J. Phys. Chem.*, 1979, **33**, 1061).

[266] R. Sangiergi, A. Passerone, and V. Lorenzelli, *J. Phys. Chem.*, 1977, **81**, 1851.

[267] L. M. Martynova and E. G. Gatilova, *Fiz. Khim. Kondens. Faz., Sverkhtverdykh Mater. Ikh Granits Razdela*, 1975, 113 (*Chem. Abstr.*, 1976, **85**, 37 489).

[268] E. G. Gatilova and L. M. Martynova, *Poverkhn. Yavleniya Dispersnykh Sist.*, 1975, **4**, 63 (*Chem. Abstr.*, 1978, **88**, 42 098).

[269] L. M. Martynova, *Poverkhn. Yavleniya Dispersnykh Sist.*, 1977, **4**, 65 (*Chem. Abstr.*, 1978, **88**, 79 644).

[270] T. Ya. Zharinova, L. N. Yakubenko, P. I. Sanin, V. V. Sher, and N. N. Kishchuk, *Ukr. Khim. Zh.* (*Russ. Ed.*), 1978, **44**, 683 (*Chem. Abstr.*, 1978, **89**, 186 461).

[271] A. V. Belaya, F. N. Kaputskii, and T. L. Yurkshtovich, *Kolloid. Zh.*, 1980, **42**, 332 [*Colloid J. (USSR*), 1980, **42**, 276].

[272] Ya. I. Korenman, A. T. Alymova, and E. I. Polumestnaya, *Kolloid. Zh.*, 1979, **41**, 1007 [*Colloid J. (USSR*), 1979, **41**, 854].

Heropolitanski[273] shows that the uptake is first decreased by the addition of alkali halides and calcium chloride, but at concentrations greater than 3% adsorption is enhanced. Ignatov[274] has studied the adsorption of methyl orange, congo red, and indigo carmine from methanolic solutions by several ion-exchangers in the temperature range 243—293 K. In most cases the limiting capacity decreases as the temperature is lowered, and the adsorption constant increases.

8 Conclusions

Several important aspects of adsorption from solution have been omitted from this review and will be dealt with subsequently. Among these are the statistical mechanics of adsorption, and especially the problems of polymer adsorption, where much progress has been made in recent years. The phenomena associated with structuring at solid surface have been further studied, and increasing attention is being paid to the kinetics of adsorption from solution. One of the areas of greatest significance in colloid science is that of surfactant adsorption, and work in this area also continues and expands.

[273] R. Heropolitanski, *Przem. Chem.*, 1979, **58,** 491 (*Chem. Abstr.*, 1980, **92,** 11 630).
[274] Yu. I. Ignatov, *Zh. Fiz. Khim.*, 1980, **54,** 406 (*Russ. J. Phys. Chem.*, 1980, **54,** 232).

4
Statistical Mechanics of Colloidal Suspensions

BY E. DICKINSON

1 Introduction

The role of statistical mechanics in describing colloidal structure and phase equilibria was recognized by Langmuir[1] and Onsager[2] over 30 years ago. Exploitation of statistical mechanics began properly, however, only a few years ago following the application of ideas from the theory of simple liquids to the problem of concentrated colloidal dispersions.[3] Reviewing the state of inter-molecular forces in 1977, Israelachvili and Ninham recalled[4] the historical dichotomy between short- and long-range forces and suggested that statistical mechanics provides the cement to bind together these two traditional strands of colloid science.

To relate macroscopic observables to forces acting between particles is the objective. The relevant measurable properties are: (a) the phase diagram, (b) scattering of photons, neutrons, and X-rays, (c) the osmotic equation of state, and (d) rheological behaviour. This Report covers (a), (b), and (c) with emphasis on the transition between ordered and disordered states. Discussion of (b) is limited to light-scattering. Current problems in relation to (d) are set out (up to mid-1980).

Five types of forces between colloidal particles may be identified: (i) repulsive forces, from the overlap of electrical double-layers; (ii) dispersion forces, from long-range van der Waals attraction between molecules in neighbouring particles; (iii) 'steric' forces, from interaction of macromolecules adsorbed at the particle surface; (iv) 'structural' and Brownian forces, from interaction with solvent molecules of the dispersion medium; and (v) hydrodynamic forces.

In an electrostatically-stabilized dispersion in which the particles do not approach within a distance of about 10 or 15 solvent diameters, the *equilibrium* behaviour depends only on the double-layer repulsion and van der Waals attraction. The interaction energy between pairs of spherical particles i and j a distance r apart is split into two parts, arising respectively from repulsive and

[1] I. Langmuir, *J. Chem. Phys.*, 1938, **6**, 873.
[2] L. Onsager, *Ann. N.Y. Acad. Sci.*, 1949, **51**, 627.
[3] E. Dickinson, *Chem. Br.*, 1980, **16**, 146.
[4] J. N. Israelachvili and B. W. Ninham, *J. Colloid Interface Sci.*, 1977, **58**, 14.

attractive forces,

$$u_{ij}(r) = u_R(r) + u_A(r). \tag{1}$$

The repulsive potential takes the form[5]

$$u_R(r) = \begin{cases} 4\pi\epsilon_r\epsilon_0 a^2\psi^2 e^{-\kappa d}/r, & (\kappa a \ll 1) \\ 2\pi\epsilon_r\epsilon_0 a\psi^2 \ln\{1 + e^{-\kappa d}\}. & (\kappa a \gg 1) \end{cases} \tag{2}$$

In equation (2), ϵ_r is the relative dielectric constant of the continuous phase, ϵ_0 is the absolute permittivity of free space, a is the particle radius, ψ is the surface potential, and d is the closest surface-to-surface separation. The quantity κ, the reciprocal of the Debye length, is defined by

$$\kappa^2 = 2\nu^2 e^2 c N_A/\epsilon_r\epsilon_0 kT, \tag{3}$$

where ν is the ionic valency of the (symmetrical) electrolyte, e is the electronic charge, c is the electrolyte concentration, N_A is Avogadro's constant, k is Boltzmann's constant, and T is the absolute temperature.

The (unretarded) attractive potential takes the form[6]

$$u_A(r) = -A_H[2a^2/(r^2 - 4a^2) + 2(a/r)^2 + \ln\{1 - (2a/r)^2\}]/6, \tag{4}$$

where A_H is the Hamaker coefficient.

This Report places emphasis on the time-averaged behaviour of assemblies with pair-wise DLVO-type potentials given by equations (1)—(4). This reflects the strong current interest in the phase equilibria of electrostatically-stabilized dispersions of spherical latex particles. Pair-wise additivity, in which the total excess[2] potential energy Φ of N particles is given by

$$\Phi = \sum_{i<j}^{N} u_{ij}, \tag{5}$$

is a widespread, convenient, but unproven, assumption. It seems likely that additivity holds for $u_A(r)$, but may be in substantial error for $u_R(r)$, especially where double-layers overlap in concentrated dispersions.[7]

There are no expressions of equivalent reliability and generality for the contributions to $u_{ij}(r)$ from 'steric' interactions. A full understanding of steric stabilization requires further progress in the theory of physical adsorption of macromolecules.[8] The problem is essentially a molecular one, and will not be discussed in this Report which is concerned with statistical mechanics at the particulate level. For the same reason, we are not concerned here with short-range solvent-mediated structural forces,[9] which can become important when particles are very close together.

The time-dependent statistical mechanics of a colloidal dispersion is complicated by the fact that there is as yet no satisfactory method for incorporating

[5] E. J. W. Verwey and J. Th. G. Overbeek, 'Theory of the Stability of Lyophobic Colloids', Elsevier, Amsterdam, 1948.
[6] J. Mahanty and B. W. Ninham, 'Dispersion Forces', Academic Press, London, 1976.
[7] S. Levine, *Discuss. Faraday Soc.*, 1978, **65**, 134.
[8] E. Dickinson and M. Lal, *Adv. Mol. Interact. Relaxation Processes*, 1980, **17**, 1.
[9] B. W. Ninham, *J. Phys. Chem.*, 1980, **84**, 1423.

hydrodynamic interactions into a concentrated system. Expressions for the diffusion coefficient and shear viscosity at low-volume fractions have been obtained by Batchelor[10,11] and Felderhof,[12] but the results are not easily extended to high-volume fractions.[13] The rheology of colloidal dispersions is normally dominated by the hydrodynamic properties of the continuous medium,[13,14] so that the problem is one of fluid dynamics rather than statistical mechanics. However, in certain special cases, for example ordered dispersions, a statistical mechanical approach to the rheology has proved to be fruitful.

2 Formalism and Methodology

Liquid-state Background.—The theory of simple dense fluids composed of spherically-symmetric molecules is now reasonably well-understood.[15,16] The structure is determined by the short-range repulsive forces,[17] and attractive forces can be allowed for in a perturbation scheme based on a reference fluid of 'hard' particles whose structure is known. This was recognized by van der Waals,[18] but then neglected for about 50 years.

The hard-sphere fluid has had a seminal influence on the development of liquid-state theory. It is the prototype reference system, and good semi-empirical expressions are now available for its equation of state,[19]

$$p/\rho kT = (1 + \eta + \eta^2 - \eta^3)/(1 - \eta)^3, \tag{6}$$

diffusion coefficient,[20]

$$D = (kT/\pi m)^{1/2}\sigma\{1.3891 - 2.3758\rho\sigma^3 + 1.0019(\rho\sigma^3)^2\}, \tag{7}$$

and shear viscosity.[21] In equations (6) and (7), p is the pressure, ρ is the number density, $\eta = \pi\rho\sigma^3/6$ is the packing fraction (equal to the volume fraction ϕ), m is the particle mass, and σ is the particle diameter. The hard-sphere fluid freezes at a density of about two-thirds the close-packed density.[16]

In extending liquid-state ideas to colloids, we imagine replacing the inter-molecular potential by a potential of mean force between dispersed particles.[3] The properties of the homogeneous solvent then appear as parameters in the potential of mean force. Figure 1 shows a schematic potential for two spherical particles with $u_{ij}(r)$ given by equations (1)—(4). If the height u_{max} of the

[10] G. K. Batchelor, *J. Fluid Mech.*, 1976, **74**, 1.
[11] G. K. Batchelor, *J. Fluid Mech.*, 1977, **83**, 97.
[12] B. U. Felderhof, *J. Phys. A*, 1978, **11**, 929.
[13] R. C. Ball and P. Richmond, *Phys. Chem. Liq.*, 1980, **9**, 99.
[14] W. B. Russel, *J. Rheol.*, 1980, **24**, 287.
[15] K. Singer (ed.), 'Statistical Mechanics', (Specialist Periodical Reports) The Chemical Society, London, 1973, Vol. 1.
[16] J. A. Barker and D. Henderson, *Rev. Mod. Phys.*, 1976, **48**, 587.
[17] J. S. Rowlinson, *Discuss. Faraday Soc.*, 1970, **49**, 30.
[18] J. D. van der Waals, 'Die Kontinuität des gasförmigen und flüssigen Zustandes', Barth, Leipzig, 1900, Vol. 2.
[19] N. F. Carnahan and K. E. Starling, *J. Chem. Phys.*, 1969, **51**, 635.
[20] D. Chandler, *J. Chem. Phys.*, 1975, **62**, 1358.
[21] J. H. Dymond, *Chem. Phys.*, 1976, **17**, 101.

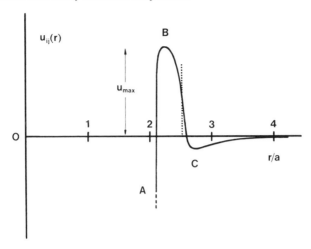

Figure 1 *Pair-wise-interaction energy between spherical particles i and j in an electrostatically-stabilized dispersion. The potential $u_{ij}(r)$ is plotted against r/a, where r is the centre-to-centre separation and a is the particle radius. Three regions of the potential are identified: primary minimum (A), primary maximum (B), (height u_{max}), and secondary minimum (C). The dotted line represents an effective hard-sphere potential*

primary maximum B is sufficiently large ($\geqslant 15kT$) to prevent coagulation into the primary minimum A, only the section of the potential to the right of B in Figure 1 affects the properties of the stable dispersion. The relevant part of $u_{ij}(r)$ has a steeply repulsive short-ranged section, qualitatively analogous to the potential between molecules in a simple liquid. The steepness of $u_R(r)$, together with the shallowness of the secondary minimum C, justifies using the hard-sphere fluid as reference system for a concentrated dispersion. In the perturbation scheme of Barker and Henderson,[22] an effective particle diameter d_{ij} is defined by

$$d_{ij} = \int_0^\infty (e^{-u_R(r)/kT} - 1)\, dr. \tag{8}$$

Distribution and Correlation Functions.—We consider a single spherical particle with position r' and velocity v' at time t' in a concentrated dispersion of mean number density ρ. The *distribution function* measures the probability of finding a particle (the same or another particle) with position r'' and velocity v'' at time t''. The osmotic equation of state is related to a time-averaged distribution function that depends on r alone, whereas the dynamic behaviour depends on time-dependent functions. A basic premise of statistical mechanics is that a time-average is equivalent to an ensemble average at fixed time; the ensemble average is denoted by angular brackets $\langle \ldots \rangle$.

[22] J. A. Barker and D. Henderson, *J. Chem. Phys.*, 1967, **47**, 2856, 4714; *Phys. Rev. A*, 1971, **4**, 806.

The radial (or static) distribution function $g(r)$ describes the equilibrium probability for two different particles a distance r apart. If $\rho(\mathbf{r})$ is the number density at position $\mathbf{r}$, then

$$g(r) = \rho^{-2}\langle \rho(\mathbf{r})\rho(\mathbf{O})\rangle. \tag{9}$$

In an isotropic disordered system, $g(r)$ approaches unity for large r.

Correlation functions describe deviations of distribution functions from their behaviour at large $\mathbf{r}$ or large t. The so-called total correlation function is defined by

$$h(r) = g(r) - 1. \tag{10}$$

The generalization of $g(r)$ and $h(r)$ to include time-dependence requires the definition of two more functions:

$$g^*(\mathbf{r}, t) = \rho^{-1}\langle \rho(\mathbf{r}, t)\rho(\mathbf{O}, 0)\rangle, \tag{11}$$

$$G(\mathbf{r}, t) = g^*(\mathbf{r}, t) - \rho. \tag{12}$$

The density–density correlation function $G(\mathbf{r}, t)$, generally known as the van Hove correlation function,[23] describes the complete equilibrium and dynamic behaviour of the dispersion. Unlike $h(r)$, $G(\mathbf{r}, t)$ also contains the correlation in the position of a *single* particle at time 0 and time t. This is called the van Hove *self*-correlation function $G_s(r, t)$, to distinguish it from the *distinct* part $G_d(r, t)$, which refers to correlations with other particles in the system. The behaviour of $G_s(r, t)$ and $G_d(r, t)$ in a dense fluid system is shown in Figure 2. At very short times, $G_d(r, t)$ approaches $h(r)$, and $G_s(r, t)$ approaches a delta function. The functions may be normalized by dividing by ρ.

The osmotic pressure of a monodisperse suspension of spherical particles interacting with the pairwise-additive potential $u(r)$ is related to $g(r)$ *via* the so-called pressure equation,[24]

$$p = \rho kT - (2\pi\rho^2/3)\int_0^\infty r^3 g(r)\{du(r)/dr\}\,dr. \tag{13}$$

Alternatively, p may be obtained from the compressibility equation,[25]

$$kT(\partial\rho/\partial p)_{V,T} = 1 + 4\pi\rho\int_0^\infty r^2 h(r)\,dr, \tag{14}$$

which makes no assumption about pairwise additivity.

The diffusion coefficient is given by

$$D = (1/6)\lim_{t\to\infty}\{d\langle r^2(t)\rangle/dt\}, \tag{15}$$

where $\langle r^2(t)\rangle$ is the mean-square particle displacement at time t, defined as

$$\langle r^2(t)\rangle = \int_0^\infty r^2 G_s(\mathbf{r}, t)\,dr. \tag{16}$$

[23] L. van Hove, *Phys. Rev.*, 1953, **89**, 1189.
[24] M. Born and H. S. Green, *Proc. R. Soc. London, Ser.* A, 1947, **191**, 168; T. L. Hill, 'An Introduction to Statistical Mechanics', Addison-Wesley, Reading, MA, 1960, Ch. 17 and 19.
[25] L. S. Ornstein and F. Zernike, *Proc. Acad. Sci. (Amsterdam)*, 1914, **17**, 793.

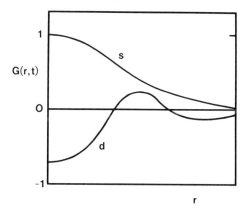

Figure 2 *Self* (s) *and distinct* (d) *parts of the van Hove correlation function in a dense fluid-like system. The functions $G_s(r, t)$ and $G_d(r, t)$ are plotted against separation r at fixed time t. Both functions are normalized*

The velocity autocorrelation function $\langle v(t)v(0)\rangle$ provides an alternative route to the diffusion coefficient:

$$D = \int_0^\infty \langle v(t)v(0)\rangle \, dt. \tag{17}$$

The scattering of radiation by particles is described in terms of an experimental structure factor $S(K, \omega)$, where K and ω denote the changes in wave-vector and angular frequency, respectively. The structure factor $S(K, \omega)$ is linked to the van Hove correlation function by a Fourier transformation:[26]

$$S(K, \omega) = (2\pi)^{-2} \iint G(r, t) e^{i(rK - \omega t)} \, dr \, dt. \tag{18}$$

For time-averaged light-scattering in the Rayleigh–Debye approximation, equation (18) reduces to[27]

$$S(K) = 1 + 4\pi\rho \int_0^\infty r^2 h(r) \frac{\sin(Kr)}{Kr} \, dr. \tag{19}$$

From equations (14) and (19) we see that

$$S(K \approx 0) = kT(\partial\rho/\partial p)_{V,T}, \tag{20}$$

which provides the route for determing osmotic virial coefficients from low-angle light-scattering experiments.

Computer Simulation.—The exact numerical solution of a many-body problem in classical statistical mechanics requires the use of a high-speed computer.

[26] J. S. Rowlinson and M. W. Evans, *Annu. Rep. Prog. Chem., Sect. A*, 1975, **72**, 5.
[27] M. Kerker, 'The Scattering of Light and Other Electromagnetic Radiation', Academic Press, New York, 1969.

There are two main types of computational procedure which have been used to simulate colloidal dispersions: the *Monte Carlo* method and the *Brownian dynamics* method. A Monte Carlo simulation gives equilibrium behaviour only, whereas a Brownian dynamics simulation gives both equilibrium and time-dependent behaviour.

An essential feature of all simulations of bulk systems is the use of periodic boundary conditions. Each 'cell' contains a small number of particles ($\sim 10^2$) that interact amongst themselves and with their periodic images. In particular, particle i interacts only with that periodic image of particle j lying closest to i (nearest-image distance convention). The original simulation cell—usually, but not always, a cube—is surrounded by an infinite set of replicas on each of its six sides. Particles may pass freely through the walls between the cells, and any particle leaving a cell through, say, the left-hand wall is deemed to re-enter at the corresponding point on the right-hand wall. By this means, the number of particles in the ensemble (canonical or microcanonical) remains constant. Interactions of an effective range greater than half the cell-width (*e.g.*, electrostatic or hydrodynamic) are difficult to handle in a computer simulation, because of the danger of particles becoming artificially correlated with their periodic images (*i.e.*, themselves).

By 'Monte Carlo method' we mean the 'importance sampling' procedure of Metropolis *et al.*,[28] in which a chain of configurations is generated in the (N, V, T) canonical ensemble such that the probability of a particular configuration of energy U appearing in the chain is proportional to $e^{-U/kT}$. The details of the Monte Carlo method are set out by Barker and Henderson.[16] Its principal exploiters in the field of colloidal dispersions are Snook and van Megen.[29—33]

The theory of Brownian motion was developed[34] to describe the dynamics of a massive particle in a medium of small particles. Ermak and McCammon have formulated[35] a Brownian dynamics simulation method suitable for concentrated dispersions. The method follows the traditional deterministic molecular dynamics algorithm used to simulate fluids,[36] except that the set of $3N$ Newtonian equations is replaced by a set of $3N$ coupled Langevin equations:[37]

$$m\dot{v}_i = F_i + \sum_j \alpha_{ij}f_j - \sum_j \zeta_{ij}v_j. \quad (1 \leqslant i, j \leqslant 3N) \tag{21}$$

[28] N. Metropolis, A. W. Rosenbluth, M. N. Rosenbluth, A. H. Teller, and E. Teller, *J. Chem. Phys.*, 1953, **21**, 1087.
[29] W. van Megen and I. K. Snook, *Chem. Phys. Lett.*, 1975, **33**, 156.
[30] I. K. Snook and W. van Megen, *J. Chem. Soc., Faraday Trans. 2*, 1976, **72**, 216.
[31] W. van Megen and I. K. Snook, *J. Colloid Interface Sci.*, 1976, **57**, 40, 47.
[32] W. van Megen and I. K. Snook, *J. Chem. Phys.*, 1977, **66**, 813.
[33] W. van Megen and I. K. Snook, *Discuss. Faraday Soc.*, 1978, **65**, 92.
[34] A. Einstein, *Ann. Phys.* (*Leipzig*), 1905, **17**, 549 (English translation in R. Furth (ed.), 'Albert Einstein, Investigations on the Theory of Brownian Movement', Dover, New York, 1956).
[35] D. L. Ermak and J. A. McCammon, *J. Chem. Phys.*, 1978, **69**, 1352.
[36] B. J. Alder and T. E. Wainwright, *J. Chem. Phys.*, 1957, **27**, 1208.
[37] J. M. Deutch and I. Oppenheim, *J. Chem. Phys.*, 1971, **54**, 3547.

In equation (21) the change in particle momentum is equated to the sum of three forces acting on a particle: (1) a systematic force F_i due to interparticle and external field interactions in direction i, (2) a randomly fluctuating force exerted on the particle by the surrounding medium (tending to increase the energy), and (3) a frictional force (tending to decrease the energy). The quantities f_i are described by a Gaussian distribution. The coefficients α_{ij} are related to a hydrodynamic configuration-dependent friction tensor ζ_{ij} by

$$\zeta_{ij} = (kT)^{-1} \sum_l \alpha_{il}\alpha_{jl}. \tag{22}$$

In a Brownian dynamics machine simulation, the configuration-space trajectories are composed of successive displacements $r_i^0 \to r_i$ over a short time-step Δt. The equation of motion is[35]

$$r_i = r_i^0 + \sum_j (\partial D_{ij}^0/\partial r_j)\Delta t + (kT)^{-1} \sum_j D_{ij}^0 F_j^0 \Delta t + R_i(\Delta t), \tag{23}$$

where the superscript 0 refers to the beginning of the time-step. In equation (23) D_{ij} is the configuration-dependent symmetric diffusion tensor related to ζ_{ij} by

$$\sum_j \zeta_{ij}D_{jl} = \sum_j D_{ij}\zeta_{jl} = kT\delta_{il}, \tag{24}$$

and $R_i(\Delta t)$ is a random displacement having a zero-mean Gaussian distribution with variance–covariance

$$\langle R_i(\Delta t)R_j(\Delta t)\rangle = 2D_{ij}^0\Delta t. \tag{25}$$

The set of displacements $\{R_i\}$ is generated stochastically. Numerical accuracy requires Δt to be sufficiently small that F_j and $(\partial D_{ij}^0/\partial r_j)$ are essentially constant during the interval.

The method of Ermak and McCammon is consistent with the Fokker–Planck equation,[38] whose solution yields the complete phase-space distribution function directly. Because the sets of particle velocities $\{v(t)\}$ and $\{v(t+\Delta t)\}$ are effectively uncorrelated for $\Delta t \gg m_i/\zeta_{ij}$, the Brownian dynamics algorithm gives no information about *instantaneous* velocities (the mean square particle velocity $\langle v^2(t)\rangle$ over time intervals $\geq \Delta t$ is the equilibrium value, $3kT/m_i$).

3 Order–Disorder Phase Transitions

Periodic structures are common in colloidal systems,[39] and order–disorder transitions are observed in colloidal dispersions of such diverse particle types

[38] T. J. Murphy and J. L. Aguirre, *J. Chem. Phys.*, 1972, **57**, 2098.
[39] I. F. Efremov and O. G. Us'yarov, *Russ. Chem. Rev.*, 1976, **45**, 435 (also, with some changes, under I. F. Efremov, in E. Matijević (ed.), 'Surface and Colloid Science', Wiley, New York, 1976, Vol. 8, p. 85).

158   Colloid Science

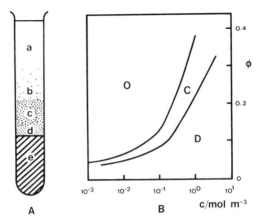

Figure 3 *Phase transition observed in emulsifier-free dispersions of polystyrene latex particles. (A) Schematic representation of sedimentation equilibrium (particle diameter 0.26 μm, ionic strength 8 mol m⁻³) with distinct regions, which are (a) transparent, (b) turbid, (c) milky-white, (d) diffusely coloured, and (e) iridescent. (B) Phase diagram (particle diameter 0.17 μm) with regions of order (O), disorder (D), and coexistence (C). The volume fraction φ is plotted against the electrolyte concentration c on a logarithmic scale*
(Redrawn from Refs. 41 and 45).

as latex spheres,[40—46] clay platelets,[1] and tobacco mosaic viruses.[47,48] Gem opals consist of a regular array of silica spheres,[49,50] whose structure is believed to arise from ordering in the colloidal state.[51] In spite of the diversity of these systems, it seems likely that statistical mechanics provides the unifying framework for understanding the physical principles underlying these ordering phenomena.[52]

Emulsifier-free, 'monodisperse' suspensions of spherical latex particles (diameter ~10^2 nm, ~10^4 charged groups) exhibit[41—46] separation into two distinct phases at univalent salt concentrations in the range 10^{-3}—10 mol m⁻³. Visual observation distinguishes a milky-white, disordered region from an iridescent, ordered region (see Figure 3A). The interparticle spacing in the

[40] P. A. Hiltner and I. M. Krieger, *J. Phys. Chem.*, 1969, **73**, 2386.
[41] S. Hachisu, Y. Kobayashi, and A. Kose, *J. Colloid Interface Sci.*, 1973, **42**, 342.
[42] A. Kose, M. Ozaki, K. Takano, Y. Kobayashi, and S. Hachisu, *J. Colloid Interface Sci.*, 1973, **44**, 350.
[43] A. Kose and S. Hachisu, *J. Colloid Interface Sci.*, 1974, **46**, 460.
[44] K. Takano and S. Hachisu, *J. Colloid Interface Sci.*, 1976, **55**, 487, 499.
[45] K. Takano and S. Hachisu, *J. Chem. Phys.*, 1977, **67**, 2604.
[46] K. Takano and S. Hachisu, *J. Phys. Soc. Jpn.*, 1977, **42**, 1775.
[47] G. Oster, *J. Gen. Physiol.*, 1949, **33**, 445.
[48] V. A. Parsegian and S. L. Brenner, *Nature (London)*, 1976, **259**, 435.
[49] J. V. Sanders, *Nature (London)*, 1964, **204**, 1151.
[50] J. V. Sanders, *Acta Crystallogr., Sect. A*, 1968, **24**, 427.
[51] J. V. Sanders and M. J. Murray, *Nature (London)*, 1978, **275**, 201.
[52] P. A. Forsyth, jun., S. Marčelja, D. J. Mitchell, and B. W. Ninham, *Adv. Colloid Interface Sci.*, 1978, **9**, 37.

iridescent phase is of the order of the wavelength of visible light, and so the forces producing the ordering must be of very long range. The experimental phase diagram for particles of diameter 170 nm is shown in Figure 3B. Under certain circumstances the Bragg diffraction from the ordered phase can be used as a technique of particle-size analysis.[53]

Numerical Results and Approximate Theories.—The phase diagram and structure of a model colloidal dispersion has been studied by computer simulation.[29—33,54] The Monte Carlo results have been supplemented by predictions from two approximate analytic theories: cell theory[55] and perturbation theory.[16]

In an attempt to imitate the latex dispersions in which phase transitions have been observed,[40—46] Snook and van Megen simulated an assembly of between 32 and 108 particles. The pair potentials were of the DLVO form given in equations (1)—(4). In a Monte Carlo simulation, the radial distribution function is calculated from

$$g(r) = \langle N(r, r + \delta r) \rangle / 2\pi\rho r^2 \, dr, \tag{26}$$

where $N(r, r + \delta r)$ is the number of particles found at a distance between r and $r + \delta r$ from a specified particle. The osmotic pressure is calculated from the virial equation,

$$p/\rho kT = 1 - (3NkT)^{-1}\left\langle \sum_{i<j} r\{du_{ij}(r)/dr\}\right\rangle. \tag{27}$$

The structural difference between ordered and disordered states is shown in Figure 4 for particles of radius 0.595 μm at a univalent electrolyte concentration of 0.1 mol m^{-3} ($\kappa a \gg 1$). At volume fractions of 0.15 and 0.25, $g(r)$ resembles that for a simple liquid; at a volume fraction of 0.35, $g(r)$ resembles that for a crystalline solid.

The properties of a concentrated, disordered dispersion may be described by perturbation theory.[31,33] In the approach of Barker and Henderson,[22] the pair potential is assumed to be of the form

$$u(r) = u_0(r) + u_1(r), \tag{28}$$

where $u_0(r)$ and $u_1(r)$ are reference and perturbation parts, respectively,

$$u_0(r) = \begin{cases} u(r), & (r \leq \sigma) \\ 0, & (r > \sigma) \end{cases}; \quad u_1(r) = \begin{cases} 0, & (r \leq \sigma) \\ u(r), & (r > \sigma). \end{cases} \tag{29}$$

The Helmholtz free energy is given by

$$A = \sum_{n=0}^{\infty} (\epsilon/kT)^n A_n, \tag{30}$$

where ϵ is some parameter that measures the attractive strength of the

[53] J. W. Goodwin, R. H. Ottewill, and A. Parentich, *J. Phys. Chem.*, 1980, **84**, 1580.
[54] K. Gaylor, W. van Megen, and I. K. Snook, *J. Chem. Soc., Faraday Trans. 2*, 1979, **75**, 451.
[55] J. A. Barker, 'Lattice Theories of the Liquid State', Pergamon, Oxford, 1963.

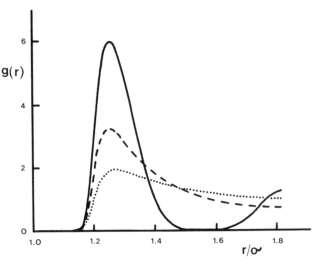

Figure 4 *Structure in Monte Carlo-simulated dispersion (particle diameter* 1.190 μm, *ionic strength* 0.1 mol m^{-3}*). The radial distribution function* g(r) *is plotted against the reduced separation* r/σ *for volume fractions of* 0.15 (····), 0.25 (---), *and* 0.35 (——)
(Redrawn from ref. 30).

potential (*e.g.*, depth of the secondary minimum), and A_0 is the free energy of of the reference system taken to be a fluid of hard spheres of diameter

$$d = \int_0^\sigma (e^{-u(r)/kT} - 1) \, dr. \tag{31}$$

In the theory of Weeks, Chandler, and Andersen (WCA),[56] equation (29) is replaced by

$$u_0(r) = \begin{cases} u(r) + \epsilon, & (r \leqslant r_m) \\ 0, & (r > r_m) \end{cases}; \quad u_1(r) = \begin{cases} -\epsilon, & (r \leqslant r_m) \\ u(r), & (r > r_m), \end{cases} \tag{32}$$

where r_m is the value of r for which $u(r)$ is a minimum. The free energy is given to first order by

$$A = A_0 + 2\pi N\rho \int_0^\infty r^2 u_1(r) g_0(r) \, dr. \tag{33}$$

The radial distribution function of the WCA reference system is approximated by

$$g_0(r) = g_{HS}(r) e^{-u_0(r)/kT}, \tag{34}$$

[56] J. D. Weeks, D. Chandler, and H. C. Andersen, *J. Chem. Phys.*, 1971, **54**, 5237; *ibid.*, 1971, **55**, 5422.

where $g_{HS}(r)$ is the function for hard spheres. The diameter d is defined by

$$\int_0^{r_m} r^2 g_{HS}(r) \, dr = \int_0^{r_m} r^2 g_0(r) \, dr. \tag{35}$$

The WCA predictions for $g(r)$ in the concentrated, disordered region agree well with the Monte Carlo results.[30]

The cell theory provides an approximate theory for the ordered phase.[54] Each particle is constrained to a 'cell' of radius r_c and free volume

$$v_f = 4\pi \int_0^{r_c} r^2 e^{-\{\psi(r)-\psi(0)\}/kT} \, dr. \tag{36}$$

The function $\psi(r)$ measures the particle potential energy a distance r from the centre of the cell. The time-averaged properties of the dispersion are calculated *via* the configurational partition function,

$$Q = v_f^N e^{-E_0/kT}, \tag{37}$$

where E_0 is the static energy of the lattice.

Snook and van Megen compared their computed phase diagram[30—33] with the available experimental data on polystyrene latex dispersions.[41,42] In agreement with experiment, they found that as the electrolyte concentration decreases so does the difference between the volume fractions of the coexisting phases (and the volume fraction at which the ordered phase first appears). The combined approximate methods of cell theory and perturbation theory agree satisfactorily with a complete Monte Carlo determination of coexisting volume fractions at $c = 0.1$ mol m^{-3} using the procedure of Hoover and Ree.[57]

The correspondence between a concentrated dispersion and an assembly of hard spheres has been pursued by several authors.[40,58—60] The Kirkwood–Alder hard-sphere transition[61] is in qualitative agreement with experiment, but the coexistence region is in general too narrow. Introduction of attractive forces, in the Monte Carlo simulations and approximate perturbation–cell theories, leads to improvement at high salt concentrations and large volume fractions.[33] But at low salt concentrations there remains the fundamental problem that the particles are not in proper thermodynamic equilibrium with bulk electrolyte; as Ninham and coauthors put it:[52] 'the diffuse double-layers of the particles fill up the entire volume of the system, and there is no place to be regarded as "bulk"'.

Wigner Lattice Model.—Marčelja *et al.* have considered[62] a Wigner lattice model in which each particle moves about in a sphere containing an equal and opposite charge centred on the lattice site. The surface potential of each particle in the dispersion is determined by solving the Poisson–Boltzmann

[57] W. G. Hoover and F. H. Ree, *J. Chem. Phys.*, 1967, **47**, 4873.
[58] M. Wadati and M. Toda, *J. Phys. Soc. Jpn*, 1972, **32**, 1147.
[59] W. van Megen and I. K. Snook, *Chem. Phys. Lett.*, 1975, **35**, 399.
[60] S. L. Brenner, *J. Phys. Chem.*, 1976, **80**, 1473.
[61] B. J. Alder, W. G. Hoover, and D. A. Young, *J. Chem. Phys.*, 1968, **49**, 3688.
[62] S. Marčelja, D. J. Mitchell, and B. W. Ninham, *Chem. Phys. Lett.*, 1976, **43**, 353.

equation subject to the constraint that the degree of dissociation of surface charge groups is determined self-consistently.

There is no statistical theory of the solid–liquid transition, even for hard spheres. So, to estimate volume fractions of coexisting phases, Marčelja *et al.* resorted to an intuitive approach based on Lindemann's empirical rule,[63] which states that a solid will melt when the root mean-square displacement of particles about their equilibrium positions exceeds some characteristic fraction f_L of the lattice spacing. For potentials of functional form r^{-n} ($n \geq 4$) it is found that $f_L \sim 0.10$. For particles of charge zq occupying volume $4\pi a^3/3$ on a f.c.c. lattice with spacing b, the Lindemann ratio is

$$f_L = \langle r^2 \rangle^{1/2}/b = (3/2)^{5/6}\pi^{-1/3}\Gamma^{-1/2}, \qquad (38)$$

where

$$\Gamma = (zq)^2/\epsilon_r\epsilon_0 akT. \qquad (39)$$

Monte Carlo studies show[64,65] that the Wigner lattice melts when the parameter Γ is about 155 ± 10 (corresponding to $f_L \sim 0.08$). Thus, whatever the form of the effective repulsive interaction, from very soft (unscreened Coulomb) to hard (say r^{-12}), the empirical Lindemann rule provides melting curves that are, at worst, qualitatively correct.

Application of the Wigner lattice model to a low-density latex dispersion gives a phase diagram in reasonable agreement with the experimental points,[41] whose large scatter exaggerates the width of the coexistence region. The model predicts[62] a coexistence region that is much too narrow, and at volume fractions above about 0.2 the ordering is better described by a Kirkwood–Alder transition.[45]

The Onsager Transition.—A dispersion ($\sim 2\%$) of California bentonite clay platelets (thickness ~ 2 nm, diameter 0.1–1 μm) will separate into an isotropic phase and a birefringent phase.[1] The birefringent phase (interparticle spacing ~ 0.1 μm) has no mechanical strength indicative of translational order; rather it seems to have orientational order, as in a nematic liquid crystal of disc-shaped micelles.[66] A similar type of ordering occurs in dilute 'solutions' of tobacco mosaic virus (diameter ~ 18 nm, length 0.28 μm) at low ionic strengths. The birefringent phase contains an ordered array of rod-shaped particles with large interparticle spacing.

It has been pointed out[52,67] that ordering in colloidal dispersions can be explained *via* the classic work of Onsager[2] by treating the particles as hard discs or hard rods. The Onsager transition is really a description of liquid-crystal formation: with increasing concentration there is a transition from an isotropic fluid phase to a nematic fluid phase. The physical driving force for the transition is entirely entropic. With increasing concentration, hard rods (or discs) lose entropy owing to hindrance of free rotation, but if some are

[63] F. A. Lindemann, *Z. Phys.*, 1910, **11**, 609.
[64] N. H. March and M. P. Tosi, *Phys. Lett. A*, 1974, **50**, 224.
[65] C. M. Care and N. H. March, *Adv. Phys.*, 1975, **24**, 101.
[66] N. Boden, P. H. Jackson, K. McMullen, and M. C. Holmes, *Chem. Phys. Lett.*, 1979, **65**, 476.
[67] M. A. Carter, *Phys. Rev. A*, 1974, **10**, 625.

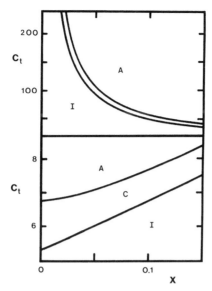

Figure 5 *Onsager transition for hard rods (top) and hard discs (bottom) with minimum and maximum dimensions d_{min} and d_{max} respectively. The transition concentration c_t (ρd_{max}^3 for rods, ρd_{min}^3 for discs) is plotted against the anisotropy ratio $x = d_{min}/d_{max}$: I, isotropic region; A, anisotropic region; C, coexistence region* (Redrawn from ref. 52).

orientated in a single direction they become more effectively packed (ordered phase), with the result that the remainder (disordered phase) have greater freedom and entropy. When the gain in entropy of the disordered phase outweighs the loss in entropy of the ordered phase, spontaneous phase separation takes place.

Figure 5 shows the Onsager transition concentration c_t as a function of the anisotropy ratio x (ratio of smallest to largest dimension) for rods and discs.[52] As the ionic strength is reduced, the range of the electrostatic forces decreases, and so the particles 'appear' more isotropic ($x \to 1$). There are qualitative differences between the behaviour of rods and discs. For instance, dc_t/dx is negative for rods and positive for discs; for rods c_t is infinite for $x = 0$ (completely anisotropic), whereas for discs it is finite.

4 Structure from Light Scattering

Time-averaged Light Scattering.—In a conventional light-scattering experiment the integrated intensity is measured as a function of scattering angle, θ, at constant wavelength, λ. Let us consider the case of vertically-polarized light incident on a dilute, structureless system of monodisperse, spherical particles. The reduced scattering intensity of the dispersion over that of the solvent, the

Rayleigh ratio $R_0(K)$, is given in the Rayleigh–Debye approximation by[27]

$$R_0(K) = \mathcal{K}wMP(K), \tag{40}$$

where M is the molar particle mass, and w is the weight concentration. The quantities K and $\mathcal{K}$ are defined by

$$K = (4\pi n/\lambda_0)\sin(\theta/2), \tag{41}$$

$$\mathcal{K} = 2n^2\pi^2(\mathrm{d}n/\mathrm{d}c)^2/\lambda_0^4 N_A, \tag{42}$$

where n is the refractive index of the medium, and λ_0 is the wavelength *in vacuo*. The single-particle form-factor $P(K)$ is approximated by[27]

$$P(K) \approx 1 - K^2 R^2/5. \tag{43}$$

In order to allow for correlations in the positions of the particles, equation (40) is supplemented by

$$R(K) = R_0(K)S(K), \tag{44}$$

where $S(K)$ is the structure factor defined in equation (19). The total correlation function is obtained from the experimental structure factor by a Fourier transformation:

$$h(r) = (2\pi^2\rho)^{-1}\int_0^\infty K^2\{S(K)-1\}\frac{\sin(Kr)}{Kr}\,\mathrm{d}r. \tag{45}$$

In principle, then, through equation (45), light-scattering studies of concentrated colloidal dispersions yield structural information in the form of $h(r)$. In practice, however, the usefulness of equation (45) is severely limited by the fact that it is valid only in the absence of multiple scattering. With concentrated aqueous dispersions of latex particles, this means studying very small sample volumes in which structure could be influenced by the surface properties of the container.[68] Even so, no measurements of this sort have yet been reported due to the experimental difficulties involved.

The multiple-scattering problem has so far restricted the study of structure to two types of dispersion: (i) dilute latex dispersions at very low salt concentration ($\kappa a \ll 1$), where structure results from the repulsion between long-range double-layers,[69—71] and (ii) polymer latex particles dispersed in an organic solvent of closely matched refractive index.[68,72]

In aqueous dispersions of small polystyrene latex particles (diameter ~ 50 nm) at very low ionic strength, Brown *et al.* observed[69] maxima in $R(K)$ similar to those found in the structure factors of simple liquids as determined by X-ray scattering. Figure 6a shows the developing structure in a sample as counter ions are removed by ion-exchange resin; Fourier transformation yields

[68] E. A. Nieuwenhuis and A. Vrij, *J. Colloid Interface Sci.*, 1979, **72**, 321.
[69] J. C. Brown, P. N. Pusey, J. W. Goodwin, and R. H. Ottewill, *J. Phys. A*, 1975, **8**, 664.
[70] J. C. Brown, J. W. Goodwin, R. H. Ottewill, and P. N. Pusey, in 'Colloid Interface Science', Academic Press, New York, 1976, Vol. IV.
[71] D. W. Schaefer, *J. Chem. Phys.*, 1977, **66**, 3980.
[72] A. Vrij, E. A. Nieuwenhuis, H. M. Fijnaut, and W. G. M. Agterof, *Discuss. Faraday Soc.*, 1978, **65**, 101.

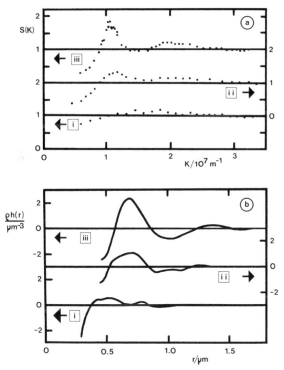

Figure 6 *Development of structure in aqueous dispersion of polystyrene latex particles (diameter ~50 nm) on removing counter ions by ion-exchange resin. (a) The measured structure factor S(K) is plotted against the wave-number K after (i) 2 h, (ii) 48 h, and (iii) over 300 h. (b) The function ρh(r) obtained by Fourier transformation is plotted against separation r*
(Redrawn from ref. 69).

the correlation functions shown in Figure 6b. The authors estimate[69] that the height of the first peak in $g(r)$ may be in error by up to 20%. The two main sources of error in the analysis are the assumption of Rayleigh–Debye scattering theory and the lack of data at small wave-number.

In analysing $S(K)$ data for poly(methyl methacrylate) latices in benzene, Nieuwenhuis and Vrij found[68] it necessary to take a different $P(K)$ function, with a smaller magnitude and smaller angular dependence, in the lower range of K. This was attributed to the ease of interpenetration of these 'micro-gel' particles. Figure 7a shows an 'adjusted' structure factor for a dispersion of concentration 5.4×10^{-2} g cm^{-3} at 6.5 °C, and Figure 7b shows the resulting total correlation function from an extrapolation of $S(K)$ to $K = 0$. The spurious peak in $h(r)$ at small r arises from truncation errors and/or inaccuracies in $S(K)$.

Information about the form of the pair potential $u(r)$ may be derived from

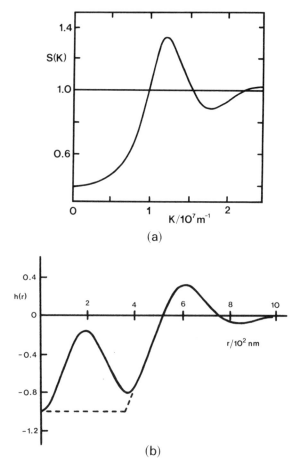

Figure 7 *Structure of dispersion of poly(methyl methacrylate) latex particles (diameter 0.53 μm) in benzene. (a) The measured structure factor is plotted against the wave-number K. (b) The total correlation function h(r) obtained by Fourier transformation is plotted against separation r. The dashed line represents the expected behaviour of h(r) at low r*
(Redrawn from ref. 68).

$h(r)$ through the Ornstein–Zernike equation,[73]

$$h_{ij}(r) = c_{ij}(r) + \rho \int c_{ij}(r_1)h_{kj}(|\mathbf{r}_1 - \mathbf{r}|)\,\mathrm{d}\mathbf{r}, \qquad (46)$$

where $c(r)$ is the so-called *direct* correlation function. To solve equation (46) an additional (approximate) relationship between the direct and indirect correlation functions is required. A successful closure relation for particles with steep

[73] R. J. Baxter in D. Henderson (ed.), 'Physical Chemistry: An Advanced Treatise', Academic Press, New York, 1971, Vol. VIIIA.

repulsive interactions is the (now) ubiquitous Percus–Yevick approximation:[74]

$$c(r) = \{1 + h(r)\}\{1 - e^{u(r)/kT}\}. \tag{47}$$

Nieuwenhuis and Vrij[68] have used equation (47) directly to determine $u(r)$ from the total correlation function shown in Figure 7b. The direct correlation function was determined by Fourier transformation of $S(K)$:

$$c(r) = (2\pi^2\rho)^{-1} \int_0^\infty K^2 \frac{\{S(K) - 1\}}{S(K)} \frac{\sin(Kr)}{Kr} \, dK. \tag{48}$$

The resulting pair potential indicates a strong repulsive interaction between cores at $r \sim 0.45$ μm. The 'secondary minimum' is thought to be spurious.

Light-scattering Studies of Floc Structure.—The light scattered from a monodisperse coagulating sol can give information on structure in flocs.[75,76] Representing the scattering contribution from non-randomness as a Rayleigh–Gans–Debye summation of phase differences between geometric centres, the mean floc structure factor is written as

$$S(K) = 1 + (2/n_0) \sum_{j=2} A_j n_j(t). \tag{49}$$

In equation (49), $n_j(t)$ is the number density of the j-fold aggregates at time t, and n_0 is the total number density at $t = 0$. Lips *et al.*[76] have evaluated the configurational factors A_j ($2 \leqslant j \leqslant 13$) for linear and close-packed aggregates.

The feasibility of deriving $g(r)$ by transforming $S(K)$ for a coagulating sol has been tested[75] by inverting an artificial set of $S(K)$ data, generated by the Smoluchowski–Benoit model[77] for linear and close-packed aggregates (particle diameter $4\lambda_0/\pi$). Direct Fourier inversion produced large oscillations in $g(r)$ due to truncation of the data set characterized by a cut-off parameter proportional to particle size and wavelength. By choosing carefully the cut-off parameter and a special modification function, House was able in principle to determine $g(r)$ from measurements of $S(K)$.[75]

Giles and Lips[78] have shown that the initial change in structure factor in a dilute coagulating system is

$$\{dS(K, E)/dE\}_{E \to 0} = 2 \sin(Kd)/Kd, \tag{50}$$

where d is the hard-core diameter, and E is a reduced coagulation time. It was found that rapid coagulation in the presence of electrolyte and nonionic flocculant produced a close-packed, ordered structure at low ionic strength, but a liquid-like, disordered structure at high ionic strength.

Photon Correlation Spectroscopy.—In the type of quasi-elastic light-scattering known as photon correlation spectroscopy (PCS), measurements are made of

[74] J. K. Percus and G. J. Yevick, *Phys. Rev.*, 1958, **110**, 1.
[75] W. A. House, *J. Chem. Soc., Faraday Trans. 1*, 1978, **74**, 1045, 1112.
[76] A. Lips, C. Smart, and E. Willis, *Trans. Faraday Soc.*, 1971, **67**, 2979.
[77] H. Benoit, R. Ullman, A. J. de Vries, and C. Wippler, *J. Chim. Phys.*, 1962, **59**, 889.
[78] D. Giles and A. Lips, *J. Chem. Soc., Faraday Trans. 1*, 1978, **74**, 733.

the autocorrelation function $G(\boldsymbol{K}, t)$ of the scattered electric field E:[79]

$$G(\boldsymbol{K}, t) = \langle E(\boldsymbol{K}, 0)E^*(\boldsymbol{K}, t)\rangle. \tag{51}$$

The function $G(\boldsymbol{K}, t)$ is the Fourier transform of $S(\boldsymbol{K}, \omega)$ with respect to ω. In a system of non-interacting particles, $G(\boldsymbol{K}, t)$ is simply a self-correlation function, and the absolute value of the normalized field autocorrelation function takes the form[80]

$$|g^1(K, t)| \propto e^{-D_0 K^2 t}, \tag{52}$$

where D_0 is the single-particle diffusion coefficient. Equation (52) is the theoretical basis of the widespread use of PCS in determining diffusion coefficients and particle dimensions in dispersions of high dilution. Biological macromolecules, especially proteins,[81,82] are also receiving increasing attention. (This is outside the scope of this Report. Aspects of PCS in strongly absorbing and non-dilute protein solutions have been discussed recently by Hall et al.[83])

Multiple scattering is a considerable problem in extracting structural information from the dynamic light-scattering of dispersions of interacting particles.[84] The effect of particle interactions on $|g^1(K, t)|$ has been studied by several authors, most notably Pusey[85] and Ackerson.[86] Experimentally, $g^1(\boldsymbol{K}, t)$ is found no longer to be of a simple exponential form.[70,87,88] The difficulty theoretically is to separate out the effects of particle interactions, polydispersity, and multiple scattering.

In a dilute dispersion it can be assumed that the instantaneous particle velocity $\boldsymbol{v}$ is separable into two parts: a rapidly-fluctuating Brownian component, and an interaction component with a fluctuation time of the order of the time for the particle to move a distance K^{-1}. With this assumption, the *initial* decay of $g^1(K, t)$ is characterized by an effective, $\boldsymbol{K}$-dependent diffusion coefficient related to the structure in the system:[89]

$$D = D_0 H/S(K). \tag{53}$$

The coefficient D is obtained by integrating the autocorrelation function of the Brownian component of $\boldsymbol{v}$ [see equation (17)]. In very dilute dispersions the correction H for hydrodynamic interactions is close to unity,[86] and equation (53) provides a viable route to $g(r)$. This approach has been used[70] to obtain $S(K)$ for very dilute aqueous latex dispersions at low ionic strengh; the curves are qualitatively consistent with those obtained from static experiments (Figure 6a). With index-matched poly(methyl methacrylate) particles in benzene,

[79] H. Z. Cummins and E. R. Pike, 'Photon Correlation and Light-beating Spectroscopy', Plenum, New York, 1974.
[80] R. Pecora, J. Chem. Phys., 1964, **40**, 1604.
[81] J. M. Schurr, CRC Crit. Rev. Biochem., 1977, **4**, 371.
[82] S. C. Lin, W. I. Lee, and J. M. Schurr, Biopolymers, 1978, **17**, 1041.
[83] R. S. Hall, Y. S. Oh, and C. S. Johnson, jun., J. Phys. Chem., 1980, **84**, 756.
[84] F. Grüner and W. Lehman, J. Phys. A, 1980, **13**, 2155.
[85] P. N. Pusey, J. Phys. A, 1975, **8**, 1433.
[86] B. J. Ackerson, J. Chem. Phys., 1976, **64**, 242.
[87] D. W. Schaefer and B. J. Berne, Phys. Rev. Lett., 1974, **32**, 1110.
[88] G. D. J. Phillies, J. Chem. Phys., 1974, **60**, 976.
[89] B. J. Berne and R. Pecora, 'Dynamic Light Scattering', Wiley, New York, 1976.

Nieuwenhuis and Vrij find that the extrema in D_0/D coincide with those in $S(K)$ from conventional light-scattering, even at high concentrations.[68]

5 Osmotic Equation of State

Information on the equation of state comes from two experimental sources: light-scattering at low ($\leqslant 1\%$) volume fractions *via* equation (20), and direct compressibility studies at high ($\geqslant 10\%$) volume fractions. For testing statistical theories of concentrated dispersions, we therefore look towards compressibility studies on well defined systems. The available compression data on lattices[90-92] show only qualitative agreement with the Monte Carlo results;[30,31] they are insufficiently precise to test the pairwise-additivity assumption employed in the computer simulations.

The presence of a primary maximum in the potential of Figure 1 should imply an inflexion in the osmotic pressure *versus* volume fraction plot indicative of coagulation. The effect is indeed found in the computer simulations[30,31] and to differing extents in the osmotic compression experiments.[91,92] Meijer *et al.*[93] have used the cell model[55] to predict critical coagulation pressures to compare with their experimental results on the centrifugation of monodisperse polystyrene latices (diameter $\sim 0.5 \, \mu$m). In the cell theory, the osmotic pressure is given by

$$p = kT(\partial \ln Q/\partial V)_{T,N}, \tag{54}$$

where Q is the configuration integral defined in equation (37). Using a DLVO-type pair potential, the agreement between theory and experiment is good at high salt concentrations, but at low ionic strength the experimental critical coagulation pressures exceed the theoretical ones by an order of magnitude. Two reasons are suggested[93] for the discrepancy: (i) a breakdown in the pairwise-additivity assumption for $\kappa a \leqslant 1$, and (ii) the presence of an additional repulsive interaction from macromolecular chains at the surface of 'hairy' latex particles.

Evans and Napper have proposed[94] a simple expression for the equation of state of a concentrated, hexagonally close-packed dispersion. The osmotic pressure is given by

$$p = \{(a\gamma^2\kappa/\nu^2)e^{-\kappa d} - (A_H a/12d^2)\}/2^{1/2}(a + d/2)^2, \tag{55}$$

where

$$\gamma = \{e^{(\nu e\psi_\delta/2kT)} - 1\}/\{e^{(\nu e\psi_\delta/2kT)} + 1\}, \tag{56}$$

and ψ_δ is the Stern potential. It is assumed in deriving equation (55) that the particles are rigidly located at their lattice points (*i.e.*, there is no Brownian motion), and that nearest neighbours interact *via* DLVO-type potentials. It is

[90] L. Barclay, A. Harrington, and R. H. Ottewill, *Kolloid Z. Z. Polym.*, 1972, **250**, 655.
[91] A. Homola and A. A. Robertson, *J. Colloid Interface Sci.*, 1976, **54**, 286.
[92] E. Dickinson and A. Patel, *Colloid Polym. Sci.*, 1979, **257**, 431.
[93] A. E. J. Meijer, W. van Megen, and J. Lyklema, *J. Colloid Interface Sci.* 1978, **66**, 99.
[94] R. Evans and D. H. Napper, *J. Colloid Interface Sci.*, 1978, **63**, 43.

found[95] that osmotic pressures from the Evans–Napper model agree well with Monte Carlo pressures for ordered phases with $p/\rho kT \gtrsim 20$. This result validates the model at high-volume fractions, and is independent of any uncertainties about pairwise additivity. At phase points close to the order–disorder transition, however, the cell theory is to be preferred.

6 Polydispersity

The particles in a real colloidal dispersion are not identical, and any complete statistical theory must take this into account. The problem of polydispersity seems first to have been recognized by Onsager[2] in his study of orientational order–disorder transitions. By a polydisperse system, we mean one in which there is a *continuous* distribution $f(x)$ of some characteristic (single-particle) variable x. According to this definition, a binary mixture is not an example of a polydisperse system. The variable x can refer to any characteristic of the particle, but in practice this is almost always the particle size.

Thermodynamics of Polydisperse Systems.—The simplest polydisperse system is perhaps that composed of hard spheres (or discs) with a continuous distribution $f(\sigma)$ of particle diameter σ. For a multicomponent mixture of hard spheres in the Percus–Yevick approximation, the compressibility equation of state is[96]

$$p = (6kT/\pi)\{\xi_0(1-\xi_3)^{-1} + 3\xi_1\xi_2(1-\xi_3)^{-2} + 3\xi_2^3(1-\xi_3)^{-3}\}, \qquad (57)$$

where

$$\xi_n = (\pi/6) \sum_i \rho_i \sigma_i^n, \qquad (n = 0, 3) \qquad (58)$$

and ρ_i and σ_i are the number density and particle diameter of species i. The generalization of equation (57) to a polydisperse system of hard spheres $(i \rightarrow \infty)$ is achieved by replacing summations by integrations,[97]

$$p = (6kT/\pi)\{\xi(1-\xi m_3)^{-1} + 3\xi^2 m_1 m_2(1-\xi m_3)^{-2} + 3\xi^3 m_2^3(1-\xi m_3)^{-3}\}, \qquad (59)$$

where $\xi = \pi\rho/6$, and m_n is the nth moment of $f(\sigma)$:

$$m_n = \int_0^\infty f(\sigma)\sigma^n \, d\sigma. \qquad (60)$$

In the disordered state, hard spheres of different sizes pack together slightly more efficiently than identical hard spheres.[17] At constant number density, therefore, the osmotic pressure of the polydisperse hard-sphere fluid is lower than that of the monodisperse system with the same volume (packing) fraction.[98] Figure 8 shows the effect on the compressibility factor as a function of

[95] E. Dickinson, *J. Chem. Soc., Faraday Trans.* 2, 1979, **75**, 466.
[96] R. J. Baxter, *J. Chem. Phys.*, 1970, **52**, 4559.
[97] J. J. Salacuse, Ph.D. Thesis, SUNY, Stony Brook, 1978.
[98] E. Dickinson, *Chem. Phys. Lett.*, 1978, **57**, 148.

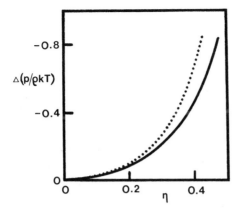

Figure 8 *Difference in compressibility factor between a polydisperse hard-sphere system and the equivalent monodisperse system. The function $\Delta(p/\rho kT)$ is plotted against the packing fraction η for polydisperse hard spheres with a rectangular distribution of diameters and a degree of polydispersity $\Delta = 0.28$:* ———, *Percus–Yevick;* · · · ·, *van der Waals one fluid*
(Redrawn from ref. 98).

packing fraction $\eta = \xi_3$ for a polydisperse system with a rectangular distribution of diameters:

$$f(\sigma) = \begin{cases} (2\Delta\bar{\sigma})^{-1}, & (1-\Delta < \sigma/\bar{\sigma} < 1+\Delta) \\ 0. & (1-\Delta \geqslant \sigma/\bar{\sigma}; \; 1+\Delta \leqslant \sigma/\bar{\sigma}) \end{cases} \tag{61}$$

In equation (61), $\bar{\sigma}$ is the most probable diameter, and Δ measures the degree of polydispersity. The difference in $p/\rho kT$ between a polydisperse system having $\Delta = 0.28$ and its equivalent monodisperse system is obtained from a generalization of the Percus–Yevick mixture result of Lebowitz and Rowlinson.[99,100] At low densities $\Delta(p/\rho kT)$ is negligible; but in a concentrated dispersion it could amount to several per cent.

The pressure change calculated from the so-called van der Waals one-fluid model is also shown in Figure 8. In this model[101] the polydisperse system is characterized by a single size parameter

$$\langle\sigma\rangle = \left\{ \iint f(\sigma_i)f(\sigma_j)\sigma_{ij}^3 \, \mathrm{d}\sigma_i \, \mathrm{d}\sigma_j \right\}^{\frac{1}{3}}, \tag{62}$$

where $\sigma_{ij} = \frac{1}{2}(\sigma_i + \sigma_j)$ for additive hard spheres. For the rectangular distribution of equation (61), it is found that[98]

$$\langle\sigma\rangle^3 = \bar{\sigma}^3\{1 + \Delta^2/2 + \mathcal{O}(\Delta^4)\}. \tag{63}$$

The van der Waals one-fluid approximation agrees with the Percus–Yevick result, except at very high-volume fractions.

[99] J. L. Lebowitz, *Phys. Rev. A*, 1964, **133**, 895.
[100] J. L. Lebowitz and J. S. Rowlinson, *J. Chem. Phys.*, 1964, **41**, 133.
[101] T. W. Leland, J. S. Rowlinson, and G. A. Sather, *Trans. Faraday Soc.*, 1968, **64**, 1447.

The one-fluid theory is extended to non-additive particles by defining

$$\sigma_{ij} = \tfrac{1}{2}(1 + \zeta_{ij})(\sigma_i + \sigma_j), \tag{64}$$

where ζ_{ij} characterizes the non-additivity of the i–jth pair. In hard-core systems, non-additivity enables us to take account of the fact that the volume excluded by two unlike particles in a mixture may be larger or smaller than expected: if larger the unlike particles tend to repel one another, but if smaller they are preferentially attracted to one another. (This is quite a different concept from the non-additivity implied by deviations from equation (5).) It has been suggested [102] that the non-additive hard-sphere system is appropriate for modelling a dispersion with *generalized polydispersity*, that is, one in which a supplementary physical variable (surface charge, for instance) is correlated with the size distribution. For substantial non-additivity of the positive kind ($\zeta_{ij} > 0$), large deviations from ideality may lead to phase separation. Evidence for this effect has been provided by a computer simulation in two dimensions of a multicomponent system with a symmetrical non-additivity distribution.[102]

Blum and Stell have proposed [103] a *permeable*-sphere model of polymer colloids (to be distinguished from the *penetrable*-sphere model of Widom and Rowlinson [104]). The pair potential between permeable particles of types i and j is of the same range as the hard-sphere potential, so that

$$c_{ij}(r) = 0, \qquad u_{ij}(r) = 0; \qquad (r \geq \sigma_{ij}) \tag{65}$$

the probability of interpenetration of any two spheres is expressed in terms of the total correlation function,

$$h_{ij}(r) = -\epsilon, \qquad (r < \sigma_{ij}) \tag{66}$$

where ϵ is the permeability parameter ($\epsilon = 1$ for hard spheres, $\epsilon = 0$ for an ideal gas). In the Percus–Yevick approximation, the model exhibits an appealing scaling relation with simple hard spheres: the osmotic pressure of polydisperse permeable spheres at density ρ is equal to that for hard spheres with the same distribution function $f(\sigma)$ at density $\epsilon\rho$.

Whereas the effect of polydispersity on the osmotic pressure of a disordered dispersion is small and negative, the effect on the osmotic pressure of an ordered dispersion appears to be both large and positive.[95] Using the model of Evans and Napper under conditions where repulsive forces dominate,[94] the effect is always to increase the pressure and free energy of the close-packed phase, in agreement with Monte Carlo results for a polydisperse system with a Gaussian distribution of particle sizes.[105] As expected intuitively, then, the overall influence of polydispersity on the order–disorder transition is always towards the disruption of long-range order. With large particles at moderate ionic strengths ($\kappa a \gg 1$) the osmotic pressure of the ordered phase is very sensitive to the parameter Δ, but the converse holds for $\kappa a \ll 1$. We expect therefore that the stability of dilute ordered dispersions ($\phi \lesssim 0.01$) at low salt

[102] E. Dickinson, *Chem. Phys. Lett.*, 1979, **66**, 500.
[103] L. Blum and G. Stell, *J. Chem. Phys.*, 1979, **71**, 42.
[104] B. Widom and J. S. Rowlinson, *J. Chem. Phys.*, 1970, **52**, 1670.
[105] E. Dickinson, *Discuss. Faraday Soc.*, 1978, **65**, 127.

concentration ($\sim 10^{-3}$ mol m^{-3}) will be insensitive to the degree of polydispersity.

Light Scattering from Polydisperse Systems.—The scattering from a polydisperse hard-sphere system in the Percus–Yevick approximation has been considered independently by Vrij[106,107] ($K = 0$) and Blum and Stell[103] (all K).

For spherical particles, the excess Rayleigh ratio $\Delta R(K)$ over that of the continuum solvent is given in the Rayleigh–Debye limit by[108]

$$\Delta R(K) = \sum_{ij} F_i F_j B_i(K) B_j(K) (\rho_i \rho_j)^{1/2} \{\delta_{ij} + H_{ij}(K)\}, \tag{67}$$

where δ_{ij} is the Dirac function, and

$$H_{ij}(K) = 4\pi (\rho_i \rho_j)^{1/2} \int_0^\infty r^2 h_{ij}(r) \frac{\sin(Kr)}{Kr} \, dr. \tag{68}$$

In equation (67) F_i is the excess zero-angle scattering amplitude (over the solvent amplitude) for a particle of diameter σ_i, and B_i is the intraparticle interference factor[109] defined by

$$B_i(K) = \int_0^\infty r^2 \alpha_i(r) \frac{\sin(Kr)}{Kr} \, dr \Big/ \int_0^\infty r^2 \alpha_i(r) \, dr, \tag{69}$$

where $\alpha_i(r)$ is the spherosymmetric distribution of scattering material inside a particle of diameter σ_i. In the continuous distribution limit, equation (67) becomes

$$\Delta R(K) = \rho \int f(\sigma_i) \{F_i B_i(K)\}^2 \, d\sigma_i$$
$$+ \rho \int \int f(\sigma_i) f(\sigma_j) F_i F_j B_i(K) B_j(K) H_{ij}(K) \, d\sigma_i \, d\sigma_j. \tag{70}$$

The extension of the Percus–Yevick solution for hard-sphere mixtures[96] to the polydisperse limit yields[103] an explicit expression for $H_{ij}(K)$, which is a complicated functon of the second and third moments of $f(\sigma)$.

The influence of polydispersity on the analysis of quasi-elastic light-scattering data is considerable. For non-interacting particles in the Stokes–Einstein régime, the effective diffusion coefficient is[69]

$$D = kT/3\pi\eta\{m_5 - (K^2/5)m_7\}\{m_6 - (K^2/5)m_8\}, \tag{71}$$

where η is the viscosity, and m_n ($n = 5$—8) is defined by equation (60). In experiments with interacting latex particles, the separate effects of polydispersity, structure, and hydrodynamics are difficult to separate.[69,110,111] Recently, it

[106] A. Vrij, *Chem. Phys. Lett.*, 1978, **53**, 144.
[107] A. Vrij, *J. Chem. Phys.*, 1978, **69**, 1742.
[108] A. Guiner and G. Fournet, 'Small Angle Scattering of X-rays', Wiley, New York, 1955.
[109] S. R. Aragon and R. Pecora, *J. Chem. Phys.*, 1976, **64**, 2395.
[110] P. N. Pusey, *J. Phys. A*, 1978, **11**, 119.
[111] P. S. Dalberg, A. Boe, K. A. Strand, and T. Sikkeland, *J. Chem. Phys.*, 1978, **69**, 5473.

has been suggested[112] that PCS experiments at small wave-vectors are dominated by effects of polydispersity in the scattering cross-section, thus providing an explanation why a slow 'tracer-like' relaxation mode of refractive-index fluctuations is found in dilute, ordered latex dispersions but not in protein solutions.[113] As indicated in equation (53), the faster mode correlates with the inverse structure factor yielding a *mutual* diffusion coefficient D.

According to Weissman,[112] the scattering intensity I in a polydisperse system arises from size fluctuations (proportional to $\langle a^6 \rangle - \langle a^3 \rangle^2$, where a is the particle radius) and number concentration fluctuations (proportional to $\langle a^3 \rangle^2$), but the division is significant only when interactions are present. Particles of similar size and surface charge will have almost identical self-diffusion coefficients but appreciably different scattering powers (proportional to $\langle a^6 \rangle$). The size distribution fluctuations will therefore be relatively little affected by interactions between the colloidal particles, since this part of the dynamic light-scattering signal is sensitive to the interparticle forces only through the diffuse electrostatic background contribution to the (Stokes-like) single-particle hydrodynamic friction factor.[114] The polydispersity scattering contribution I_p satisfies the inequality

$$(1 - \langle a^3 \rangle^2 / \langle a^6 \rangle) I < I_p < (1 - \langle a^3 \rangle^2 / \langle a^6 \rangle) \{S(K)\}^{-1}, \tag{72}$$

which is consistent with the magnitude of the slow mode measured in dilute, ordered latex dispersions.[110,111]

7 Time-dependent Behaviour

Neglect of Hydrodynamic Interactions.—The coupling of hydrodynamic flow exerts a major influence on the dynamics of colloidal dispersions.[13,14] In certain special cases, however, it has proved reasonable or expedient to neglect the hydrodynamic interactions. One such instance is the very dilute, electrostatically-stabilized dispersion in which particles interact *via* a screened Coulomb potential, that is, equation (2) with $\kappa a \ll 1$.

Using the algorithm of Ermak and co-workers[35,115,116] in the absence of hydrodynamic interactions, Gaylor *et al.* have simulated[117] a system of spherical particles ($a = 23$ nm, $\psi = 0.22$ V, $T = 300$ K, $c = 10^{-3}$ mol m^{-3}) at volume fractions up to 7×10^{-4}. Particles were displaced according to equation (23) with D^0 taking the Stokes–Einstein value, so that the term in $(\partial D_{ij}^0 / \partial r_j)$ is trivially zero. The particle self-diffusion coefficient, as derived from the mean-square displacement, is as much as 30% below the infinite dilution value at a volume fraction as low as 5×10^{-5}. The effect on D in this system arises from the extremely long range of the double-layer forces; if c is increased, then, as expected, D also increases. The reduction in D from the repulsive interparticle

112 M. B. Weissman, *J. Chem. Phys.*, 1980, **72**, 231.
113 M. B. Weissman, R. C. Pan, and B. R. Ware, *J. Chem. Phys.*, 1978, **68**, 5069.
114 G. D. J. Phillies, *J. Chem. Phys.*, 1977, **67**, 4690.
115 D. L. Ermak and Y. Yeh, *Chem. Phys. Lett.*, 1974, **24**, 243.
116 D. L. Ermak, *J. Chem. Phys.*, 1975, **62**, 4189, 4197.
117 K. J. Gaylor, I. K. Snook, W. van Megen, and R. O. Watts, *Chem. Phys.*, 1979, **43**, 233; *J. Phys. A*, 1980, **13**, 2513.

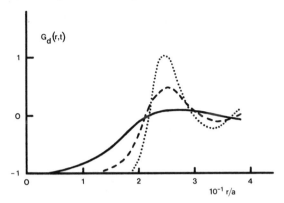

Figure 9 *Relaxation of pair correlations in Brownian dynamics simulated dispersion (particle diameter 46 nm, volume fraction 3×10^{-4}, ionic strength 10^{-3} mol m^{-3}). The van Hove distinct correlation function $G_d(r, t)$ at time t is plotted against r/a, where r is the centre-to-centre separation and a is the particle radius:* $\cdots\cdots$, $t = 0$; $---$, $t = 1$ ms; ———, $t = 4$ ms
(Redrawn from ref. 117).

forces appears to contradict hydrodynamically-based theories in the low volume fraction limit.[10,118,119] It is suggested[117] that the discrepancy may arise from different definitions of self- and mutual-diffusion coefficients.

The two parts of the van Hove correlation function have been extracted from the Brownian dynamics simulation of Gaylor *et al.*[117] With increasing volume fraction, the van Hove self-correlation function deviates strongly from the free diffusion behaviour:

$$G_s(r, t) = (4\pi D_0 t)^{-3/2} e^{-(r^2/4D_0 t)}. \tag{73}$$

The distinct function $G_d(r, t)$ is illustrated in Figure 9 for a system of volume fraction 3×10^{-4}. The system relaxes towards the limiting value of $G_d(r, \infty) = 0$.

Evans and Watts[120] have simulated an assembly of particles with repulsive pair potentials under the influence of homogeneous shear using a non-equilibrium molecular dynamics method.[121] The shear viscosity is given by

$$\eta = (2\pi\rho^2/15) \int_0^\infty r^3 \{du(r)/dr\}\lambda(r)\,dr. \tag{74}$$

The derivation of equation (74) involves an amalgam of rheology and statistical mechanics. Simply stated, it is based on an integral equation of the same general form as equation (13), but involving a distribution function $g(r)$, which now depends on the strain-rate tensor, and with the stress tensor replacing

[118] J. L. Aguirre and T. J. Murphy, *J. Chem. Phys.*, 1973, **59**, 1833.
[119] J. L. Anderson and C. C. Reed, *J. Chem. Phys.*, 1976, **64**, 3240.
[120] D. J. Evans and R. O. Watts, *Chem. Phys.*, 1980, **48**, 321.
[121] D. J. Evans, *Mol. Phys.*, 1979, **37**, 1745.

pressure. The shear viscosity is the coefficient relating corresponding off-diagonal elements of the stress- and strain-rate tensors. The potential was taken to be of the form

$$u(r) = Ae^{-\alpha r}/r, \tag{75}$$

where A and α are constants. The function $\lambda(r)$ is the first-order correction to the equilibrium radial distribution function $g(r)$. By assuming that the application of strain at rate γ is equivalent to a fixed strain-rate $\tau\gamma$ for some relaxation time τ, $\lambda(r)$ can be written as[120]

$$\lambda(r) \approx -\tau r\{dg(r)/dr\}. \tag{76}$$

Using equations (74)—(76), a reduced shear viscosity,

$$\eta^* = \eta/k^2(mA)^{1/2}, \tag{77}$$

was extracted from the simulation for several values of reduced shear-rate γ^*. Table 1 shows that as γ^* increases the reduced viscosity decreases and the

Table 1 *Reduced shear viscosity η^* and reduced pressure p^* as a function of reduced shear-rate γ^* in non-equilibrium molecular dynamics simulation*[120]

$\gamma^*/10^{-2}$	$\eta^*/10^{-3}$	$p^*/10^{-3}$
0.31	3.09	3.024
0.62	2.67	3.026
1.24	2.30	3.028
2.44	2.28	3.035

reduced pressure p^* increases. Evans and Watts describe[120] the system as exhibiting shear dilatancy in conjunction with shear thinning. With respect to rheological measurements on real dispersions, two comments are relevant. First, the simulated effect occurs at high shear rate in the absence of any interparticle hydrodynamic forces. And, secondly, we should not expect the reduced *osmotic* pressure p^* to correlate with the *total* volume of the dispersion (particles + continuum solvent).

The viscosity of disordered, 'monodisperse' polystyrene latices increases as the order–disorder transition is approached.[122] This behaviour has been successfully interpreted in terms of the free-volume Cohen–Turnbull theory[123] of molecular transport in a liquid.

High-frequency Hookean shear moduli of ordered latices have been compared with predictions based on pairwise-additivity of DLVO-type potentials.[124] A simple static lattice model gives a simple expression for the modulus,[125]

$$G = k_p r^{-1}\{d^2u(r)/dr^2\}, \tag{78}$$

[122] K. Okano and S. Mitaku, *J. Phys.*, 1980, **41**, 585.
[123] M. H. Cohen and D. Turnbull, *J. Chem. Phys.*, 1959, **31**, 1164.
[124] R. Buscall, *Discuss. Faraday Soc.*, 1978, **65**, 133.
[125] J. W. Goodwin and A. M. Khidher, in 'Colloid Interface Science', Academic Press, New York, 1976, Vol. IV.

where k_p is a packing constant. Equations (2) ($\kappa a \ll 1$) and (78) provide a good fit of the experimental results over a range of volume fraction ($0.1 < \phi < 0.4$) and particle size ($25 < a < 100$ nm).

Bohlin has developed[126,127] a statistical theory, applicable to colloidal dispersions, in which macroscopic flow is the consequence of co-operative rearrangements of microscopic elements. The model is based on a lattice microstructure whose elements can exist in 'stressed' or 'relaxed' states. Stress relaxation at constant strain is co-operative owing to the presence of an energy of interaction between neighbouring elements in different states. In common with all lattice models, the predictions are sensitive to the (somewhat arbitrary) choice of co-ordination number.

Effect of Hydrodynamic Forces between Particles.—Consider a pair of spheres of radii a_1 and a_2 acted on by external forces $\boldsymbol{F}_1$ and $\boldsymbol{F}_2$, respectively. If inertial forces are negligible, and there is zero torque on each sphere, their velocities are given by[10]

$$\left.\begin{array}{l} \boldsymbol{v}_1 = b_{11}\boldsymbol{F}_1 + b_{12}\boldsymbol{F}_2, \\ \boldsymbol{v}_2 = b_{21}\boldsymbol{F}_1 + b_{22}\boldsymbol{F}_2, \end{array}\right\} \tag{79}$$

where

$$b_{ij} = [A_{ij}(r)(\boldsymbol{rr}/r^2) + B_{ij}(r)\{\boldsymbol{I} - (\boldsymbol{rr}/r^2)\}]/3\pi\eta(a_i + a_j). \quad (i,j = 1,2) \tag{80}$$

In equation (80), $r = |\boldsymbol{r}|$ is the distance between the particle centres and $\boldsymbol{I}$ is the unit tensor. The mobility tensor b_{ij} contains four independent coefficients, A_{11}, A_{12}, B_{11}, and B_{12}, each of which depends on r and a_2/a_1. The behaviour of these coefficients has been discussed by Batchelor[10,128] in relation to the estimation of first-order corrections to the sedimentation velocity and diffusion coefficient. The mean settling velocity is

$$\bar{\boldsymbol{U}} = (1 - K\phi)\boldsymbol{U}_0, \tag{81}$$

where $\boldsymbol{U}_0$ is the velocity from Stokes' Law. For a statistically random distribution where $g(r) \equiv 1$ for $r \geq 2a$ in a dispersion of identical hard spheres, the sedimentation coefficient K is 6.55.[128]

In a dilute monodisperse suspension the equilibrium pair distribution function is given by

$$g(r) = e^{-u(r)/kT}. \tag{82}$$

In a dispersion with repulsive interparticle forces, the effect of using $g(r)$ from equation (82) is to increase the sedimentation coefficient K.[129] This reduction in the settling velocity arises from an exclusion of certain pair configurations from

[126] L. Bohlin and J. Kubát, *Solid State Commun.*, 1976, **20**, 211.
[127] L. Bohlin, *J. Colloid Interface Sci.*, 1980, **74**, 423.
[128] G. K. Batchelor, *J. Fluid Mech.*, 1972, **52**, 245.
[129] E. Dickinson, *J. Colloid Interface Sci.*, 1980, **73**, 578.

the hydrodynamic averaging process. A full analysis of the effect of interparticle forces on $\bar{U}$, however, requires allowance for the change in $g(r)$ due to flow itself. On the other hand, we note that *sedimentation equilibrium* of colloidal particles is independent of any coupling between hydrodynamic and Brownian forces. Vrij[130] has obtained an expression for the concentration–distance profile for a polydisperse hard-sphere system in the Percus–Yevick approximation.

The mobility tensor b_{ij} refers to an isolated pair of particles. According to Batchelor,[10] 'the determination of the hydrodynamic functions occurring in the expressions for the mobility of one particle in the presence of two other particles would be a formidable task'. Since there is little evidence that hydrodynamic forces are to any extent additive, a rigorous treatment of the dynamics of concentrated dispersions is not yet in sight.

An intuitive approch to the problem is to assume an approximate form for the diffusion tensor D_{ij}, and then 'see what happens' in the computer simulation. Ermak and McCammon suggest[35] the Oseen tensor[131]

$$D_{ij} = (kT/8\pi\eta r_{ij})\{\mathbf{I} + (\mathbf{r}_{ij}\mathbf{r}_{ij}/r_{ij}^2)\}, \qquad (i \neq j) \tag{83}$$

or the Rotne–Prager tensor[132]

$$D_{ij} = (kT/8\pi\eta r_{ij})[\{\mathbf{I} + (\mathbf{r}_{ij}\mathbf{r}_{ij}/r_{ij}^2) + (2a^2/r_{ij}^2)\{(\mathbf{I}/3) - (\mathbf{r}_{ij}\mathbf{r}_{ij}/r_{ij}^2)\}]. \quad (i \neq j) \tag{84}$$

Both tensors have the convenient property that

$$\sum_i (\partial D_{ij}/\partial r_{ij}) \equiv 0, \tag{85}$$

enabling the gradient term to be dropped from equation (23). So far, only dimers and trimers have been studied.[35] The diffusion coefficients of a harmonic trimer, determined from the mean-square displacement of the centre of mass, was found to increase by up to 100% due to hydrodynamic interactions. Changes in trimer orientation were studied[35] through the zero-adjusted autocorrelation function of $\cos \theta(t)$,

$$A_\theta(t) = \langle \cos \theta(t) \cos \theta(0) \rangle - \langle \cos \theta(t) \rangle^2, \tag{86}$$

where θ is the angle formed by the three-particle centres. Hydrodynamic interactions slow the decay rate of $A_\theta(t)$. Equations (83) and (84) give similar results within the statistical uncertainties. Simulations of small groups of particles are of interest in their own right for comparison with observations of floc motion by quasi-elastic light-scattering[133] and direct microscopy.[134]

[130] A. Vrij, *J. Chem. Phys.*, 1980, **72**, 3735.
[131] H. Yamakawa, 'Modern Theory of Polymer Solutions', Harper and Row, New York, 1971.
[132] J. Rotne and S. Prager, *J. Chem. Phys.*, 1969, **50**, 4831.
[133] D. R. Bauer, *J. Phys. Chem.*, 1980, **84**, 1592.
[134] R. M. Cornell, J. W. Goodwin, and R. H. Ottewill, *J. Colloid Interface Sci.*, 1979, **71**, 254.

There are considerable logistical problems in incorporating hydrodynamic interactions into a computer simulation of a concentrated colloidal dispersion. In the close-touching régime ($r_{ij} \rightarrow a_i + a_j$) many otherwise satisfactory approximations break down.[135] Dickinson has suggested[129] that, far from being a complicating factor, the presence of repulsive interparticle forces will help to validate the simple hydrodynamic approximations by excluding close-touching configurations. Finally, there is the problem of the boundary conditions: how to allow for the long-range hydrodynamic interactions without introducing anomalous periodic correlations?

[135] J. Happel and H. Brenner, 'Low Reynolds Number Hydrodynamics', Prentice-Hall, New York, 1965.

5
Micellar Structure and Catalysis

BY J. M. BROWN AND R. L. ELLIOTT

1 Introduction

The review follows a similar pattern to that presented in Colloid Science, Volume 3, Chapter 6 with coverage of what the authors judge to be the most important developments between mid-1976 and mid-1980. In order to keep reportage within manageable proportions, some important areas, including reverse micelles in non-polar media and catalysis by polymers or in association complexes have been deliberately omitted. The former subject has been reviewed recently.[1]

Interest in the structure and chemistry of micelles has been sustained and the period under consideration has been marked by many novel developments in excited-state reactions, and by a much clearer understanding of the detailed mechanism of catalysis of ionic reactions. Ben-Naim has written a book summarizing the current picture of hydrophobic interactions[2] and a volume concerned with micelle and enzyme catalysis includes an excellent summary of physicochemical techniques.[3] Two sets of conference reports edited by Mittal[4,5] include a substantial body of short reviews of all aspects of micelle chemistry.

2 Micellar Structure and Solubilization

Computational Aspects.—Theoretical studies on the micellar aggregation phenomenon follow the basic premises of Tanford, and of Israelachvili and Ninham described some time ago.[6a] Their basic hypothesis is that the shapes

[1] A Kitahara, *Adv. Colloid Interface Sci.*, 1980, **12,** 109.
[2] 'The Hydrophobic Bond', ed. A. Ben-Naim, John Wiley and Sons, New York, 1978.
[3] B. Lindmann and H. Wennerstrom, *Top. Curr. Chem.*, 1980, **87**.
[4] 'Micellization, Solubilization and Microemulsions', Vol. 1 and 2, ed. K. L. Mittal, Plenum Press, 1977. Proceedings of the International Symposium on Micellization, Solubilization and Microemulsions held at the 7th Northeastern Regional Meeting of the Am. Chem. Soc., Albany, N.Y. 1976.
[5] 'Solution Chemistry of Surfactants', Vol. 1 and 2, ed. K. L. Mittal, Plenum Press, 1979. Proceedings of the 52nd Colloid and Surface Science Symposium, Division of Colloid and Surface Chemistry, Am. Chem. Soc. held in Knoxville, Tennessee, 1978.
[6] (a) J. N. Israelachvili and B. W. Ninham, *J. Colloid Interface Sci.*, 1977, **58,** 17; (b) J. N. Israelachvili, D. J. Mitchell, and B. W. Ninham, *J. Chem. Soc., Faraday Trans. 1*, 1976, **72,** 1525.

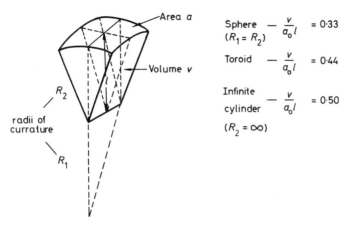

Sphere $-\dfrac{v}{a_0 l}$ = 0·33
$(R_1 = R_2)$

Toroid $-\dfrac{v}{a_0 l}$ = 0·44

Infinite
cylinder $-\dfrac{v}{a_0 l}$ = 0·50
$(R_2 = \infty)$

Figure 1 *Packing model used by Ninham and Israelachvili.[6a] A monomer of volume v and chain length l packs at surface of area a described by radii of curvature R_1 and R_2*

and structures of micelles may be explained by considering constraints introduced by packing forces. Using a very simple geometrical model for headgroup and alkyl-chain self and solvent interactions, they predict spherical and rod-like micelles as well as vesicle formation (Figure 1). The latter authors have reviewed their contribution[6b] and a discussion of more general aspects of hydrophobic solutes in water has been presented by Ninham and co-workers in the most recent volume of Franks' treatise.[7] Interactions in aqueous solution, with particular reference to small peptides appear in an account by Sheraga.[8]

A very detailed discussion of free-energy changes in micellization, which includes an explicit dissection into translational, rotational, electrostatic, and dipolar contributions, has appeared.[9] The statistical mechanical treatment permits a reasonably accurate estimation of critical micelle concentrations and aggregation numbers for simple surfactants and the boundary between micelle and vesicle formation for two-chain surfactants, the latter being preferred at high ionic strength. Electrostatic repulsion is greater for vesicles than for micelles at both inner and outer surfaces. The size distribution in micelles has been computed using a 'shell model' modification of the law of mass action.[10] Calculations on the size, shape, and degree of hydration of polyoxyethyleneglycol monoethers[11] suggest a non-spherical aggregate, possibly an oblate ellipsoid, in accord with the disc-like structure suggested by sedimentation equilibria.[12] Tanford's oblate ellipsoidal model for non-spherical micelles has

[7] D. Y. C. Chan, D. J. Mitchell, B. W. Ninham, and B. A. Pailthorpe, in 'Water, A Comprehensive Treatise', ed. F. Franks, Plenum Press, NY, Vol. 6, p. 239.
[8] H. Scheraga, *Acc. Chem. Res.*, 1979, **12**, 7.
[9] R. Nagarajan and E. Ruckenstein, *J. Colloid Interface Sci.*, 1979, **71**, 580.
[10] G. Kegeles, *J. Phys. Chem.*, 1979, **83**, 1728.
[11] R. J. Robson and E. A. Dennis, *J. Phys. Chem.*, 1977, **81**, 1075.
[12] C. Tanford, Y. Nozaki, and M. F. Rohde, *J. Phys. Chem.*, 1977, **81**, 1555.

been criticised[13] on the grounds that a capped cylinder has greater surface area for a given volume, but this view is unlikely to gain general currency.

Critical Micelle Concentrations.—Lipophilic crown-ether derivatives promise many interesting applications. Surface-tension measurements on the annelides (1a), (1b), and (2) demonstrate that micelles are formed in dilute aqueous solution;[14] no protonation of (1a) occurs in neutral solution above the c.m.c. In the latter two cases the metal ion is completely bound to the macrocycle and evaluation of the tensiometric data demonstrates that the surface area of annelides increases on metal-ion complexation. The optically active surfactant (3) exhibits increased circular dichroism above its c.m.c.,[15] and the c.m.c. of racemic (4) is higher than that of its pure enantiomers.[16]

Merocyanines (5) undergo a red shift of their electronic absorption maximum on micellization, the degree of which increases with the hydrophobicity of the alkyl group.[17] This implies solubilization in a water-poor region in both cationic and anionic co-micelles. Pyridinium iodide (6) has a charge-transfer absorption band in its u.v. spectrum, but only above the c.m.c., and the frequency suggests a polarity in the environment of the head-group similar to that of ethanol.[18]

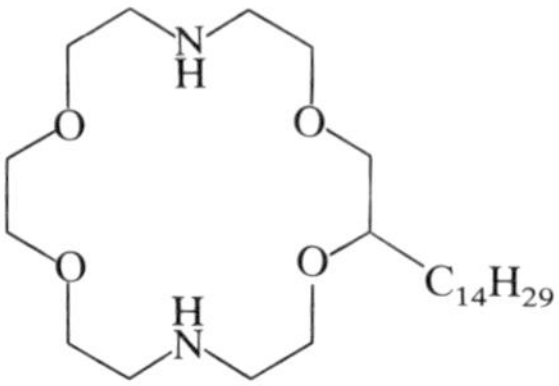

(1) a; uncomplexed
    b; Ag$^+$ complex

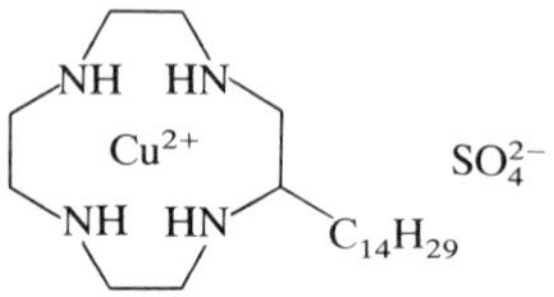

(2)

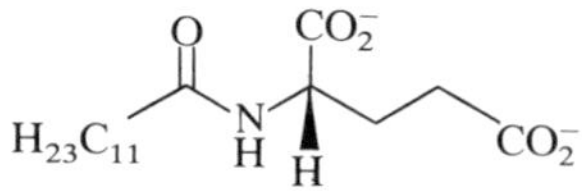

(3)

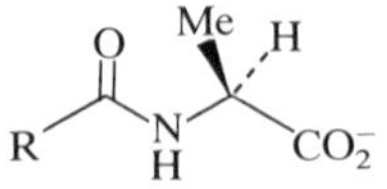

(4)

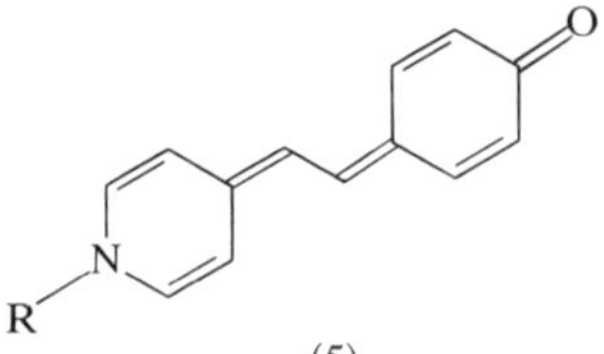

(5)

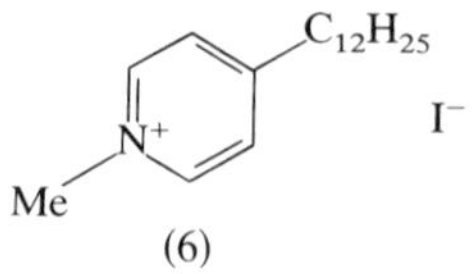

(6)

[13] J. E. Leibner and J. Jacobus, *J. Phys. Chem.*, 1977, **81**, 131.
[14] Y. Moroi, E. Pramauro, M. Grätzel, E. Pelizzetti, and P. Tundo, *J. Colloid Interface Sci.*, 1979, **69**, 341.
[15] K. Sakamoto and M. Hatano, *Bull. Chem. Soc. Jpn.*, 1980, **53**, 339.
[16] S. Miyagishi and M. Nishida, *J. Colloid Interface Sci.*, 1978, **65**, 380.
[17] M. J. Minch and S. S. Shah, *J. Org. Chem.*, 1979, **44**, 3253.
[18] E. J. R. Sadhotter and J. B. F. N. Engberts, *J. Phys. Chem.*, 1979, **83**, 1854.

Hamman has criticised over simplifications in equations relating the c.m.c, micelle size, and the free-energy change on micellization and derived the correct formula, together with permissible approximations.[19] Ruckenstein and Nagarajan[20] point out that a critical micelle concentration to some extent depends on the method by which it is measured, since this may depend either on the total number of molecules (osmotic pressure, freezing point depression) or on weight averages (light scattering). They define a concentration C_{crit} below which the size distribution of aggregates decreases monotonically with their size but above which it exhibits two extrema, corresponding to micelles and monomers or small oligomers. This concentration is normally the lower bound of phenomenological c.m.c. values.

Positron annihilation, namely the decay of 'positronium atoms' e^+e^- formed by positron substrate collisions, provides a sensitive probe of critical micelle concentration[21] since the intensity of the slow decay changes abruptly at the c.m.c. Although many workers have been critical of perturbing probes in c.m.c. determination, and particularly of the dye solubilization method, the fluorescence of 1-ammonio-8-naphthalene sulphonate give correct c.m.c. values for anionic surfactants.[22]

A detailed study reveals how the c.m.c. of alkylphosphates varies with chain length and anionic state.[23]

Properties of Micelles.—The partial molar volume $\bar{V}_L$, of surfactants in aqueous solution increases above the critical micelle concentration. A careful study[24] involving a combination of equilibrium ultracentrifugation and isopiestic distillation leads to accurate data for anionic surfactants. For example $\bar{V}_L = 248$ for micellated sodium lauryl sulphate, and $\bar{V}_L = 236$ for the monomer. The results obtained may be elaborated to provide estimates of the micelle charge (0.356 for sodium lauryl sulphate in 0.1 M-NaCl) and aggregation number ($N = 91$ in 0.1 M-HCl). As might be expected, the c.m.c. increases monotonically with pressure up to 5.5×10^3 bars, and recent results obtained by monitoring naphthalene solubilization appear to negate earlier conclusions from conductimetric experiments.[25] Volume changes on micellization have been determined for simple surfactants by densimetry,[26] as have the volume changes on protonation of n-alkylcarboxylate and n-alkylamine micelles.[27] Densimetric and conductimetric measurements have been applied to solutions of decyltrimethylammonium alkylcarboxylate suggesting greatly increased water ordering consequent on hydrophobic interaction,[28] and to the structure

[19] S. D. Hamman, *Aust. J. Chem.*, 1978, **31**, 919.
[20] R. Nagarajan and E. Ruckenstein, *J. Colloid Interface Sci.*, 1977, **60**, 221; E. Ruckenstein and R. Nagarajan, *ibid.*, 1978.
[21] Y.-C. Jean and H. J. Ache, *J. Am. Chem. Soc.*, 1978, **100**, 984.
[22] K. S. Birdi, H. N. Singh, and S. U. Dalsager, *J. Phys. Chem.*, 1979, **83**, 2733; K. S. Birdi, T. Krag, and J. Klausen, *J. Colloid Interface Sci.*, 1977, **62**, 562; *cf.* K. S. Birdi, *J. Phys. Chem.*, 1977, **81**, 935; H. C. Chiang and A. Luxton, *ibid.*, 1977, **81**, 936.
[23] J. Arakawa and B. A. Pethica, *J. Colloid Interface Sci.*, 1980, **75**, 441.
[24] D. A. Doughty, *J. Phys. Chem.*, 1979, **83**, 2621.
[25] S. Rodriguez and H. Offen, *J. Phys. Chem.*, 1977, **81**, 47.
[26] K. M. Kale and R. Zana, *J. Colloid Interface Sci.*, 1977, **61**, 312.
[27] R. Zana, *J. Phys. Chem.*, 1977, **81**, 1817.
[28] D. Oakenfull and D. E. Fenwick, *Aust. J. Chem.*, 1977, **30**, 741.

(7)

of bile-salt aggregates,[29] the latter not without controversy.[30] The ion-pair association constants of alkylammonium carboxylates (7) have been measured conductimetrically and show strong chain–steroid hydrophobic interactions, with $\delta\Delta G^0 = -2.77$ kJ mol^{-1} per additional CH_2 group in the side-chain.[31]

The ion-association coefficient of ionic micelles is important in current interpretations of micellar catalysis. It has been determined for sodium alkyl-sulphates using a cation selective electrode, and increases with added sodium chloride concomitant with a substantial increase in micelle molecular weight.[32] There is a simple equation (1) for its estimation:

$$1000S_2 = (1-\beta)^2 \eta^{2/3}(1000S_1 - \Lambda_x) + (1-\beta)\Lambda_x, \qquad (1)$$

where Λ_x is the molar conductivity of the counterion; η is the micelle size; S_1 the slope of specific conductivity *versus* concentration above the c.m.c., and S_2 the slope of specific conductivity *versus* concentration below the c.m.c.

It has been demonstrated[33] that β increases and then decreases with increasing sodium chloride concentration for sodium nonanoate solutions although the reverse is true of the viscosity. The limiting equivalent conductance Λ_2 permits estimation of the kinetic charge of micelles.[34]

Scattering Technique.—Annelides (8) *cf.* ref 14 have been studied by a variety of physicochemical techniques[35] and light-scattering experiments demonstrate an average aggregation number of about 100 for the parent but 370 for its lithium salt. Large and small aggregates co-exist for the barium nitrate complex with $\bar{N}_A$ around 6.7×10^4 for the former. Micelles of HCl-protonated dodecyldimethylamine are spherical at low sodium chloride concentration but rod-shaped at high salt concentration.[36] A detailed study of quasielastic light scattering by sodium lauryl sulphate as a function of temperature and sodium chloride concentration concludes, by comparison of the hydrodynamic radius

[29] L. R. Fisher and D. G. Oakenfull, *J. Phys. Chem.*, 1977, **83**, 1838.
[30] R. Zana, *J. Phys. Chem.*, 1978, **82**, 2441; D. G. Oakenfull and L. R. Fisher, *ibid.*, 1978, **82**, 2442.
[31] D. Oakenfull and D. E. Fenwick, *Aust. J. Chem.*, 1977, **30**, 355.
[32] T. S. Brun, H. Høilvand, and E. Vikingstad, *J. Colloid Interface Sci.*, 1978, **63**, 590.
[33] E. Vikingstad, A. Skauge, and H. Høiland, *Acta. Chem. Scand.*, Ser. A 1979, **33**, 235.
[34] D. Stiger, *J. Phys. Chem.*, 1979, **83**, 1670.
[35] J. de Moigne and J. Simon, *J. Phys. Chem.*, 1980, **84**, 177.
[36] S. Ikeda, S. Ozeki, and M. A. Tsunoda, *J. Colloid Interface Sci.*, 1980, **73**, 127.

$$C_{12}H_{25}$$

$$(CH_2)_2\overset{+}{N}Me_3\bar{N}O_3$$
(8)

$\bar{R}_h$ and the mean radius of gyration $\bar{R}_g$, that its micelles are rod-shaped in salt solution.[37] The aggregate size increases further at high salt and high surfactant concentrations due to microgel formation, causing anomalous light scattering.[38] Small-angle X-ray scattering by non-ionic Triton X-100 solutions has been employed to derive a radius of 38.2 Å at 293 K and a mean molecular weight of 10^5, the micelle size increasing with temperature.[39] The aggregation number of sodium lauryl sulphate exhibits a minimum at 100 mPa., where it is claimed the c.m.c. is maximum, observed by laser light scattering.[40] These results may complicate the interpretation of naphthalene solubilization data (ref. 25).

A critical assessment[41] of methods available for the measurement of aggregation number concludes that isopiestic distillation coupled with sedimentation equilibria provides accurate data, but that the results of quasi-elastic light scattering are difficult to interpret and that luminescence quenching data are not soundly evaluated.

Solubilization.—Mukherjee[42] has provided a general review of micellar aggregation and solubilization and discussed[43] the microenvironments of benzene and naphthalene in Triton X-100 micelles in terms of a two-state model involving competition of core and surface solubilization. The partition coefficients of *p*-nitrophenylacetate and methyl *p*-toluenesulphonate between water and cetyltrimethylammonium bromide micelle have been measured using dialysis by an artificial kidney.[44] Since monomeric surfactant may be transported across the regenerated cellulose membrane, such experiments require careful attention to detail and caution in their interpretation. In the present case, K values obtained accord well with the literature.

Mixed solutions of homologous surfactants show ideal mixing behaviour, suggesting random co-micellization but mixtures of anionic and non-ionic

[37] C. Y. Young, P. J. Missel, N. A. Mazer, G. B. Benedek, and M. C. Carey, *J. Phys. Chem.*, 1978, **82**, 1375.
[38] S. Hayashi and S. Ikeda, *J. Phys. Chem.*, 1980, **84**, 744.
[39] H. H. Paradies, *J. Phys. Chem.*, 1980, **84**, 599.
[40] M. Niskikido, M. Shinozaki, G. Sugihara, M. Tanaka, and S. Kaneshina, *J. Colloid Interface Sci.*, 1980, **75**, 271.
[41] J. P. Kratohvil, *J. Colloid Interface Sci.*, 1980, **75**, 271.
[42] P. Mukherjee, *Ber. Bunsenges. Phys. Chem.*, 1978, **82**, 931.
[43] P. Mukherjee and J. R. Cardinal, *J. Phys. Chem.*, 1978, **82**, 1621.
[44] G. B. van de Langkruis and J. B. F. N. Engberts, *Tetrahedron Lett.*, 1979, 3991.

surfactants do not.[45] In the latter case a non-ionic micelle is formed at low concentrations, which is capable of incorporating the anionic surfactant below the c.m.c. Fluorocarbon–hydrocarbon surfactant mixtures show some mutual solubility; the phase diagram is rather complex and exhibits a critical solution temperature.[46]

The free energy of transfer of alkylphenols and alkylphenoxides to cationic micelles has been measured.[47] For the neutral parent compound $\delta\Delta G_0 = -5.680$ kcal mol^{-1} and for its anion $\delta\Delta G_0 = -6.870$ kcal mol^{-1}. Incremental methyl substitution changes this value by -0.300 kcal mol^{-1} in both cases and it was concluded that the environment of solubilized molecule and anion are very similar. The size of micelles can be quite subtly dependent on counterion however, since sodium p-toluenesulphonate and sodium p-toluate increase the viscosity of cetyltrimethylammonium bromide solutions sharply but sodium benzenesulphonate and disodium phenylphosphate do not.[48] The anthra-quinone sulphonate (9) is absorbed very slowly $k_1 = 0.37$ s^{-1} by cationic micelles, the process being accelerated by added KBr.[49]

(9)

3 Magnetic Resonance

Electron Spin Resonance.—Nitroxide radicals of varying structure have been employed in studies of micelle structure and mobility. The basic spectrum is a triplet due to ^{14}N-electron coupling, which may show hyperfine coupling to β-C–H in appropriate cases. On micelle formation or incorporation of the probe the spectrum normally broadens because of reduction in the rotational correlation time and shows enhanced broadening and change in positions of the high-field line. Cationic micelles incorporate the probe (10) with an association constant of 3×10^5 M^{-1} and at low surfactant concentrations there is a broadened spectrum superimposed on the normal micelle–monomer-averaged spectrum, due to multiple incorporation of (10) in a single micelle.[50] This disappears at higher surfactant concentrations. Similar studies have been

[45] C. P. Kurzendorfer, M. J. Schwuger, and H. Lange, *Ber. Bunsenges. Phys. Chem.*, 1978, **82**, 962.
[46] N. Funasaki and S. Hada, *Chem. Lett.*, 1979, 717; *J. Chem. Soc., Chem. Commun.*, 1980, 253.
[47] C. A. Bunton and L. Sepulveda, *J. Phys. Chem.*, 1979, **83**, 680.
[48] J. W. Larsen and L. B. Tepley, *J. Org. Chem.*, 1976, **41**, 2969.
[49] Y. Miyashita and S. Hayano, *Chem. Lett*, 1978, 987.
[50] C. L. Kwan, S. Atik, and L. A. Singer, *J. Am. Chem. Soc.*, 1978, **100**, 4783.

(10) $n = 9, 11, 13,$ or 15   (11)   (12)

(13) a; $x = 1, y = 14$
     b; $x = 12, y = 3$ (14)

carried out for (11) in cationic micelles and (12) in anionic micelles.[51] Spin-probes (13a) and (13b) have been widely used in the determination of phospholipid structure, and in sodium laurylsulphate micelles the ratio of their mobilities is similar to that in water. In basic media (where the micelle-bound carboxylic acid is fully ionized) the rotational correlation time of the former shortens, whereas that of the latter lengthens appreciably.[52] A set of spin probes of widely varying hydrophobicity has been employed to derive rotational correlation times as a function of structure.[53,54] To the unpractised eye, the data all appear similar, but hydrophilic probes have a higher enthalpy of rotation in micelles than in water, whereas hydrophobic probes show the reverse behaviour. The probe (14) resides in the polyoxyethylene group of a alkylpolyoxyethylene glycol monoether, on the basis of e.s.r. studies. The interpretation of temperature-dependent broadening in the e.s.r. spectrum of (10) is now suggested to be caused by differential effectiveness of micelles and monomers in Heisenberg spin-exchange.[55] The factors affecting the e.s.r. spectra of nitroxide probes partitioned between water and lipid phases have been analysed in detail.[56] The mobility of nitroxide probes in bile-salt micelles is much lower than in sodium laurylsulphate micelles.[57]

[51] M. Aizawa, T. Komatsu, and T. Nagagawa, *Bull. Chem. Soc. Jpn*, 1980, **53,** 957.
[52] H. Yoshioka, *J. Colloid Interface Sci..*, 1978, **66,** 352; H. Yoshioka, *Chem. Lett.*, 1977, 1477.
[53] H. Yoshioka, *J. Am. Chem, Soc.*, 1979, **101,** 28.
[54] H. Yoshioka, *J. Colloid Interface Sci.*, 1978, **63,** 378.
[55] K. K. Fox, *J. Chem. Soc., Faraday Trans. 1*, 1978, **74,** 220.
[56] C. F. Polnaszek, S. Schreier, K. W. Butler, and I. C. P. Smith, *J. Am. Chem. Soc.*, 1978, **100,** 8223.
[57] L. R. Fisher and D. Oakenfull, *Aust. J. Chem.*, 1979, **32,** 31.

A combination of n.m.r. paramagnetic broadening and e.s.r. has been utilized to locate N-methylphenazenium radical cations in sodium laurylsulphate micelles.[58] The conclusion that 'the site of attachment of NMPH$^{\cdot+}$ to the micelle is in the hydrocarbon region a few angstroms below the head-groups' is at variance with expectation and may reflect a superficial analysis of the data, which neglects intrinsic differences in T_2 values at different sites.

Proton Magnetic Resonance.—CIDNP has been observed in micellar solution during the photolysis of substituted bibenzyls (ref. 301). The relative enhancement of intensities in starting material and product suggest that cleavage of bond α in (15) occurs preferentially and that the lifetime of the bibenzyl radical pair in the micelle is insufficiently long to allow significant relaxation of nuclear-spin polarization before coupling.[59] O-Alkyltyrosines (16) change conformational preference on micelle formation with the rotamer shown most dominant, based on analysis of the ABC spin-coupling system observed in 100 MHz spectra.[60] In mixed micelles with Triton X-100, but not otherwise, the heterotopic α-CH$_2$ groups of phospholipids show distinct chemical shifts at 220 MHz.[61]

(15)   (16)

Carbon Magnetic Resonance.—The characteristic shape of a plot of ^{13}C chemical shifts *versus* the inverse of surfactant concentration permits calculation of the aggregation number for small micelles, with sodium octanoate giving values of 5—10 and nonyltrimethylammonium bromide a value of 33.[62] Incorporation of sodium sorbate into dodecylammonium chloride micelles affords marked changes with C(2), and to a lesser extent C(4) and C(5) deshielded but other carbon nuclei, particularly C(5) quite strongly shielded, implying a change in *sym-cis* and *sym-trans* rotamer populations.[63] In an attempt to assess the degree of water penetration into micelles, Menger and co-workers have examined the ^{13}C chemical shifts of carbonyl groups incorporated into surfactant chains and of simple carbonyl compounds solubilized in cationic micelles. Octanal, 1-naphthaldehyde, and 7-oxotridecane all show deshielding of *ca.* 5 p.p.m. on transfer from water to micelle, whereas (17) is little affected.[64] This was taken to demonstrate water penetration into the

[58] C. A. Evans and J. R. Bolton, *J. Am. Chem. Soc.*, 1977, **99**, 4502.
[59] R. S. Hutton, H. D. Roth, B. Kraeutler, W. R. Cherry, and N. J. Turro, *J. Am. Chem. Soc.*, 1979, **101**, 2227.
[60] F. M. Menger and J. M. Jerkunica, *Tetrahedron Lett*, 1977, 4569.
[61] M. F. Roberts and E. A. Dennis, *J. Am. Chem. Soc.*, 1977, **99**, 6142.
[62] B. O. Persson, T. Drakenberg, and B. Lindman, *J. Phys. Chem.*, 1979, **83**, 3011.
[63] D. L. Reger and M. M. Habib, *J. Phys. Chem.*, 1980, **84**, 77.
[64] F. M. Menger, J. M. Jerkunica, and J. C. Johnston, *J. Am. Chem. Soc.*, 1978, **100**, 4676.

(17)

(18)

micellar core but the interpetation has been criticised because of the lack of independent evidence on the distribution of the carbonyl group in the micelle. Wennerstrom and Lindman point out that the stability range of amphipathic phases is not consistent with marked water penetration,[65] although Menger's ideas on micelle structure are consistent with assembled CPK models of micelles.[66]

The annelide (18) forms micelles of a tractable size for ^{13}C n.m.r. spectroscopy only in 33% aqueous tetrabutylammonium hydroxide. Under these conditions Ba^{2+} exchange with free ligand is considerably faster than for the analogous compound in which methyl replaces octadecyl.[67]

There have been several attempts to rationalize ^{13}C spin–lattice relaxation times in micelles. It was observed that all the T_1 values in sodium octanoate above the c.m.c. are dependent on the spectrometer frequency so that slow molecular motions (overall tumbling of the micelle) must contribute to relaxation.[68] A detailed model based on a density-matrix formulation was constructed, which gave a good fit with experiment, and suggested that the micellar interior is very fluid but that carbons close to the polar group are relaxed predominantly by overall motion of the micelle. A detailed treatment of relaxation in a six-carbon chain incorporating correlated *gauche–trans* isomerizations has been applied to micelles and vesicles.[69] Some aspects of the work and particularly the use of $T_2^* \equiv T_2$ suggest that further experimentation is required. Spin–lattice relaxation times of ω-phenylalkanoates[70] (19) show marked decrease within the aromatic ring on micelle formation and the R-parameter [defined as $T_1(ortho)T_1^{-1}(para)$] increases concomitantly. This suggests that off-axis motions of the phenyl ring are suppressed on micellization. The $^{13}C(F_2)$ spectrum of (20) is the A portion of an AX_2 multiplet but with the centre line considerably sharper than the wings.[71] A theoretical explanation is provided in terms of the cross-correlation spectral density, and

[65] H. Wennerstrom and B. Lindman, *J. Phys. Chem.*, 1979, **83**, 2931.
[66] F. M. Menger, *Acc. Chem. Res.*, 1979, **12**, 111.
[67] J. Le Moigne, P. Gramain, and J. Simon, *J. Colloid Interface Sci.*, 1977, **60**, 565.
[68] H. Wennerstrom, B. Lindman, O. Söderman, T. Drakenberg, and J. B. Rosenholm, *J. Am. Chem. Soc.*, 1979, **101**, 6860.
[69] R. E. London and J. Avitabile, *J. Am. Chem. Soc.*, 1977, **99**, 7765.
[70] F. M. Menger and J. M. Jerkunica, *J. Am. Chem. Soc.*, 1978, **100**, 688.
[71] J. H. Prestegard and D. M. Grant, *J. Am. Chem. Soc.*, 1978, **100**, 4664.

$(CH_2)_9CO_2H$ $CH_3(CH_2)_7CF_2(CH_2)_4CO_2^-\overset{+}{Na}$

(19)   (20)

with some assumptions this leads to a correlation time for molecular reorientation.

Magnetic Resonance of Other Nuclei.—The relaxation times of ^{35}Cl and ^{81}Br as counterions to cationic micelles suggest that the former is spherical at high concentrations but the latter rod-like, results supported by quasi-elastic light scattering.[72] In mixtures of the two surfactants quadrupole spin–spin relaxation of ^{35}Cl and ^{81}Br demonstrate that two types of aggregate co-exist with high counterion specificity. The first conclusion is entirely in accord with ^{14}N n.m.r. studies of cetyltrimethylammonium chloride and bromide; for the former $T_1 \equiv T_2$ at all concentrations, whereas T_2 decreases rapidly for the latter above 0.25 M.[73] The chemical shifts and quadrupole relaxation rates of $^{23}Na^+$ and chemical shifts of $^{133}Cs^+$ as counterions to anionic micelles give useful information on ion-hydration changes with variation of head-group.[74]

4 Dynamics of Micelle Formation and Breakdown

The previous report (Colloid Science, Vol. 3, Ch. 6) was written at a time when the theoretical description introduced by Aniansson and Wall appeared to explain the results of fast kinetic experiments provided by several groups of workers. In the past four years this work has been consolidated and extended in a most impressive way. Two reviews,[75] which summarize the present state of the field, and which contain much useful data, have appeared.

Theory.—The basis of current ideas is the observation that micellar solutions derived from carefully purified surfactants show two relaxation times in perturbation spectroscopy, which vary with the chain length and head-group and the ratios between which vary from 10 to 10^5. The fast relaxation time relates to exchange of surfactant monomers between micelles, and the slow process to breakdown of a micelle. Consider a hypothetical aggregate-size-distribution curve represented by Figure 2 in system (a), which suffers a sudden perturbation, as in a pressure-jump experiment. This will result in a rapid re-equilibration by monomer association and dissociation to the distribution (b), in which the average size is at equilibrium but the total number of micelles is not. There then follows a much slower process involving a change in the total number of micelles, resulting in the final state (c). Aniansson has re-stated the basic theory,[76] which gives a simple algebraic representation of the fast

[72] J. Ulmius, B. Lindman, G. Lindblom, and T. Drakenberg, *J. Colloid Interface Sci.*, 1978, **65**, 88.
[73] U. Henriksson, L. Odberg, J. C. Eriksson, and L. Westman, *J. Phys. Chem.*, 1977, **81**, 76.
[74] H. Gustavsson and B. Lindman, *J. Am. Chem. Soc.*, 1978, **100**, 4647.
[75] M. Kahlweit and M. Teubner, *Adv. Colloid Interface Sci.*, 1980, **13**, 1; H. Hoffmann, *Prog. Colloid Polym. Sci.*, 1978, **65**, 140.
[76] E. A. G. Aniansson, *Ber. Bunsenges. Phys. Chem.*, 1978, **82**, 981.

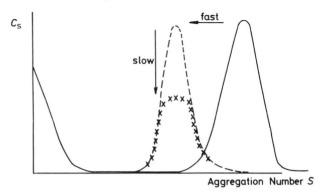

Figure 2 *Response of a micellar solution to a rapid perturbation. The fast process corresponding to τ_1 (———→ ----) represents a change in micelle size and the slow process corresponding to τ_2 (----→ × × × × ×), a change in micelle distribution*

relaxation process τ_1 as:

$$\frac{1}{\tau_1} = \frac{k^-}{\sigma^2} + \frac{k^-}{\eta} \frac{A_{\text{tot}} - \bar{A}_1}{\bar{A}_1}, \tag{2}$$

where A_{tot} is the total surfactant concentration, $\bar{A}_1$ the monomer concentration, σ the half-width of the distribution curve, η the aggregation number, and k^- the rate of exit of a surfactant ion-pair from the micelle. Thus k^- (and hence k^+, the rate of association of monomer with micelle) and σ^2 may be obtained from a plot of $1/\tau_1$ *versus* A_{tot}, well above the c.m.c.

Similar theoretical considerations give the slow relaxation time as:

$$\frac{1}{\tau_2} = \frac{\eta^2}{R\bar{A}_1} \left(1 + \frac{\sigma^2}{\eta} \frac{A_{\text{tot}} - \bar{A}_1}{\bar{A}_1}\right)^{-1}, \tag{3}$$

where $R = \Sigma \left(1/\bar{k}_s - \bar{A}_s\right)$ with $\bar{A}_s$ the concentration of aggregates at the distribution minimum and k_s^- the rate constant for dissociation of a monomer from these aggregates. Plotting $1/\tau_2$ *versus* A_{tot} above the c.m.c. therefore gives information on the minimum in the size-distribution curve, not readily accessible by other means. A consolidating paper[77] represents the current state of the theory of micellar dynamics, and explains the amplitudes of relaxation processes and their time development. Consideration of exit rates k^- to partial dissociation of monomers[78] leads to a model with pronounced head-group and chain protrusion from the micellar surface with re-equilibrations on a 20 ps time-scale.

Experiment.—Several research groups have contributed exprimental relaxation data, and interpreted it in terms of the Aniansson and Wall model. A selection of results are recorded in Table 1. Hoffmann and co-workers conducted a

[77] S. N. Wall and E. A. G. Aniansson, *J. Phys. Chem.*, 1980, **84,** 727.
[78] E. A. G. Aniansson, *J. Phys. Chem.*, 1978, **82,** 2805.

Table 1 *Fast and slow relaxation times for cationic surfactants*[83]

Surfactant	conc/M	c.m.c.	τ_1/s	τ_2/s	$t/°C$
$C_{12}H_{25}\overset{+}{N}H_3Cl^-$	1.6×10^{-2}	1.46×10^{-2}	—	1.1×10^{-2}	25
$C_{12}H_{25}\overset{+}{N}H_3Br^-$	1.6×10^{-2}	1.2×10^{-2}	—	4.9×10^{-1}	25
$C_{12}H_{25}\overset{+}{N}H_3I^-$	1.6×10^{-2}	1.1×10^{-2}	2×10^{-5}	—	20
$C_{14}H_{29}\overset{+}{N}H_3Cl^-$	5.0×10^{-3}	2.8×10^{-3}	10^{-4}	4	25
$C_{14}H_{29}\overset{+}{P}yCl^-$	8.4×10^{-3}	4×10^{-3}	2.9×10^{-6}	—	25
$C_{16}H_{33}\overset{+}{N}H_3Cl^-$	1.5×10^{-3}	1.03×10^{-3}	4.1×10^{-4}	—	40

detailed early study of alkylpyridinium halides[79] and related work is recorded by Inoue and co-workers.[80,81] The latter authors generally observe only one relaxation time corresponding to τ_2 for dodecylpyridinium halides in T-jump or P-jump experiments although τ_1 is slow enough to be measured (η being large) for the iodide in the presence of excess KI. They propose a slight modification of the Aniansson and Wall model to take account of counterion dissociation from the micelle. Hoffmann[82] records relaxation results for dodecyl- and tetradecyl-pyridinium perfluoroalkylcarboxylates. Although the aggregation numbers derived or used seem erroneous there is a clear trend in k^-, which decreases with increasing chain length of either surfactant or its counterion. The comparison between alkylammonium and alkylpyridinium halides is very interesting.[83] For a given chain length and counterion τ_1 is about an order of magnitude longer and τ_2 up to three orders of magnitude longer in the former case. This means that the residence times of individual molecules in the micelle is longer for ammonium salts and also that the formation of micelles is much slower (sharper minimum in the size-distribution curve) or the distribution width is much greater. Micelle-structure and -distribution curves must be strongly temperature dependent since the slow relaxation time τ_2 for tetradecylammonium chloride, 5×10^3 M, is 4 seconds at 25 °C ($\tau_1 = 100$ μsec) but 15 msec at 45 °C.

The theoretical treatment of the model of Aniansson and Wall is not directly applicable to the time constants obtained in ultrasonic experiments, but this gap has been bridged.[84] The amplitude is predicted to be zero at the c.m.c. then increases with concentration to a broad maximum and thence slowly decreases, as observed for sodium dodecyl sulphate.[85] An analysis of amplitudes in P-jump kinetics[86] implies the possibility of a third relaxation process due to a change in electrolyte properties. This counterion binding equilibrium may have been observed in ultrasonic studies of sodium decyl sulphate.[87] Attempts by the former authors to modify the Aniansson and Wall theory

[79] H. Hoffmann, R. Nagel, G. Platz, and W. Ulbricht, *Colloid Polym. Sci.*, 1976, **254**, 812.
[80] T. Inoue, R. Tashiro, Y. Shibuya, and R. Shimozawa, *J. Phys. Chem.*, 1978, **82**, 2037.
[81] T. Inoue, R. Tashiro, Y. Shibuya, and R. Shimozawa, *J. Colloid Interface Sci.*, 1980, **73**, 105.
[82] H. Hoffmann, B. Tagesson, and W. Ulbricht, *Ber. Bunsenges. Phys. Chem.*, 1979, **83**, 148.
[83] H. Hoffmann, R. Lang, D. Pavlovic, and W. Ulbricht, *Croat. Chem. Acta*, 1979, **52**, 87.
[84] M. Teubner, S. Diekmann, and M. Kahlweit, *Ber. Bunsenges. Phys. Chem.*, 1978, **82**, 1278.
[86] S. K. Chan. U. Hermann, W. Ostner, and M. Kahlweit, *Ber. Bunsenges. Phys. Chem.*, 1977, **81**, 60; 396; *ibid.*, 1978, **82**, 380.
[87] S. Diekmann, *Ber. Bunsenges. Phys. Chem.*, 1979, **83**, 528.

led to a brief scuffle.[88] A further study of alkylsulphates by P-jump[89] and an extensive kinetic investigation of divalent cation alkylsulphates[90] are reported. The fast relaxation time is very similar for cobalt or nickel laurylsulphates to its sodium salt, but $1/\tau_2$ increases rapidly with surfactant concentration suggesting a strongly bound counter-ion. Theory has also been correlated with stopped-flow relaxation studies.[91,92]

The kinetics of binding acridine orange and other dyes to sodium laurylsulphate micelles has been studied.[93] The lower the charge on the dye the faster it is absorbed into the micelle, and the rate constant of reaction increases with pH, being $430 \, s^{-1}$ at pH 12.4, the micellar pK'_a of the dye.

Ultrasonic relaxations of tetradecyl- and hexadecyl-trimethylammonium bromide in the presence of n-pentanol are reported.[94] Only one relaxation time is observed, assumed to be dissociation of n-pentanol from the micelle. The rate constant at zero concentration is $1.8 \times 10^7 \, s^{-1}$, two orders of magnitude smaller than the value obtained for an ion of comparable size.

5 Reactions in Micelles

Micellar Effects on Chemical Equilibria.—A few studies have been made of acid–base equilibria in micelles. Hydronium ion activity in anionic micelles has been measured conductimetrically using hydrophilic indicators, it being found[95] that a plot of $m^s_{H^+}$ *versus* $[H^+]+[Na^+]$ is linear with a slope of 0.82. The quantity $m^s_{H^+}$ is defined as the number of micellized hydrogen ions per surfactant head group, namely $m^s_{H^+} = [H^+]_{tot} - [H^+]_w/\{[D]_{tot} - c.m.c.\}$, where $[D]_{tot}$ is the total catalyst concentration. The use of fluorescent indicators (21a) and (21b) in anionic, neutral, and cationic surfactants[96] permitted the evaluation of the electrical potential at the micellar surface as a function of added electrolytes. Indicator pK_a values for mixed micelles[97] and pK_a values of weak

(21) a; X = OH
b; X = NMe$_2$

(22)

[88] E. A. G. Aniansson and S. N. Wall, *Ber. Bunsenges. Phys. Chem.*, 1977, **81**, 1293; S. K. Chan and M. Kahlweit, *ibid.*, 1977, **81**, 1294.
[89] T. Inoue, Y. Shibuya, and R. Shimozawa, *J. Colloid Interface Sci.*, 1978, **65**, 370.
[90] W. Bäumuller, H. Hoffmann, W. Ulbricht, C. Tondre, and R. Zana, *J. Colloid Interface Sci.*, 1978, **64**, 430.
[91] K. Baumgart, G. Klar, and R. Strey, *Ber. Bunsenges. Phys. Chem.*, 1979, **83**, 1222.
[92] C. Tondre and R. Zana, *J. Colloid Interface Sci.*, 1978, **66**, 544.
[93] A. D. James, B. H. Robinson, and N. C. White, *J. Colloid Interface Sci.*, 1977, **59**, 328.
[94] S. Yiv and R. Zana, *J. Colloid Interface Sci.*, 1978, **65**, 286.
[95] C. A. Bunton, K. Ohmenzetter, and L. Sepulveda, *J. Phys. Chem.*, 1977, **81**, 2000.
[96] M. S. Fernandez and P. Fromherz, *J. Phys. Chem.*, 1977, **81**, 1755.
[97] N. Funasaki, *J. Colloid Interface Sci.*, 1977, **62**, 189.

acids and bases in non-ionic surfactant solution are reported.[98] The degree of hydration of (22) is unaffected by micellation, but is lower in aqueous acetonitrile. This suggests that the activity of water at the micellar surface is little different from bulk solution.[99]

Reactions Catalysed by Simple Cationic Micelles.—There have been important recent developments in our understanding of nucleophilic catalysis in micelles. The main credit in this must go to Romsted, whose PhD thesis[100a] contains the analysis now generally accepted to be the best available description, and which is more readily accessible in a review article.[100b] It is a pseudo-phase model, which is superior to earlier approaches because it makes specific allowance for ion-dissociation and ion-exchange at the micelle surface. The derived equation for a second-order reaction is:

$$k_2 = \frac{k_m \beta P_a (c_t - \text{c.m.c.})}{[K_a(c_t - \text{c.m.c.}) + 1][I_t + KX_t]} + \frac{k_w}{K_a(c_t - \text{c.m.c.}) + 1}, \quad (4)$$

where P_a is the partition constant of substrate between aqueous and micellar phases, and I_t, X_t the total concentrations of reactive and non-reactive counterions, respectively. The overall second-order rate constant is k_2 and the rate-constants for reaction in the aqueous and micellar phases k_w and k_m, respectively. c_t is the total concentration of surfactant and β the fractional counterion binding to the Stern layer.

The closeness of fit may be gauged from the experimental and theoretical rate vs. concentration curves for hydrolysis of p-nitrophenyl carboxylates catalysed by quaternary ammonium surfactant micelles (Figure 3). The shape of the curve is satisfactorily explained for unimolecular, bimolecular, and termolecular reactions. An alternative speculative model[101] is effectively superseded by this work. Romsted's approach has been extended in a set of model calculations relating to salt and buffer effects on ion-binding,[102] acid-dissociation equilibria, reactions of weakly basic nucleophiles, first-order reactions of ionic substrates in micelles, and second-order reactions of ionic nucleophiles with neutral substrates. In like manner the reaction between hydroxide ion and p-nitrophenyl acetate has been quantitatively analysed for unbuffered cetyltrimethylammonium bromide solutions. This permits the derivation of a micellar rate constant $k_m^{OH^-} = 6.5 \text{ M}^{-1}\text{s}^{-1}$ compared to the bulk rate constant of $k_{aq}^{OH^-} = 10.9 \text{ M}^{-1}\text{s}^{-1}$. The equilibrium constant for ion-exchange at the surface of the micelle $K_m(\text{Br}^- \rightleftharpoons \text{OH}^-)$ was estimated as 40 ± 10.[103] The hydrolysis of p-nitrophenyl butyrate similarly gives $K_m(\text{Br}^- \rightleftharpoons \text{OH}^-) = 10$, and buffer effects can be accounted for.[104,105] A complementary approach gives a

[98] E. Azaz and M. Donbrow, J. Phys. Chem., 1977, **81**, 1635.
[99] J. Perez de Albrizzio and E. H. Cordes, J. Colloid Interface Sci., 1979, **68**, 292.
[100] (a) L. R. Romsted, Ph.D. Thesis, Indiana University, 1975; (b) ref. 4, p. 509.
[101] D. Piszkiewicz, J. Am. Chem. Soc., 1976, **98**, 3053; ibid., **99**, 1550, 7695.
[102] F. H. Quina and H. Chiamovich, J. Phys. Chem., 1979, **83**, 1844.
[103] M. Almgren and R. Rydholm, J. Phys. Chem., 1979, **83**, 360.
[104] N. Funasaki, J. Colloid Interface Sci., 1978, **64**, 461.
[105] N. Funasaki, J. Phys. Chem., 1979, **83**, 237.

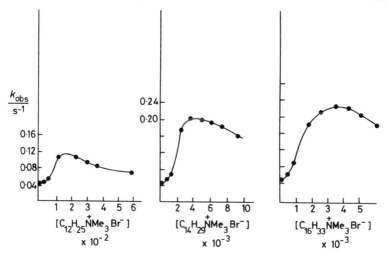

Figure 3 *Experimental (●) and predicted (———) [from equation (4)] rate constants for micelle-catalysed hydrolysis of p-nitrophenylhexanoate at 25 °C, pH 10.07*

(23)

compatible result.[106] If *N*-methyl-4-cyanopyridinium iodide (23) is hydrolysed in the presence of quaternary ammonium surfactants then reaction takes place exclusively in the intermicellar phase. Knowing that the substrate is not incorporated in the micelle, then counterion binding equilibria may be evaluated from the extent of inhibition of hydrolysis as a function of surfactant concentration. The reaction of *p*-nitrophenyl acetate with *p*-substituted thiophenoxide anions is catalysed by cetyltrimethylammonium bromide with rate accelerations up to 100-fold. The phase distribution of the anion may be evaluated spectroscopically and on this basis the acceleration is quantitatively accounted for without recourse to enhanced second-order rate constants in the micellar phase. The critical micelle concentration is 5×10^{-4} but inspection of rate concentration plots suggests appreciable catalysis at lower surfactant concentrations.[107] Hydrophobic nucleophiles may form small aggregates or even reactive ion-pairs with cationic surfactants at very low concentrations,[108]

[106] H. Chaimovich, J. B. S. Bonilha, M. J. Politi, and F. H. Quina, *J. Phys. Chem.*, 1979, **83**, 1851; F. H. Quina, M. J. Politi, I. M. Cuccovia, S. M. Martins-Franchetti, and H. Chaimovich, *J. Phys. Chem.*, 1980, **84**, 361.

[107] I. M. Cuccovia, E. H. Scroter, P. M. Monteiro, and H. Chaimovich, *J. Org. Chem.*, 1978, **43**, 2248.

[108] J. M. Brown and J. R. Darwent, *J. Chem. Soc., Chem. Commun.*, 1979, 169.

the differences being evident in a comparison between hydroperoxy and peroxycumyl anions as nucleophiles in the micelle-catalysed transacylation of p-nitrophenyl phenylacetate. In the former case reaction is appreciably catalysed at very low surfactant concentrations and the maximum rate-accelerations are, respectively, 90 and 9×10^3. Such hydrophobic ion-pair catalysis is very evident in dilute aqueous solutions of trioctylmethylammonium chloride,[109] which accelerate the reactions of hydrophobic anionic nucleophiles in a rather selective manner. The relative second-order rate accelerations in transacylation of p-nitrophenyl acetate by (24), (25), and (26) are displayed alongside the formulae, data being obtained at 7×10^{-5} M trioctylmethylammonium chloride, and compared to 1×10^{-3} M cetyltrimethylammonium bromide solutions for which values are shown in square brackets. Some specificity in the reaction of hydroxamate[110] and histidine[111] nucleophiles with anionic or neutral reactive esters is observed in the presence of cationic surfactants above their c.m.c. The oximate (27) reacts with phosphate and carboxylate esters, affording substantial rate enhancements in the presence of cetyltrimethylammonium bromide, as might be expected for a highly hydrophobic nucleophile.[112] The reaction rate constants observed under burst conditions (excess of substrate) are double those observed under pseudo-first-order conditions where the oxime is in excess.

(24) 1.2×10^3 $[1.05 \times 10^2]$

(25) 6.7×10^3 $[1.9 \times 10^3]$

(26) 18 $[8.7 \times 10^2]$

(27)

(24)—(26); k_2 values in M^{-1} s^{-1}, pH 9

Few studies of amide hydrolysis have been carried out under conditions of micellar catalysis. Gani and Viout[113] find that there is weak acceletation of the hydrolysis of activated amides (carrying electron-withdrawing alkyl groups or anion-stabilizing N-substitutents) by cetyltrimethylammonium bromide in the pH range 10—14 but the pH dependence can be quite different from the

[109] Y. Okahata, R. Ando, and T. Kunitake, *J. Am. Chem. Soc.*, 1977, **99**, 3067.
[110] R. Ueoka, K. Shimamoto, and K. Ohkubo, *J. Org. Chem.*, 1978, **43**, 1815; R. Ueoka and K. Ohkubo, *Tetrahedron Lett.*, 1978, 4131.
[111] R. Ueoka, H. Matsuura, S. Nakahata, and K. Ohkubo, *Bull. Chem. Soc. Jpn.*, 1980, **53**, 347.
[112] C. A. Bunton and Y. Ihara, *J. Org. Chem.*, 1977, **42**, 2865.
[113] V. Gani and P. Viout, *Tetrahedron*, 1978, **34**, 1337.

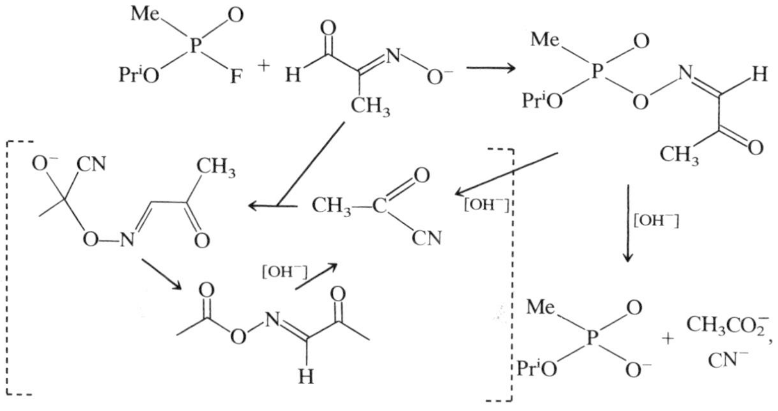

Scheme 1

uncatalysed reaction. This implies that the interplay between nucleophilicity of OH$^-$ and water activity at the micelle surface can be quite complex.

Bunton and co-workers[114] have carried out a quantitative analysis of the reaction of phenoxide and *p*-cresoxide ions with *p*-nitrophenyldiphenyl-phosphate in cetyltrimethylammonium chloride and find that the rate acceleration of up to 4×10^3 is quantitatively accounted for by the pseudophase model of micellar catalysis over most of the concentration range. The major deficiency is below the c.m.c. where pre-micellar aggregation may distort the interpretation. Hydrolysis of the same phosphate ester is catalysed by dodecyl phosphate in cationic micelles but not in homomicelles of the anionic nucleophile.[115] Alanyl adenylate polymerizes to high molecular weight polyalanines in cetyltrimethylammonium bicarbonate micelles in water, or with more efficiency in inverse micelles.[116] The hydrolysis and aminolysis of acetyl phosphate catalysed by ammonium ion–amine micelles or by alkyltetramine metal-ion complexes has been studied, and appears to exhibit a rate maximum at the pH where co-operative catalysis involving both protonated and neutral amine molecules is facilitated.[117] The reaction of fluoroisopropoxymethylphosphane oxide with oximato-2-propanone is catalysed by cationic micelles and gives rise to cyanide ion but by a different mechanisms from the reaction in water. The latter is stoicheiometric (Scheme 1) and involves nucleophilic attack of hydroxide ion at the acyl group, whereas elimination to acyl cyanide predominates in the micelle (Scheme 1, bracketed pathway). This is catalytically regenerated by reaction with oximate ion in preference to hydrolysis, and up to 30 moles of cyanide ion may arise from one mole of phosphofluoridate.[118]

[114] C. A. Bunton, G. Cerichelli, Y. Ihara, and L. Sepulveda, *J. Am. Chem. Soc.*, 1979, **101**, 2429.
[115] C. A. Bunton, S. Diaz, G. M. van Fleteren, and C. Paik, *J. Org. Chem.*, 1978, **43**, 258.
[116] D. W. Armstrong, R. Seguin, C. J. McNeal, R. D. MacFarlane, and J. H. Fendler, *J. Am. Chem. Soc.*, 1978, **100**, 4605.
[117] L. L. Melhado and C. D. Gutsche, *J. Am. Chem. Soc.*, 1978, **100**, 1850.
[118] J. Epstein, P. Cannon, jun., R. Swidler, and A. Baraze, *J. Org. Chem.*, 1977, **42**, 759.

Other Catalysed Reactions in Cationic Micelles.—Hydroxide-ion promoted eliminations may be catalysed by cationic micelles. Tagaki and co-workers found that the E_{1cb} mechanism of hydrolysis was selectively facilitated for *p*-nitrophenyl esters derived from phenylacetic acids carrying a carbanion-stabilizing *para*-substituent,[119] or from phenylthioacetic acid. This was indicated by a plateau in the pH-rate profile, corresponding to the micellar pK_a of (28) and by a linear free-energy relationship in micelle-catalysed hydrolysis of the aromatic esters between $\log k_{obs}$ and σ_{R^-} of their *para*-substituents. Presumably carbanion formation is favoured over esterolysis in the micelles because of the additional electrostatic stabilization favouring pre-equilibrium proton transfer. Competition between S_N1, S_N2, and E_2 reactions of (29) in both cationic and anionic micelles has been examined.[120] In cetyltrimethylammonium bromide micelles, elimination is accelerated to a greater extent than bimolecular substitution, and unimolecular substitution is inhibited; all reaction paths are inhibited in sodium laurylsulphate micelles. Rather more substantial catalysis is observed in hydroxyethyldimethyloctadecylammonium chloride micelles of the E_2 reactions of *para*-substituted phenethyl bromides. The isotope effect in the *p*-nitro-case increases from 7.66 (water) to 9.07 (micelles).[121] Methyl transfer from the *bis*-cationic micelle of (30) to

(28)

(29)   (30)

thiophenoxide ion is an order of magnitude faster than is observed for short-chain non-micellar analogues.[122] The non-explosive E_2 degradation of RDX is catalysed by cationic micelles,[123] as is the elimination of trichloromethide from 2,2,2-trichloro,1,1-bis-*p*-chlorophenylethanol.[124] The base-catalysed conversion of (31) into *o*-cyanophenol is weakly catalysed by

[119] W. Tagaki, S. Kobayashi, K. Kurihara, A. Kurashima, Y. Yoshida, and Y. Yano, *J. Chem. Soc., Chem Commun.*, 1976, 843.
[120] C. Lapinte and P. Viout, *Tetrahedron*, 1979, **35**, 1931.
[121] Y. Yano, Y. Yoshida, A. Kurashima, Y. Tamura, and W. Tagaki, *J. Chem. Soc., Perkin Trans. 2*, 1979, 1128.
[122] R. A. Moss and W. J. Sanders, *Tetrahedron Lett.*, 1979, 1669.
[123] M. Groce and Y. Okamoto, *J. Org. Chem.*, 1979, **44**, 2100.
[124] F. Nome, E. W. Schwingel, and L. G. Ionescu, *J. Org. Chem.*, 1980, **45**, 705.

(31)　　　　　　　　　　　　　(32)

(33)

cationic micelles, and occurs *via* intramolecular formation of benzisoxazole.[125] This is a rare example of a micelle-catalysed unimolecular reaction. Hoffmann elimination from (32) is relatively insensitive to the presence of micelles.[126] Cyclization of (33) to form a labile acylpyridinium salt (which subsequently hydrolyses) is fastest in the case of $n = 7$, although it is claimed that the rate is only about three-fold slower for the case of ($n = 33$) where the cyclized product has a 111-membered ring.[127] It was additionally claimed that no evidence for micelles could be obtained at the concentration employed (5×10^{-5} M) and that 'the contribution from intermolecular reactions can be ignored'!!!

Oxidation and Reduction.—Shinkai and co-workers have developed micellar models for flavin-dependent oxidases, which operate through the trapping of reactive carbanions. Thus 4-chlorobenzoylformic acid reacts with cyanide ion in aqueous solution to give 4,4'-dichlorobenzoin (72%) together with a small quantity of 4-chlorobenzoic acid (2.4%) formed by aerial oxidation of the carbanion. In the presence of a cationic surfactant micelle, the yield of oxidation product is greatly increased and there is a further rate acceleration in the presence of the hydrophobic flavin (34) since the mechanism shown (Scheme 2) is now strongly favoured.[128] This type of reaction path also accounts for the oxidation of aromatic aldehydes in the same reaction system, and molecular oxygen may replace the flavin but with lower efficiency. The oxidation of nitroethane to acetaldehyde is catalysed by a similar combination of hydrophobic flavin and cationic surfactant and does not occur in their absence.[129]

[125] G. Meyer, *J. Org. Chem.*, 1979, **44**, 3983.
[126] M. J. Minch, S.-S. Chen, and R. Peters, *J. Org. Chem.*, 1978, **43**, 31.
[127] M. Sisido, E. Yohikawa, Y. Imanishi, and T. Higashimura, *Bull. Chem. Soc. Jpn.*, 1978, **51**, 1464.
[128] S. Shinkai, T. Ide, and O. Manabe, *Chem. Lett.*, 1978, 583; S. Shinkai, T. Yamashita, and O. Manabe, *J. Chem. Soc., Chem. Commun.*, 1979, 301; S. Shinkai, T. Yamashita, Y. Kusano, T. Ide, and O. Manabe, *J. Am. Chem. Soc.*, 1980, **102**, 2335.
[129] S. Shinkai, T. Yamashita, and F. Yoneda, *J. Chem. Soc., Chem. Commun.*, 1976, 986.

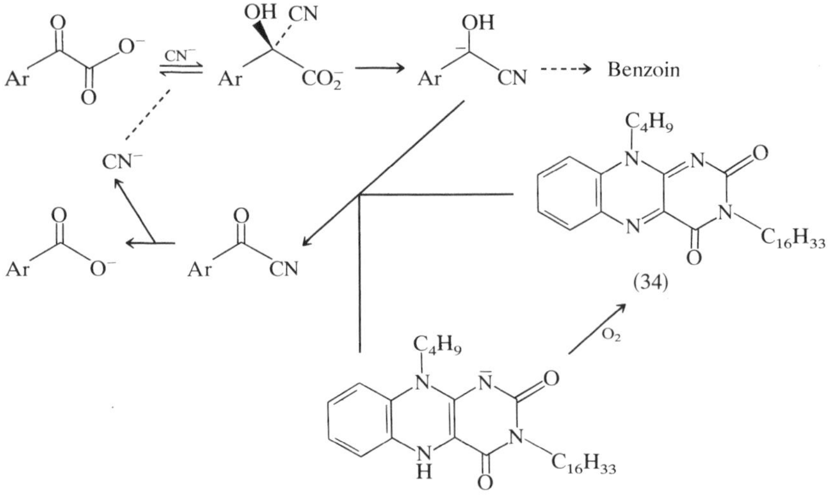

Scheme 2

Oxidation of terminal olefins to methyl ketones by aqueous palladium chloride and oxygen is very slow, but addition of micellar sodium lauryl sulphate increases the rate of formation of 2-octanone from 1-octene twenty-fold at 50 °C. There is weaker catalysis by the non-ionic surfactant Brij-35 and inhibition by cationic surfactants.[130] Oxidation of diosphenol (35) in basic aqueous tetradecyltrimethylammonium chloride is faster and more effective than in water, giving a higher yield of (36).[131] Two attempts at effecting the enantioselective reduction of aromatic ketones, one in micelles of *R*-dodecyl-dimethyl-α-phenylethylammonium bromide[132] and the other in sodium cholate micelles,[133] both give optical yields of less than 2%. Rather more success was obtained in the catalysed oxidation of L-Dopa, 3,4-dihydroxyphenyl-alanine. In the presence of the Cu[II] complex of *N*-lauroyl-L-histidine in cetyl-trimethylammonium bromide micelles reaction was 1.42 (pH 6.90, 30 °C) to

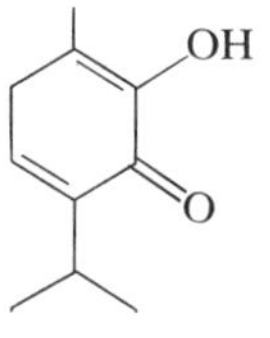

(35) 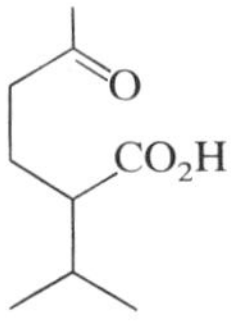   (36) 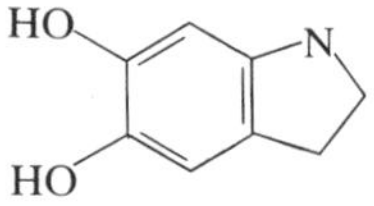   (37)

[130] C. Lapinte and H. Riviere, *Tetrahedron Lett.*, 1977, 3817.
[131] M. Utaka, S. Matushita, H. Yamasaki, and A. Takeda, *Tetrahedron Lett.*, 1980, **21**, 1063.
[132] S. I. Goldberg, N. Baba, R. L. Green, R. Pandian, J. Stowers, and R. B. Dunlap, *J. Am. Chem. Soc.*, 1978, **100**, 6768.
[133] T. Sugimoto, Y. Matsumura, T. Imanishi, S. Tanimoto, and M. Okano, *Tetrahedron Lett.*, 1978, 3431.

2.50 (pH 6.90, 30 °C) times faster than for the corresponding experiment with *N*-lauroyl-D-histidine. The product (37) is derived from an intermediate *o*-quinone, and reaction under micellar conditions is strongly catalysed.[134] Cationic micelles enhance the reactivity of L-ascorbic acid in its reduction of methylene blue, with a rate maximum around pH 8. The corresponding reduction of the dye by L-cysteine is also strongly catalysed, but the oxidation of the leuco-product back to methylene blue with oxygen is not greatly affected by micelles.[135]

Reactions Catalysed by Anionic Micelles.—The success of quantitative analysis of reactions catalysed by cationic micelles has encouraged a more detailed analysis of reactions catalysed by anionic micelles. The hydrolysis of the methyl and ethyl acetals of *p*-nitrobenzaldehyde[136] in the presence of anionic detergents may be simulated with reasonable accuracy by:

$$k_{obs} = \frac{k_w[H_3O_w^+] + k_m m_{H_3O^+} K([D] - c.m.c.)}{1 + K_s([D] - c.m.c.)}, \qquad (5)$$

where [D] is the total catalyst concentration, so that it is necessary to know the extent of hydrogen-ion binding to the cationic micelle. This was estimated from conductivity data, and the corrected second-order rate constants for micelle-based reaction are considerably lower than reaction rates in water although substantial catalysis is observed. Bunton and co-workers have additionally carried out studies of the benzidine rearrangement where specific acid catalysis may be unimolecular or bimolecular in hydronium ion concentration. The term in $[H_3O^+]^2$ for rearrangement of 1,2-diphenyl, 1,2-di-*o*-tolyl, and 1,2-di-*o*-anisylhydrazines is very strongly enhanced by sodium lauryl sulphate micelles, but the $[H_3O^+]$-dependent term is rather less spectacularly affected.[137] The general shape of the rate–surfactant concentration profile appears to be in accord with the theories of Romsted, and a later paper[138] demonstrates a quantitative analysis of the reaction of 1,2-diphenylhydrazine in aqueous sodium lauryl sulphate in which the agreement between theory and experiment is generally within 10% over a hundred-fold range of rate-constants. The treatment provides a third-order rate constant, which is eighty-fold lower than in water (despite the $>10^3$ acceleration) and fails to predict reaction rates at very low concentrations of surfactant. In a further contribution from the same laboratory,[139] hydride transfer from 1-benzyldihydronicotinamide (38) to triarylmethyl cations (39), and reaction of the latter with borohydride and hydroxide ions, was investigated. The hydride-transfer reaction is weakly catalysed by cationic micelles and more strongly catalysed by anionic micelles although quantitative analysis demonstrates that the rate constant is similar in the two environments and the difference is due to more

[134] K. Yamada, H. Shosenji, Y. Otsubo, and S. Ono, *Tetrahedron Lett.*, 1980, **21**, 2649.
[135] M. Senō, K. Kousaka, and H. Kise, *Bull. Chem. Soc. Jpn.*, 1979, **52**, 2970.
[136] C. A. Bunton, L. S. Romsted, and G. Savelli, *J. Am. Chem. Soc.*, 1979, **101**, 1253.
[137] C. A. Bunton and R. J. Rubin, *J. Am. Chem. Soc.*, 1976, **98**, 4236.
[138] C. A. Bunton, L. S. Romsted, and H. J. Smith, *J. Org. Chem.*, 1978, **43**, 4299.
[139] C. A. Bunton, N. Carrasco, S. K. Huang, C. H. Paik, and L. S. Romsted, *J. Am. Chem. Soc.*, 1978, **100**, 5420.

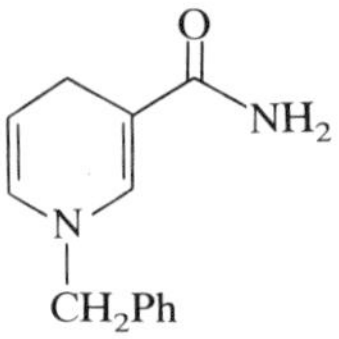

(38)

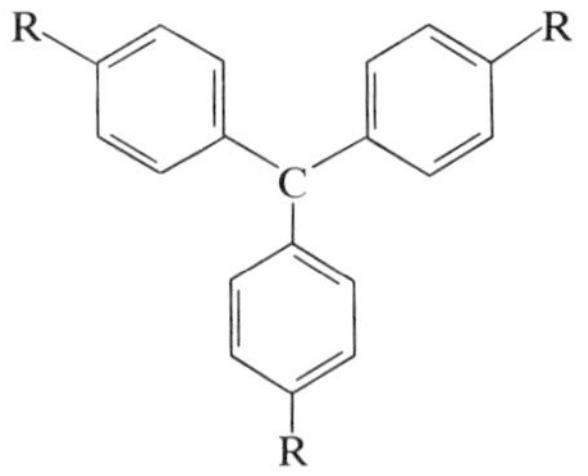

(39) R = H, OMe, or NMe$_2$

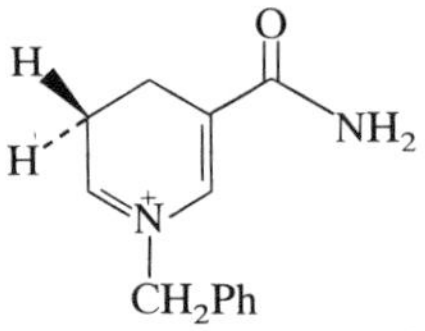

(40)

effective binding of (39) to the anionic micelle. Acid-catalysed hydration of (38) is weakly catalysed by anionic micelles and inhibited by cationic micelles,[140] and involves a rate-determining carbon protonation to give (40), which is attacked by water in a subsequent rapid step. In order to simulate the observed rate-constants it is necessary to postulate that unproductive protonation at the alternative double-bond site occurs competitively. The hydrolysis of decylphosphoric acid involves elimination of metaphosphate ion and is weakly catalysed above the c.m.c.[141] There are several competing effects because the non-micellar reaction is catalysed both by H$_3$O$^+$ and by Cl$^-$. Hydrolysis of hydroxamic acids (41) is catalysed by micelles of sodium lauryl sulphate in acidic media, and rate constants correlate well with Hansch π-parameters for the lipophilic substituent R in (41).[142]

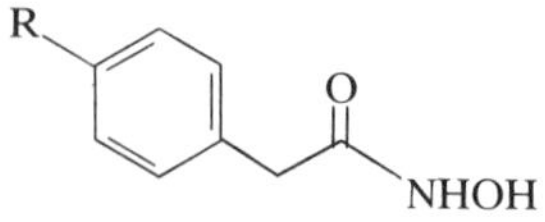

(41) 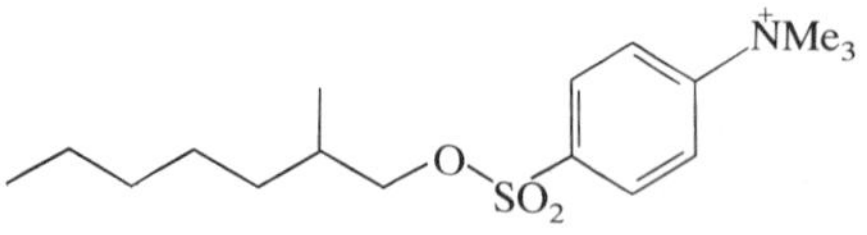   (42)

The bromination of phenyl n-pentyl ether is more *para*-selective in anionic micelles than it is in water.[143] This contrasts with the lower *para*-selectivity of nitration of bromobenzene in the *cationic* micelles formed by dissolving lauric acid in 95% H$_2$SO$_4$.[144] It is not clear whether these effects are due to substrate orientation or to micelle-induced changes in the selectivity parameter for electrophilic aromatic substitution. The rates of solvolysis of alkyl *p*-trimethyl-ammoniumbenzenesulphonate trifluoromethanesulphonates (42) are strongly inhibited by anionic micelles of sodium lauryl sulphate or sodium dodecanoate. In water, homomicelles of (42) or sodium dodecanoate micelles, undergo inversion of stereochemistry, but in sodium lauryl sulphate 22% retention of

[140] C. A. Bunton, F. Rivera, and L. Sepulveda, *J. Org. Chem.*, 1978, **43**, 1166.
[141] C. A. Bunton, S. Diaz, L. S. Romsted, and O. Valenzuela, *J. Org. Chem.*, 1976, **41**, 3037.
[142] D. C. Berndt and L. E. Sendelbach, *J. Org. Chem.*, 1977, **42**, 3305.
[143] D. A. Jaeger and R. E. Robertson, *J. Org. Chem.*, 1977, **42**, 3298.
[144] F. M. Menger and J. M. Jerkunica, *J. Am. Chem. Soc.*, 1979, **101**, 1896.

$$\text{Me}$$

(43)

configuration (56% net inversion) is observed. This and other results are consistent with competitive nucleophilic attack on substrate by the sulphate group leading to the product of double inversion.[145] It has been established by differential pulse voltammetry that the oxidation of 10-methylphenothiazine (43) to the cation radical is reduced in the presence of sodium lauryl sulphate below the c.m.c. and remains constant above that concentration. This may reflect preferential surfactant–radical cation pairing at the electrode surface.[146]

Micellar Effects on Inorganic Reactions.—Electron transfer between ferric ion and phenothiazine is inhibited by cationic micelles and accelerated by up to 10^4-fold in anionic micelles of sodium lauryl sulphate.[147] In both cases the rate–surfactant concentration profile can be simulated accurately. Anionic micelles only cause a small effect on the reactivity of ruthenium(III) tris(bipyridyl)$^{3+}$ with molybdenum(IV) octacyanide^{4-} but accelerate the reaction between ferrous ion and tris(tetramethylphenanthroline)iron(III)$^{3+}$. In the latter case a plot of surfactant concentration is linear with the reciprocal of the observed rate constant.[148] Fast outer-sphere electron-transfer reactions may decrease in rate constant by up to four orders of magnitude when one of the reactants is solubilized in an anionic micelle. When this partner is neutral the inhibition is reduced somewhat by added salt, but when it is cationic the effect may be attenuated by competitive binding of Na^+ or H_3O^+ and exclusion of reactant from the micelle.[149] Oxidation of diethyl sulphide is catalysed by micelles of sodium lauryl sulphate containing carboxylate-ions by the mechanism shown. (Scheme 3). The rate advantage is quantitatively accounted for by the entropy term, and hexanoate is forty-fold more effective than acetate.[150] Electron-transfer between the anionic *trans*-1,2-diaminocyclohexane N,N,N',N'-tetra-acetate complexes of manganese(III) and cobalt(II) is catalysed by CTAB micelles leading to maximum rate enhancements of 160-fold. The kinetics may be quantitatively analysed in terms of the Berezin–Romsted model.[151] Hydrogenation of conjugated dienes by tripotassium pentacyanocobalt hydride may be carried out more effectively in micelles of Brij 35 or Triton-X-100 [both non-ionic poly(oxyethylenephenyl ethers)], since the catalyst is more stable, although rate and product distributions are only marginally affected.[152]

[145] C. N. Sukenik and R. G. Bergman, *J. Am. Chem. Soc.*, 1976, **98**, 6613.
[146] G. L. McIntire and H. N. Blount, *J. Am. Chem. Soc.*, 1979, **101**, 7720.
[147] E. Pelizzetti and E. Pramauro, *Ber. Bunsenges, Phys. Chem.*, 1979, **83**, 996.
[148] E. Pelizzetti and E. Pramauro, *Inorg. Chem.*, 1980, **19**, 1407; *ibid.*, 1979, **18**, 882.
[149] H. Bruhn and J. Holzwarth, *Ber. Bunsenges. Phys. Chem.*, 1978, **82**, 1006.
[150] P. R. Young and K. C. Hou, *J. Org. Chem.*, 1979, **44**, 947.
[151] A. A. Bhalekar and J. B. F. N. Engberts, *J. Am. Chem. Soc.*, 1978, **100**, 5914.
[152] D. L. Reger and M. M. Habib, *J. Mol. Catal.*, 1978, **4**, 315; *Adv. Chem. Ser.*, 1978, **173**, 43.

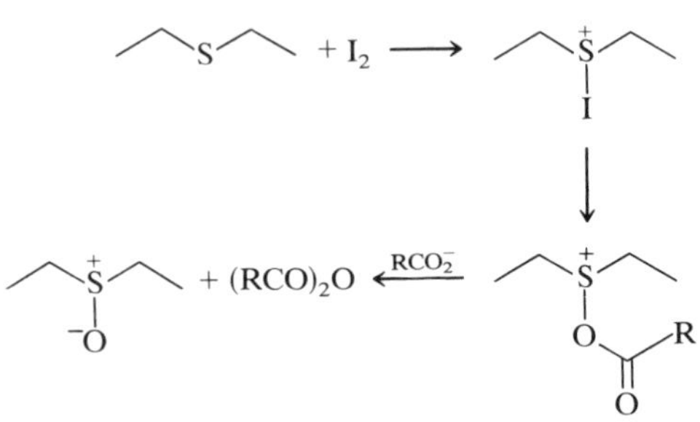

Scheme 3

Ligation of Vitamin B12a (aquocobalamin) by hydrophobic cysteineamides is inhibited in both anionic and cationic micelles, particularly the latter.[153] The biologically important methyl transfer between methylcobalamin and mercuric ion is inhibited by anionic micelles and suppressed by cationic micelles. There are considerable spectroscopic changes (λ_{max} 465 → 520 nm) on solubilization of methylcobalamin in sodium lauryl sulphate consistent with the equilibrium (44) ⇌ (45) being displaced to the right in micelles, with (45) being very much lower in reactivity.[154]

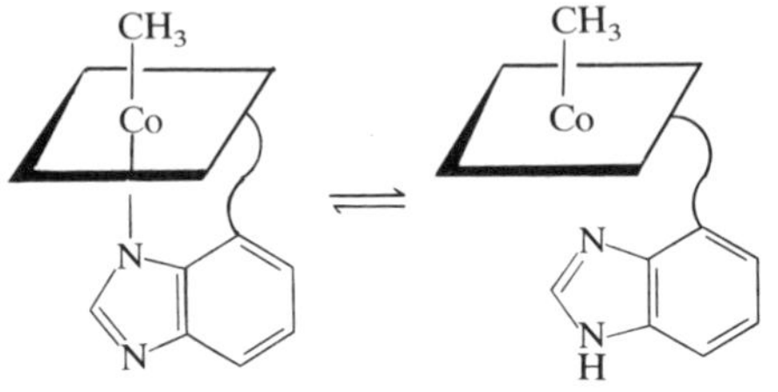

A successful quantitative analysis of the reaction between aquated nickel(II) ions and azo(2-pyridine)4'-N,N-dimethylaniline (46) in sodium lauryl sulphate leads to conclusions that catalysis occurs at the micellar surface and is due entirely to concentration of the reactants[155] (Figure 4). Addition of sodium chloride (sodium ions are claimed to be completely displaced by nickel ions at the micelle surface)[156] decreases the reaction rate marginally but tetraethylammonium chloride has a much more pronounced inhibitory effect. The reaction characteristics were found to be strongly dependent on the source of sodium lauryl sulphate. There is significant catalysis below the c.m.c. if both

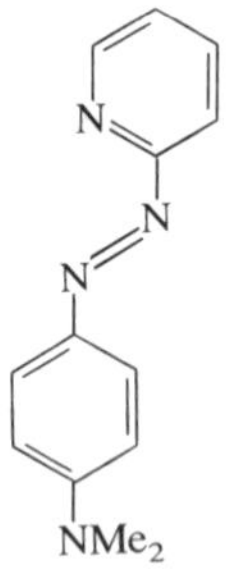

[153] F. Nome and J. H. Fendler, *J. Am. Chem. Soc.*, 1977, **99**, 1557.
[154] G. C. Robinson, F. Nome, and J. H. Fendler, *J. Am. Chem. Soc.*, 1977, **99**, 4969.
[155] A. D. James and B. H. Robinson, *J. Chem. Soc., Faraday Trans. 1*, 1978, **74**, 10; S. Dickmann and J. Frahm, *ibid.*, 1979, **75**, 2199.
[156] J. E. Newbery, *J. Colloid Interface Sci.*, 1980, **74**, 483.

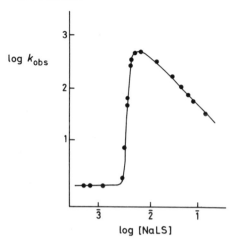

Figure 4 *Observed* (●) *and calculated* (*solid line*) *rate-constants for the reaction between* (46) *and* Ni_{aq}^{2+}

dye and metal-ion solutions contain surfactant prior to mixing, suggesting the formation of pre-micellar aggregates,[157] although the ultimate rate-enhancement is independent of the method of reactant delivery. Similar results were obtained with Mn^{2+} and (46),[157] although here the complexation constant is very much lower. Since Ni^{2+} binds very strongly to anionic micelles, and murexide (47) is a good indicator for its presence in aqueous solution, a combination of Ni^{2+} and (47), *i.e.* (48), may be used to detect micelle formation.[158] The reaction of a range of nitrogen ligands with aquated nickel(II) is catalysed by sodium lauryl sulphate with the same corrected micellar rate constant (*cf.* the analysis in ref. 155). This supports the view that in every case the rate-determining stage is dissociation of a water molecule from the hexa-aquo-ion.[159]

NH_4^+

(47)
(48) Ni^{2+} salt

[157] J. Holzwarth, W. Knocke, and B. H. Robinson, *Ber. Bunsenges. Phys. Chem..*, 1978, **82,** 1001.
[158] M. Fischer, W. Knocke, B. H. Robinson, and J. H. Maclagan-Wedderburn, *J. Chem. Soc., Faraday Trans. 1*, 1979, **75,** 119.
[159] V. C. Reinsborough and B. H. Robinson, *J. Chem. Soc. Faraday Trans. 1*, 1979, **75,** 2395.

6 Organic Reactions in Functional Micelles

The general emphasis in this work is to compare the reactivity and selectivity of function-bearing micelles in catalysis with their simple counterparts. The pattern of research has been guided by the analogy of enzymic catalysis, and surfactants carrying the common prosthetic groups of protease enzymes have been synthesized. Most studies have been concerned with acyl transfer from reactive esters, and several authors have sought to demonstrate co-operativity between different catalytic sites.

The simplest example of a functional micelle is (49), previously demonstrated to be more effective than its trimethylammonium analogue in both esterolysis and bimolecular elimination reactions. It has now been demonstrated that micelles of (49) are more effective catalysts for the hydrolysis of *p*-nitrobenzoyl phosphate dianion at high pH than non-functional surfactants.[160] 2,4-Dinitrochloro- and fluoro-benzene react with micelles of (49) at high pH 10^4 times faster than with hydroxide ion at a comparable external pH. The initial product is (50) and this in turn is hydrolysed in micelles 2.6×10^2 times faster than is 2,4-dinitrophenyl 2-(trimethylammonium)ethyl ether in water at pH 12.[161] Acyl transfer between *p*-nitrophenyl acetate and (49) gives an intermediate whose hydrolysis is not micelle catalysed.[162] In contrast to the rate acceleration observed in that case, hydrolysis of *p*-nitrophenyl acetate is inhibited by micelles of (51) since the phenoxide nucleophile is weak and at the reaction pH its micelles are zwitterionic, not cationic.[163] Synthesis of functional choline-type micelles is facilitated by the use of sulphonate (52), which is reactive towards thiophenoxide in aqueous micelles, but its water-insoluble trifluoromethanesulphonate reacts with a range of anions under phase-transfer conditions.[164]

$$C_{16}H_{33}\overset{+}{N}Me_2CH_2CH_2OH$$
(49)

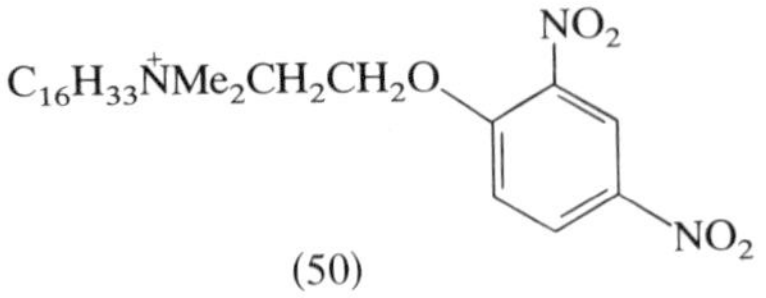

(50)

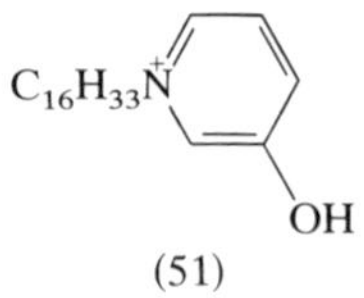

(51)

$$C_{16}H_{33}\overset{+}{N}Me_2CH_2CH_2OSO_2CH_3$$
(52)

[160] C. A. Bunton and M. McAnemy, *J. Org. Chem.*, 1977, **42**, 475.
[161] C. A. Bunton and S. Diaz, *J. Am. Chem. Soc.*, 1976, **98**, 5663.
[162] G. Meyer and P. Viout, *Tetrahedron*, 1977, **33**, 1959.
[163] V. Gani, *Tetrahedron Lett.*, 1977, 2277.
[164] R. A. Moss and W. J. Sanders, *J. Am. Chem. Soc.*, 1978, **100**, 5247.

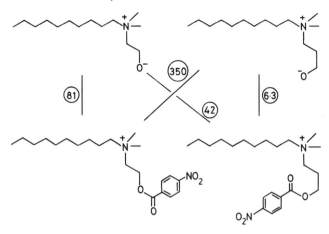

Figure 5 *Selectivity in the reaction of surfactant esters with surfactant alkoxides*

The Jerusalem group have produced a range of papers[165—169] on choline-derived cationic micelles and their selectivity in ester hydrolysis reactions. Among their conclusions are (*a*) the Brønsted β value for micellar cholines such as (49) is in the region of 0.31 to 0.36,[165] (*b*) micelles of carboxylate (53) catalyse the hydrolysis of *m*-nitrophenyl decanoate by a general base mechanism, ($k_{H_2O}/k_{D_2O} = 2.25$) but operate by a nucleophilic mechanism in the hydrolysis of 2,4-dinitrophenyl decanoate ($k_{H_2O}/k_{D_2O} = 1.15$). The lower homologue (54) operates only by a nucleophilic mechanism and is a less effective catalyst.[167] Finally they demonstrate, (*c*) that there is some selectivity (Figure 5) in acyl transfer from micellar *p*-nitrophenyl esters to micellar alkoxides.[169]

Surfactant peroxyanions are more effective nucleophiles than their alkoxy- or acyloxy-counterparts both in micelles, and below the critical micelle concentration. In an attempt to generate the micellar peracid (55), Brown and Darwent reacted the corresponding *p*-nitrophenyl ester (56) with hydrogen peroxide and observed that the production of *p*-nitrophenolate ion was far from first-order in (55).[170] It proved possible to simulate the results in terms of a scheme in which the initially formed peracid anion reacts with (56) to form a diacyl peroxide, which is then attacked by hydrogen peroxide anion to give two molecules of (55). At the sub-micellar concentrations employed it was estimated that reaction between (55) and (56) occurred 10^3 faster than might be

[165] R. Shiffman, M. Chevion, J. Katzhendler, C. Rav-Acha, and S. Sarel, *J. Org. Chem.*, 1977, **42**, 856.
[166] R. Shiffman, C. Rav-Acha, M. Chevion, J. Katzhendler, and S. Sarel, *J. Org. Chem.*, 1977, **42**, 3279.
[167] C. Rav-Acha, M. Chevion, J. Katzhendler, and S. Sarel, *J. Org. Chem.*, 1978, **43**, 591.
[168] D. Wexler, A. Pillersdorf, R. Shiffman, J. Katzhendler, and S. Sarel, *J. Chem. Soc., Perkin Trans. 2*, 1978, 479.
[169] A. Pillersdorf and J. Katzhendler, *J. Org. Chem.*, 1977, **44**, 934.
[170] J. M. Brown and J. R. Darwent, *J. Chem. Soc., Chem. Commun.*, 1979, 171.

$$C_{10}H_{21}\overset{+}{N}Me_2(CH_2)_3CO_2^-$$
(53)

$$C_{10}H_{21}\overset{+}{N}Me_2(CH_2)_2CO_2^-$$
(54)

$$C_{14}H_{29}CH(CH_2)\overset{+}{N}Me_3$$

O   OR   Cl$^-$

(55) R = OH
(56) R = p-$C_6H_4NO_2$

$$CF_3SO_3^-$$
$$C_{16}H_{33}\overset{+}{N}Me_2CH_2CH_2OOH$$
(57)

expected on the basis of simpler models, due to hydrophobic association. Hydroperoxide (57) (*cf.* ref. 108), which has pK_a of 9.6 in micelles, is a powerful esterolytic nucleophile,[171] as expected.

The nucleophilic entity in the protease enzyme papain is a cysteine residue, and thus the nucleophilic reactivity of cysteine derivatives in micelles had been investigated by Heitmann in 1968. More recently a number of thiol-bearing surfactants have been prepared and their nucleophilic reactivity investigated. Moss and co-workers have prepared (58) from the corresponding hydroxyethyl compound (49) and found it to be highly reactive in the cleavage of p-nitrophenyl esters,[172] with a rate acceleration of 10^3 in p-nitrophenyl hexanoate hydrolysis, and a kinetic advantage of 2.8×10^3 over the simple thiocholine analogue in the same reaction. The alkylation of (58) with iodoacetamide was also micelle-catalysed. Below the c.m.c. of (58) the

Cl$^-$
$$C_{16}H_{33}\overset{+}{N}Me_2CH_2CH_2SH$$
(58)

reaction-rate levels off, apparently to a higher value than is observed in buffer alone at pH 7. The reaction pathway of cysteine-derived (59) in esterolytic reactions is more complex.[173] The initial reaction with p-nitrophenyl acetate forms an S-acyl derivative, which then undergoes a slower $S \rightarrow N$-transacylation followed by a second nucleophile attack by the released thiolate to form a kinetically stable S,N-diacyl derivative (Scheme 4). Murakami and co-workers have prepared (60) and (61) by the conventional synthetic methods of peptide chemistry.[174] Since they exhibit very similar reactivities towards p-nitrophenyl hexanoate, notwithstanding the difference in structure, catalysis does not depend on the depth of location in a hydrophobic micellar core.

[171] R. A. Moss and K. W. Alwis, *Tetrahedron Lett.*, 1980, **21**, 1303.

[172] R. A. Moss, G. O. Bizzigotti, T. J. Lukas, and W. J. Sanders, *Tetrahedron Lett.*, 1978, 3661; R. A. Moss, G. O. Bizzigotti, and C. W. Huang, *J. Am. Chem. Soc.*, 1980, **102**, 754; *cf.* L. Anoardi, R. Fornasier, D. Sostero, and U. Tonellato, *Gazz. Chim. Ital.*, 1978, **108**, 707.

[173] R. A. Moss, R. C. Nahas, and T. J. Lukas, *Tetrahedron Lett.*, 1978, 507; R. A. Moss, T. J. Lukas, and R. C. Nahas, *J. Am. Chem. Soc.*, 1978, **100**, 5920.

[174] Y. Murakami, A. Nakano, K. Matsumoto, and K. Iwamoto, *Bull. Chem. Soc. Jpn*, 1979, **52**, 3573.

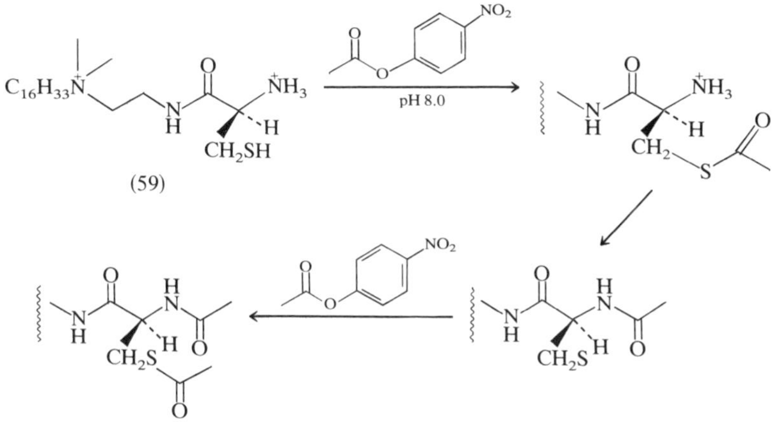

Scheme 4

These cationic surfactants behave rather differently from their anionic counterparts (62) and (63), which are effective in acyl transfer below the critical micelle concentration but relatively unreactive in homomicelles. The importance of electrostatic effects was demonstrated by the much greater reactivity of (62) and (63) in co-micelles with cetyltrimethylammonium bromide.[175]

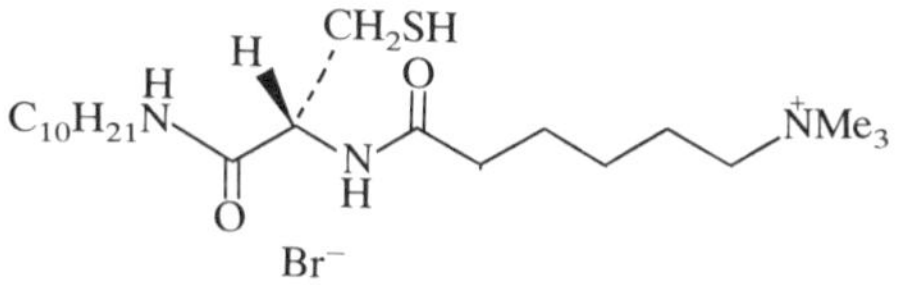

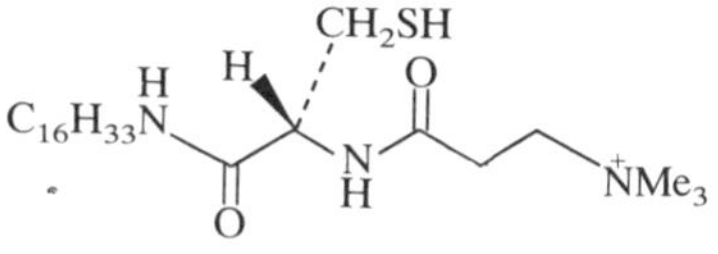

Brønsted coefficients for esterolytic reactivity of anionic nucleophiles in micelles are quite small, so that oximates and hydroxamates are much more reactive than hydroxide ion at a given pH in the weakly basic region. Thus the surfactant (64) is claimed to be the most powerful among functional micelle-forming species in promoting the transacylation of *p*-nitrophenyl acetate and hexanoate. The acyloxime is hydrolysed quite slowly at pH 8, so that catalytic turnover is feeble.[176] Oxime (65) 'catalyses' the hydrolysis of phosphate ester

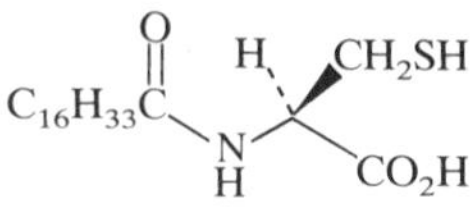

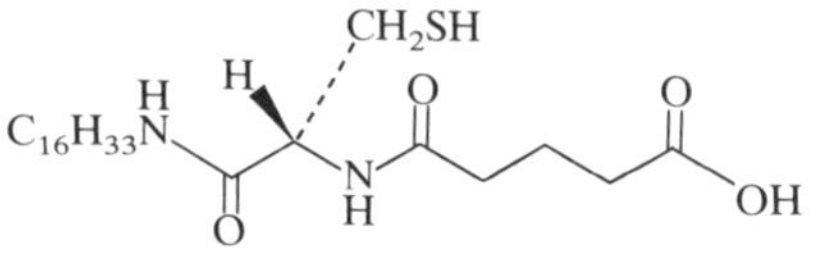

[175] Y. Murakami, A. Nakano, and K. Matsumoto, *Bull. Chem. Soc. Jpn.*, 1079, **52**, 2996.
[176] L. Anoardi, F. de Buzzacarini, R. Fornasier, and U. Tonellato, *Tetrahedron Lett.*, 1978, 3945.

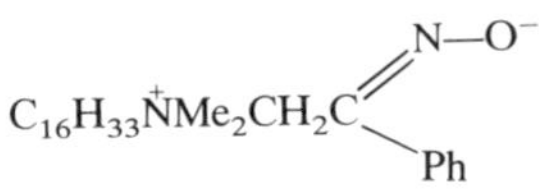

$$C_{16}H_{33}\overset{+}{N}Me_2CH_2C\overset{N-O^-}{\underset{Ph}{}}$$

(64)

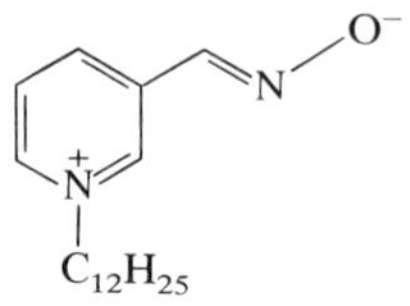

(65)

$$(EtO)_2\overset{O}{\underset{}{P}}\!-\!O\!-\!\langle\text{ring}\rangle\!-\!NO_2$$

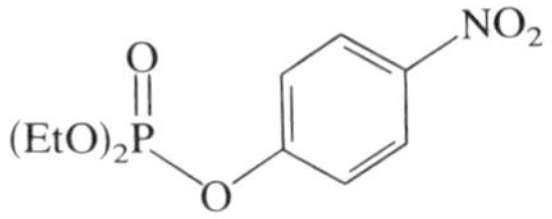

(66)

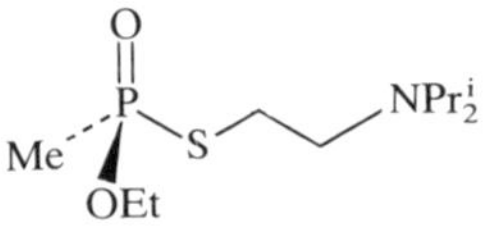

(67a)

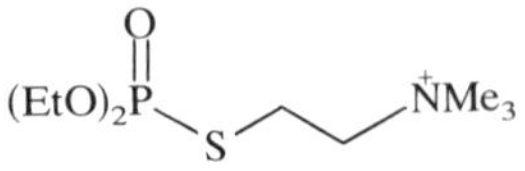

(67b)

(66) and thiophosphonate (67a), but not that of (67b) which is presumably excluded from its micelles. Shorter-chain analogoues of (65) are ineffective, and the rate acceleration observed may be accounted purely in terms of approximation.[177] Kunitake and co-workers[178] have carried out a systematic study of micelle-bound hydroxamic acids, their relative reactivities being indicated in Figure 6. The conclusion of this study is that rate enhancement is a combination of pK_a reduction brought about by the electrostatic effect of the cationic head-group, and approximation in the micelle. Dramatic catalysis of the hydrolysis of 2,4-dinitrophenyl sulphate is observed in the presence of cationic hydroxamic acid micelles.[179] Kunitake's view of the importance of hydrophobic ion-pairs in micellar catalysis is not now widely held in view of the success of alternative quantitative treatments, and other workers[180] demonstrate that the change in absolute magnitude of rate constants for hydroxamate (68)–p-nitrophenyl ester reactions is very small. Using an alternative type of micellar hydroxamic acid showing complex ionization behaviour (69) it was demonstrated that only the anionic form of the micellar hydroxamic acid[181] is effective in catalysis of p-nitrophenyl ester hydrolysis. In phosphate buffer (micellar counterion HPO_4^{2-}) the hydroxamic acids are significantly weaker than in borate buffer (micellar counterion $Cl^- \gg H_3BO_4^-$). A search for co-operative catalysis between micellar hydroxamate and histidine was unsuccessful (see later).

The study of micellar imidazoles in acyl-transfer reactions, following a classic paper of Gitler and Ochao-Solano has proved to be popular. Tagaki and

[177] J. Epstein, J. J. Kaminski, N. Bodor, R. Enever, J. Sowa, and T. Higuchi, *J. Org. Chem.*, 1978, **43**, 2816.
[178] T. Kunitake, Y. Okahata, S. Tanamachi, and R. Ando, *Bull. Chem. Soc. Jpn.*, 1979, **52**, 1967.
[179] T. Kunitake and T. Sakamoto, *Bull. Chem. Soc. Jpn.*, 1979, **52**, 2624.
[180] A. Pillersdorf and J. Katzhendler, *J. Org. Chem.*, 1979, **44**, 549.
[181] J. M. Brown and J. L. Lynn, jun., *Ber. Bunsenges. Phys. Chem.*, 1980, **84**, 70.

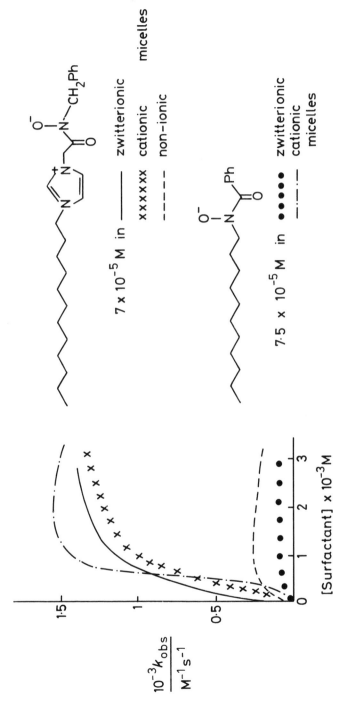

Figure 6 *Relative reactivities of zwitterionic and anionic hydroxamates in micellar catalysis*

$$C_{16}H_{33}\overset{+}{N}Me_2(CH_2)_3C \overset{O}{\underset{NHOH}{\diagdown}}$$

(68)

$$CH_3(CH_2)_x CH(CH_2)_y \overset{+}{N}Me_3 \quad Cl^-$$
$$O \diagdown \underset{Me}{\overset{OH}{\underset{N}{\diagup}}}$$

(69) $x + y = 15$

co-workers[182] have carried out a very thorough investigation of aromatic ester hydrolysis catalysed by micelles of (70) and (71). These have very different pK_a values (3.5 and *ca.* 6, respectively under kinetic conditions) and the former reacts entirely *via* an imidazolide anion[183] so that the reaction rate increases concomitantly with pH. The acylimidazole breaks down rapidly in both cases, unlike the circumstance which prevails in mixed micelles of *N*-myristoyl histidine and cetyltrimethylammonium bromide, where stable N_π-acyls are formed. Below pH 8, the neutral form of (71) is the effective nucleophile, giving rise to a characteristic pH-rate profile plateau. This comparison is brought out clearly in related work by Tonellato,[183] Figure 7.

$$C_{18}H_{37}\overset{+}{N}Me_2 \diagup\diagdown \underset{\underset{H}{N}}{\overset{N}{\diagdown}}$$

(70)

$$C_{18}H_{37}\overset{+}{N}Me_2 \diagup\diagdown\diagup \underset{N}{\overset{NH}{\diagdown}}$$

(71)

True catalysis of esterolysis, (as opposed to rapid acyl transfer and formation of a stable intermediate) requires a second rapid step in which the acyl group is transferred to water or to an oxygen nucleophile. This has presented micellar histidines and imidazoles with a second role, in experiments first carried out separately by Tagaki, Moss, and Tonellato. In the simplest experiment[184] a mixture of (49) with myristoylhistidine (72) reacts with *p*-nitrophenyl acetate first by nucleophilic attack of imidazole and then transfer of the acyl group to (49) in a rapid step. In the bifunctional surfactant (73) this process was shown to be intermolecular (but presumably intramicellar).[185] The acylimidazole can be observed spectroscopically when the substrate is *p*-nitrophenyl hexanoate. Similar conclusions are recorded in a study by Tonellato,[186] who additionally

$$C_{11}H_{23} \overset{O}{\underset{\underset{H}{N}}{\diagdown}} \overset{CO_2H}{\underset{}{\overset{H}{\diagup}}} \diagdown \underset{N}{\overset{NH}{\diagdown}}$$

(72)

$$C_{16}H_{33}\overset{\overset{Me}{|}}{N}\text{---}CH_2CH_2OH \quad Br^-$$
$$\underset{\underset{H}{N}}{\overset{N}{\diagdown}}$$

(73)

[182] W. Tagaki, D. Fukushima, T. Eiki, and Y. Yano, *J. Org. Chem.*, 1979, **44,** 555.
[183] U. Tonellato, *J. Chem. Soc., Perkin Trans. 2*, 1976, 771.
[184] W. Tagaki, S. Kobayashi, and D. Fukushima, *J. Chem. Soc., Chem. Commun.*, 1977, 29.
[185] R. A. Moss, R. C. Nahas, and S. Ramaswami, *J. Am. Chem. Soc.*, 1977, **99,** 629.
[186] U. Tonellato, *J. Chem. Soc., Perkin Trans. 2*, 1977, 821.

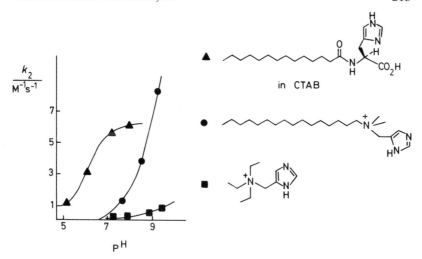

Figure 7 *Hydrolysis of p-nitrophenyl acetate by imidazole-bearing surfactants*

showed that a mixture of (49) and surfactant imidazole (71) behaved similarly to (73). In the hydrolysis of activated amides, hydroxyethylammonium surfactant micelles are very much more effective than surfactant imidazoles.[187]

Acylimidazoles are known to react rapidly with hydroxamates, and this process has been demonstrated to occur in micelles by n.m.r. monitoring of the reaction sequence between (69), (74), and p-nitrophenyl acetate. Both the acetylated derivatives are unstable to hydrolysis with that of (69) reacting by a pH-dependent mechanism around pH 7—8 with an isotope effect of 2.7, and unaffected by added (74).[188] Hydrolysis of acetylated (74) is catalysed by the

$C_{12}H_{25}CH(CH_2)_4\overset{+}{N}Me_3$ Cl^-

(74)

$C_{12}H_{25}N$ ... (75)

(76)

(77)

[187] L. Anoardi and U. Tonellato, *J. Chem. Soc., Chem. Commun.*, 1977, 401.
[188] J. M. Brown, P. A. Chaloner, and A. Colens, *J. Chem. Soc., Perkin Trans. 2*, 1979, 71.

free histidine. Kunitake has prepared surfactants (75) and (76), which contain both imidazole and hydroxamic acid moieties. These are effective catalysts in *p*-nitrophenyl ester hydrolysis but reaction occurs by sequential acylation of the hydroxamate and subsequent imidazole-catalysed hydrolysis.[189] This contrasts with the behaviour of polymer (77)[190] where the imidazole acts as a general base in deacylation of *p*-nitrophenyl acetate as shown. This is apparent both from the term for reaction due to undissociated hydroxamic acid (the polymer containing only imidazole residues is comparatively unreactive) and a solvent isotope effect of 1.6. This co-operative reactivity is very rare in micellar and other hydrophobic catalysis, and is particularly interesting since the charge-relay mechanism for chymotrypsin, which inspired the work, is being challenged.[191]

Following the initial observation of stereoselectivity in ester hydrolysis by optically active surfactant histidines, several recent observations have served to reinforce its generality. Yamada and co-workers suggest that both D- and L-myristoylhistidine show the same selectivity (1.5:1) in mixed micelles with (78) towards *N*-ethoxycarbonylphenylalanine *p*-nitrophenyl ester, and thus the

(78)

L-alanyl residue plays no part.[192] The importance of chain length in influencing both reaction rate and stereoselectivity is demonstrated for related systems.[193] The effectiveness of asymmetric micellar catalysis in amino-acid ester hydrolysis increases with increasing hydrophobicity of the side-chain[194] and Ohkubo and co-workers show that (79) is capable of a higher degree of selectivity between L- and D-amino-acid esters than is (80).[195] The rather sharp changes of optical yield with variation of amide structure (Figure 8) suggest that hydrophobic forces play a considerable part in organizing the structure and stereochemistry of the transition state.[195] The combination of a histidinyl residue and micellar catalysis is insufficient to effect asymmetric induction,[196] however. The original results were obtained with (74) as one pure diastereomer (now known to be *SS*)[197] but the species with opposite chain

[189] T. Kunitake, Y. Okahata, and T. Sakamoto, *J. Am. Chem. Soc.*, 1976, **98**, 7799.

[190] T. Kunitake and Y. Okahata, *J. Am. Chem. Soc.*, 1976, **98**, 7793.

[191] For example, W. W. Bachovchin, and J. D. Roberts, *J. Am. Chem. Soc.*, 1978, **100**, 8041.

[192] K. Yamada, H. Shosenji, and H. Ihara, *Chem. Lett.*, 1979, 491.

[193] Y. Ihari, *J. Chem. Soc., Chem. Commun.*, 1978, 984.

[194] K. Yamada, H. Shosenji, H. Ihara, and Y. Otsubo, *Tetrahedron Lett.*, 1979, 2529.

[195] K. Ohkubo, K. Sugahara, K. Yoshinaga, and R. Ueoka, *J. Chem. Soc., Chem. Commun.*, 1980, 637.

[196] R. A. Moss, T. J. Lukas, and R. C. Nahas, *Tetrahedron Lett.*, 1977, 3851.

[197] J. M. Brown, R. L. Elliot, C. G. Griggs, G. Helmchen, and G. Nill, *Angew. Chem., Int. Ed. Engl.*, 1981, **20**, 890.

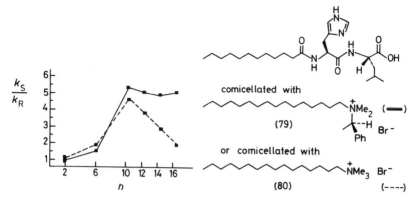

Figure 8 *Relative reactivity of the enantiomers of* $H(CH_2)_{n+1}CONHCH(CH_2Ph)$-
$COOC_6H_4NO_2$ *with surfactant histidines*

configuration is less reactive and non-stereoselective in micelle-catalysed hydrolysis of *p*-nitrophenyl *N*-acetylphenylalanines.[198] An earlier model for the transition state of asymmetric catalysis may be modified to (81), which explains the structural variations presented here.

Asymmetric induction in micelles implies the possibility that conformational changes may occur on micellation of a reacting substrate. This has been confirmed in the study of a range of micellar catalysts effecting the hydrolysis of (82).[199] As the catalytic efficiency increases, the reactivity of *SS*-(82) relative

[198] J. M. Brown, Ciba Symposium Proceedings 53, 1978, p. 149 (in memoriam Sir Robert Robinson).
[199] R. A. Moss, Y.-S. Lee, and T. J. Lukas, *J. Am. Chem. Soc.*, 1979, **101**, 2499.

to *RS*(82) is enhanced. The head-group type (thiolate or imidazole) does not appear to be important as the ratio changes from 0.62 (buffer) to 1.00 (CTAB), 2.68 (70) and 4.33 (58).

Much related work has occurred based on organized catalysis in natural and synthetic macrocycles, polymers, and modified surfaces, which can only be hinted at briefly in such a review. We highlight Lehn's application of crown-ethers (83)[200] and (84)[201] in the catalysis of hydrolysis of chiral dipeptide esters (*S*:*R* up to 50:1) and hydride transfer to ketones, respectively, Knowles's successful preparation of specifically substituted α-cyclodextrins (85)[202] and their involvement in phosphate ester hydrolysis, and Murakami's syntheses of catalytically active *p*-cyclophanes.[203,204] The long-standing contribution of

(83)   (84)

$X = CONH(CH_2)_2-N$ (with CONHBu)

$X = CONHCH$ (with CO_2Me and CH_2SH)

(85)

[200] J.-P. Behr and J.-M. Lehn, *J. Chem. Soc., Chem. Commun.*, 1978, 143.
[201] J.-M. Lehn and C. Sirlin, *J. Chem. Soc., Chem. Commun.*, 1978, 949.
[202] J. Boger, D. G. Brenner, and J. R. Knowles, *J. Am. Chem. Soc.*, 1979, **101**, 7630; J. Boger and J. R. Knowles, *ibid.*, 1979, **101**, 7632.
[203] Y. Murakami, A. Nakano, K. Matsumoto, and K. Iwamoto, *Bull. Chem. Soc. Jpn.*, 1978, **51**, 2690.
[204] Y. Murakami, J. Sunamoto, H. Kondo, and H. Okamoto, *Bull. Chem. Soc. Jpn.*, 1977, **50**, 2420; Y. Murakami, Y. Aoyama, M. Kida, and A. Nakano, *ibid.*, 1977, **50**, 3365.

Klotz and co-workers to catalysis by hydrophobic polyethyleneimines continues[205,206] and a novel application of water soluble microgels[207] containing pendant hydroxamic acid groups is reported.

7 Bicatenate Surfactants; Structure and Catalysis

The tendency of natural phospholipids to form organized structures based on two-dimensional repeating bilayers has been recognized for half a century. These aggregates, which form the basis of biological membranes are very different from micelles in their physical properties. Dissociation of monomers is slow, and electron microscopy reveals stable vesicular structures capable of entrapping added solutes, which diffuse across the bilayer quite slowly. It is rather surprising that synthetic models were not examined until three or four years ago, and gratifying to discover that their behaviour resembles the biological prototypes quite closely.

Kunitake and co-workers made the first observations[208] of single-walled vesicles in sonicated solutions of bis-dodecyldimethylammonium bromide by electron microscopy. Light-scattering suggests an average molecular weight of 7×10^5. It was later shown that multilamellar vesicles are formed at short sonication times but single-walled aggregates predominate after longer sonication. The ^{1}H n.m.r. line-width is temperature-dependent being broad at 20 °C but narrowest at 40 °C.[209] Radio-labelled glucose is entrapped by the vesicles,[210] and similar aggregates derived from dioctadecyldimethylammonium chloride have been characterized by low-angle laser scattering and photon correlation spectroscopy. As sonication progresses the average molecular weight of the vesicle decreases to a constant value[211] of 7.26×10^7 daltons, with a hydrodynamic radius calculated to be 396 Å. The data are best fitted by assuming a prolate structure. Various physicochemical techniques have been applied by Fendler and co-workers in the characterization of dioctadecyldimethylammonium chloride vesicles.[212—214] The turbidity decreased dramatically for samples sonicated at 20 °C when heated above 35 °C, with a second minor decrease at 47 °C. A change in optical absorbance at 400 nm on introducing the sonicated vesicles to 0.1 M-KCl is interpreted as due to their osmotic shrinkage, the observed half-life being 1—2 minutes. Positron annihilation (see ref. 21) confirms the phase-transition temperatures. The spin probe (86) binds strongly to vesicles $(K \sim 350 \text{ M}^{-1})$, and is scavenged by sodium

[205] R. S. Johnson, J. A. Walder, and I. M. Klotz, *J. Am. Chem. Soc.*, 1978, **100**, 5159.
[206] M. A. Hierl, E. P. Gamson, and I. M. Klotz, *J. Am. Chem. Soc.*, 1979, **101**, 6020.
[207] R. A. Weatherhead, K. A. Stacey, and A. Williams, *J. Chem. Soc,, Chem. Commun.*, 1979, 598.
[208] T. Kunitake, Y. Okahata, K. Tamaki, F. Kumamaru, and M. Takayanagi, *Chem. Lett.*, 1977, 387; T. Kunitake and Y. Okahata, *J. Am. Chem. Soc.*, 1977, **99**, 3860; *Chem. Lett.*, 1977, 1337; T. Kajiyama, A. Kumaro, M. Tagayanagi, Y. Okahata, and T. Kunitake, *Chem. Lett.*, 1979, 645.
[209] K. Deguchi and Y. Mino, *J. Colloid Interface Sci.*, 1978, **65**, 155.
[210] R. McNeil and J. K. Thomas, *J. Colloid Interface Sci.*, 1980, **73**, 522.
[211] U. Herrmann and J. H. Fendler, *Chem. Phys. Lett.*, 1979, **64**, 270.
[212] C. D. Tran, P. L. Klahn, A. Romero, and J. H. Fendler, *J. Am. Chem. Soc.*, 1978, **100**, 1622.
[213] K. Kano, A. Romero, B. Djermouni, H. J. Ache, and J. H. Fendler, *J. Am. Chem. Soc.*, 1979, **101**, 4030.
[214] Y. Y. Lim and J. H. Fendler, *J. Am. Chem. Soc.*, 1979, **101**, 4023.

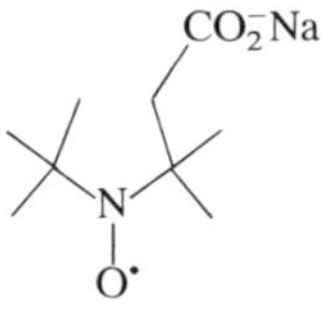

(86) 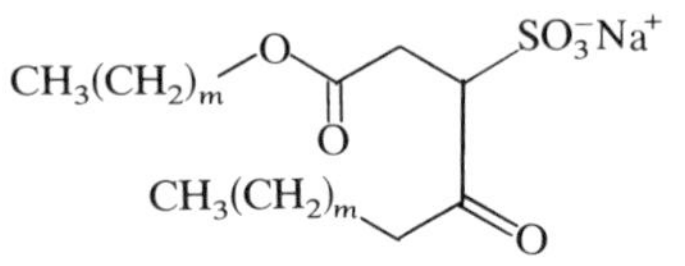   (87)

CO_2^-Na

$CH_3(CH_2)_m$—O ... $SO_3^-Na^+$

$CH_3(CH_2)_m$

ascorbate. Even in the presence of excess scavenger a small number of radicals remain, which decreases with increasing surfactant concentration. This is interpreted as partial binding of (86) on the inner surface of the vesicle, rendering it inaccessible to reductant. Anionic dialkylphosphates may form vesicles or lamellae, depending on the chain length, and dialkylsulphates derived from maleic anhydride (87) behave similarly.[215]

There is an elegant demonstration[216] that bis-quaternary ammonium salts such as (88) form a variety of large aggregates in aqueous solution. Rod-like structures were explained by chain-folding, with both terminal cationic groups exposed to the external medium. In the presence of cholesterol, very large single-walled vesicles are apparent by electron microscopy.[216] Pyridinium salts (89) and (90) also form large closed vesicles on sonication,[217] which entrap

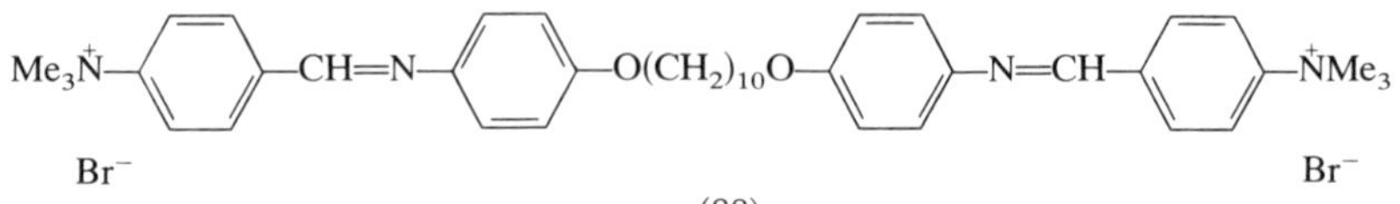

$Me_3\overset{+}{N}$—⟨ ⟩—CH=N—⟨ ⟩—O(CH$_2$)$_{10}$O—⟨ ⟩—N=CH—⟨ ⟩—$\overset{+}{N}Me_3$

Br^- Br^-

(88)

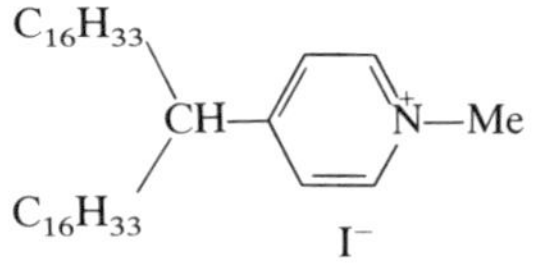

(89) 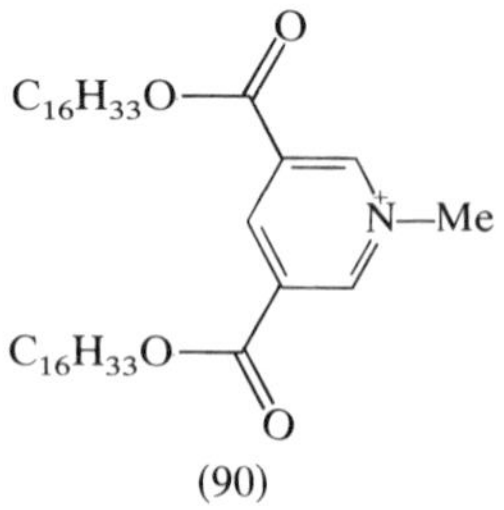   (90)

$C_{16}H_{33}$
CH—⟨ ⟩—$\overset{+}{N}$—Me I^-
$C_{16}H_{33}$

$C_{16}H_{33}O$—C(=O) ... $\overset{+}{N}$—Me
$C_{16}H_{33}O$—C(=O)

carboxyfluorescin causing efficient self-quenching of its fluorescence. The aggregate formed from (89) at 0 °C leaked the dye rapidly, whereas those formed from (90) at 50 °C are very much less permeable. Several compounds with the general formulae (91) or (92) form bilayer structures that exhibit a phase change around 30—40 °C, observable by differential scanning calorimetry. Both vesicles and lamellae were observed by electron microscopy

[215] T. Kunitake and Y. Okahata, *Bull. Chem. Soc. Jpn.*, 1978, **51**, 1877.
[216] Y. Okahata and T. Kunitake, *J. Am. Chem. Soc.*, 1979, **101**, 5231.
[217] E. J. R. Sudholter, J. B. F. N. Engberts, and D. Hoekstra, *J. Am. Chem. Soc.*, 1980, **102**, 2467.

CH$_3$(CH$_2$)$_n$—⟨benzene⟩—N=CH—⟨benzene⟩—$\overset{+}{\text{N}}$Me$_3$

Br$^-$

(91)

CH$_3$(CH$_2$)$_n$—⟨benzene⟩—N=CH—⟨benzene⟩—O(CH$_2$)$\overset{+}{\text{N}}$Me$_3$

Br$^-$

(92)

H$_{37}$C$_{18}$—$\overset{+}{\text{N}}$⟨ring⟩$\overset{+}{\text{N}}$—C$_{18}$H$_{37}$ 2Cl$^-$

(93)

depending on the structure of the aggregate, with no obvious trends.[218] The specificity of (93) in binding dianionic nucleolides in preference to mono- or tri-anions may be associated with vesicle or at least large aggregate formation.[219]

Kinetics of Catalysed Reactions.—Few examples have appeared but there is already sufficient information to make it clear that vesicles affect organic reactions in a very different manner to micelles. Kunitake and co-workers showed[220] that hydrolysis of p-nitrophenyl acetate and hexadecanoate was accelerated by vesicles of bis-dodecyldimethylammonium bromide containing the cholesteryl analogue (94). The acetate is unaffected by the method of sample preparation but the hexanoate reacts much faster if the reactants are co-sonicated (and therefore exist within the same surfactant vesicle) than on separate mixing showing that transfer of reactant and substrate between vesicles must be slow. Reaction of imidazole (95) with p-nitrophenyl hexadecanoate in the presence of bicatenate surfactants occurs predominantly by an intravesicle mechanism. The Arrhenius plot for OH$^-$ catalysed hydrolysis is linear, whereas that for the acyl transfer exhibits a transition temperature that increases with the surfactant alkyl-chain length.[221] A more detailed comparison between different types of surfacant in promoting hydroxamate and imidazole esterolyses is made in the full paper.[222] The most interesting aspect is that (94) but not (96) is considerably more reactive in vesicles than in micelles, implying that some structural specificity is possible. In the hydrolysis of p-nitrophenyl acetate catalysed by heptanethiolate in dioctadecyldimethylammonium

[218] T. Kunitake and Y. Okahata, *J. Am. Chem. Soc.*, 1980, **102,** 549.
[219] I. Tabushi, J. Imuta, N. Seko, and Y. Kobuke, *J. Am. Chem. Soc.*, 1978, **100,** 6287.
[220] T. Kunitake and T. Sakamoto, *J. Am. Chem. Soc.*, 1978, **100,** 4615.
[221] T. Kunitake and T. Sakamoto, *Chem. Lett.*, 1979, 1059.
[222] Y. Okahata, R. Ando, and T. Kunitake, *Bull. Chem. Soc. Jpn.*, 1979, **52,** 3647.

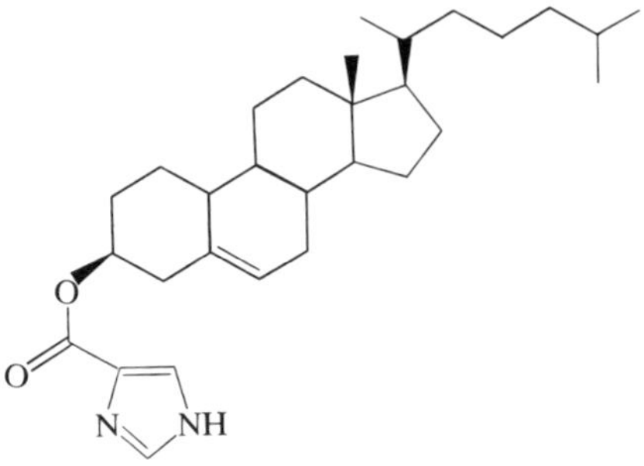

(94)

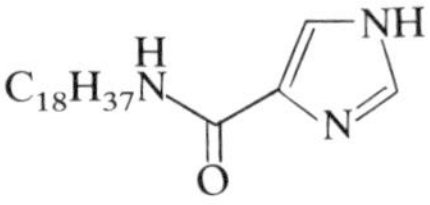

(95)

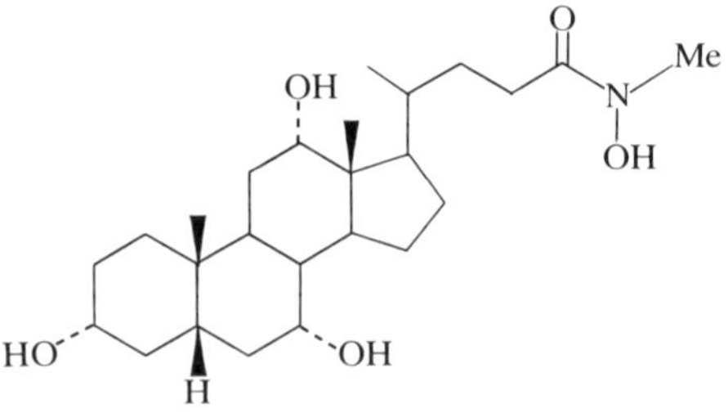

(96)

chloride (mainly!) considerable rate enhancements were observed at high dilution of surfactant, which tailed off above 8×10^{-4} M.[223]

Excited-state Phenomena.—Czarniecki and Breslow[224] provide an experimental test for structure and organization in micelles and vesicles based on inter-molecular hydrogen-atom abstraction by benzophenone triplet-state. This leads to functionalization at a particular site in the chain, which may be identified by degradation and mass spectrometry. With didodecyl phosphate and (97) or (98), the specificity is as shown (Figure 9) for sonicated surfactant, presumed to be in vesicular form. This shows a much altered specificity from the corresponding reaction in micelles. When sonication is carried out in a sodium borate buffer, a technique considered to provide multilamellar layers (opaque dispersion!) then the terminal selectivity is completely lost and C(6) is the most favoured site for attack.

The excimer fluorescence of pyrene and substituted pyrenes has been studied in natural[225] and synthetic[226] vesicles. The rate constant for electron transfer to pyrene absorbed in didodecyldimethylammonium chloride vesicles was studied by pulse radiolysis, and evaluation of these results suggests a vesicle diameter of 20 nm. Sonication gives rise to an aggregate of smaller dispersisity than does methanol injection. Irradiation of pyrene in dihexadecyl

[223] I. M. Cuccovia, R. M. X. Aleixo, R. A. Mortora, P. B. Filho, J. B. S. Bonilha, F. H. Quina, and H. Chaimovich, *Tetrahedron Lett.*, 1979, **33**, 3065.
[224] M. F. Czarniecki and R. Breslow, *J. Am. Chem. Soc.*, 1979, **101**, 3675.
[225] W. Schnecke, M. Grätzel, and A. Henglein, *Ber. Bunsenges. Phys. Chem.*, 1977, **81**, 821.
[226] A Henglein, T. Proske, and W. Schnecke, *Ber. Bunsenges. Phys. Chem.*, 1978, **82**, 956.

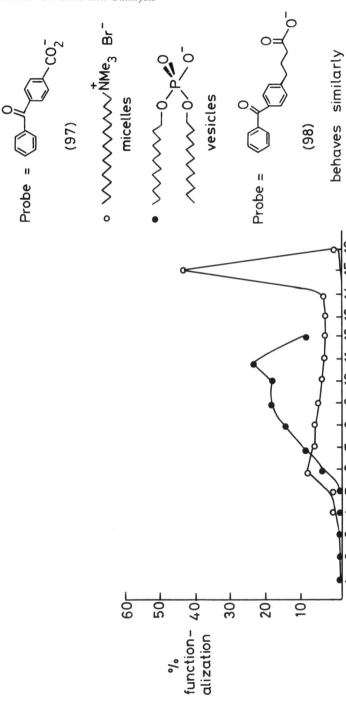

Figure 9 *Selectivity in H-atom abstraction from surfactant chains by solubilized benzophenones in their triplet excited state*[224,297]

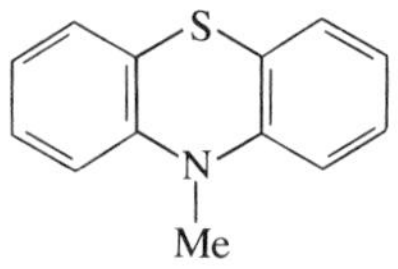

(99)

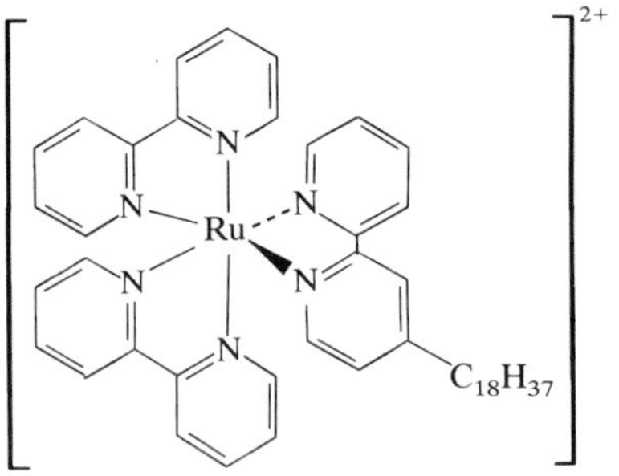

(100)

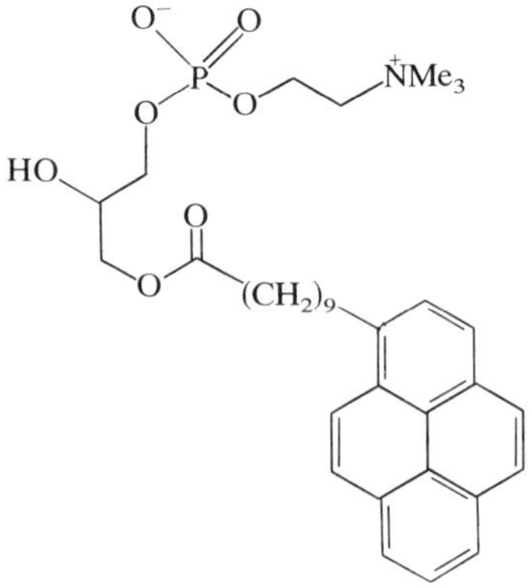

(101)

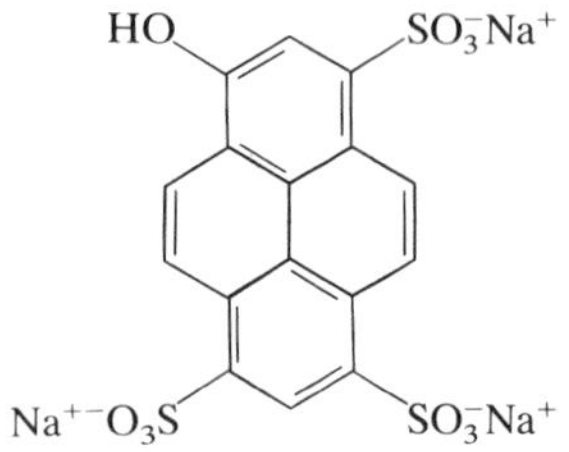

(102)

phosphate anionic vesicles leads to electron ejection from the vesicle. The resulting hydrated electron may be trapped by benzophenone giving its radical anion, which is spectroscopically observable ($t_{1/2} \sim 220$ µs).[227] Photolysis of *N*-methylphenothiazine (99) in dioctadecyldimethylammonium chloride vesicles to which (100) is bound leads to efficient electron transfer from heteroarene to ruthenium. The radical cation of (99) may then undergo geminate recombination to reform starting materials, it may escape and react in the water core of the vesicle, or it may escape into aqueous solution, where it is relatively long-lived.[228] In related work[229] it was demonstrated that fluorescence quenching of (101) in cationic vesicles by added surface-bound (102) occurred with a rate constant of 6.2×10^{11} M^{-1} s^{-1} and is the main source of energy transfer after photoexcitation. Sonication before irradiation, causing a partial redistribution of (102) to the inner surface to the vesicle, increases the efficiency of quenching when the acceptor is present at low concentrations.

[227] J. R. Escabi-Perez, A. Romero, S. Lukac, and J. H. Fendler, *J. Am. Chem. Soc.*, 1979, **101**, 2231.

[228] P. P. Infelta, M. Grätzel, and J. H. Fendler, *J. Am. Chem. Soc.*, 1980, **102**, 1479.

[229] T. Namura, J. R. Escabi-Perez, J. Sunamoto, and J. H. Fendler, *J. Am. Chem. Soc.*, 1980, **102**, 1484.

8 Excited-state Chemistry in Micelles

Micellar photochemistry, and the generation of transients in micellar media by other means, have attracted much effort in the last two or three years.[230] This is because of its potential in energy conversion, where physical separation of the oxidative and reductive parts of the catalytic cycle may permit photochemical cleavage of water. Some very promising results have already been obtained.

Radiolysis and Related Techniques.—Solvated electrons are the primary product of pulse radiolysis, and react with micellated pyrene with a rate constant of $10 \, \text{M}^{-1} \text{s}^{-1}$, the corresponding reaction with nitropyrene being twice as rapid. The resulting anion radical decays by an extramicellar bimolecular second-order reaction for which the rate constant is $3.8 \times 10^9 \, \text{M}^{-1} \, \text{s}^{-1}$.[231] Pulse radiolysis of naphthalene in micelles of sodium lauryl sulphate permits the estimation of its reactivity towards e_{aq}^-, H·, and OH·. The rate of decay of naphthalene radical anion may be analysed to reveal its binding constant (25—30 compared to 550 for the parent hydrocarbon) to anionic micelles.[232] In a more general study of the reactivity of hydrated electrons towards electron acceptors in micelles, a maximum was observed for substrates with an electron affinity in the range 1.6—2.0 eV and both higher or lower electron affinities led to a reduction in rate. This entirely accords with the predictions of Marcus theory.[233] The rate constant was substrate independent in cationic micelles and CH₃ĊHOH (from e_{aq}^- and added ethanol) reacts with electron acceptors at a constant rate constant of $3.5 \times 10^9 \, \text{M}^{-1} \, \text{s}^{-1}$, which is independent of the nature of the micellar head-group. In reactions of hydrated electrons with the cationic dye dimethylviologen absorbed in the Stern layer of sodium laurylsulphate micelles, significant perturbation of reactivity is observed which persists to sub-micellar concentrations.[234]

In the presence of nitrous oxide, the main high-energy species formed in pulse radiolysis of water is the hydroxyl radical. This reacts with sodium laurylsulphate by hydrogen-atom abstraction, which occurs at a range of possible sites in the alkyl chain. The resulting radical reacts with ferricyanide ion in an extramicellar environment, and their rate of decay permits values of $k^+ = 1.5 \times 10^9 \, \text{M}^{-1} \text{s}^{-1}$ and $k^- = 1.8 \times 10^5 \, \text{s}^{-1}$ to be derived (see pp. 190).[235] By similar techniques, binding constants for aliphatic alcohols to cationic micelles were obtained. Surfactant (103) reacts less rapidly with hydroxide radicals above its c.m.c., the main product being through attack on the aromatic ring

[230] For review of recent literature see B.A. Lindig and M. A. J. Rodgers, *Photochem. Photobiol.*, 1980, **31**, 617; K. Kalyanasundaram, *Chem. Soc. Rev.*, 1978, **1**, 453; a more recent and general review is by N. J. Turro, M. Grätzel, and A. M. Braun, *Angew. Chem., Int. Ed. Engl.*, 1980, **19**, 675.

[231] T. Proske, C. H. Fischer, M. Grätzel, and A. Henglein, *Ber. Bunsenges. Phys. Chem.*, 1977, **81**, 916.

[232] E. L. Evers, G. G. Jayson, I. D. Robb, and A. J. Swallow, *J. Chem. Soc., Faraday Trans. 1*, 1979, **75**, 528.

[233] A. J. Frank, M. Grätzel, A. Henglein, and E. Janata, *Ber. Bunsenges. Phys. Chem.*, 1976, **80**, 547.

[234] M. A. J. Rodgers, D. C. Foyt, and Z. A. Zimek, *Radiation Res.*, 1978, **75**, 296.

[235] M. Almgren, F. Grieser, and J. K. Thomas, *J. Chem. Soc. Faraday Trans. 1*, 1979, **75**, 1674.

$$\text{C}_6\text{H}_{13}$$

(103)

giving a transient absorption at 320 nm.[236] Sodium hydrogen sulphite reacts with hydrated electrons to give the radical $SO_3^{\cdot-}$, which may be trapped by addition to the double-bond of dodec-1-ene in sodium laurylsulphate solution.[237]

The radiolysis of bromine in aqueous cetyltrimethylammonium bromide gives rise to micellated $Br_2^{\cdot-}$ radical anions.[238] There is no intermicellar reaction, the main quenching process being

$$Br_2^{\cdot-}(\text{free}) + Br_2^{\cdot-}(\text{micelle}) \xrightarrow{k_2} Br^-(\text{free}) + Br^-(\text{micelle}) + Br_2, \qquad (6)$$

where $k_2 = 2 \times 10^{11} \text{ M}^{-1} \text{ s}^{-1}$.

The micellar residence time of $Br_2^{\cdot-}$ was calculated to be $1.5 \times 10^5 \text{ s}^{-1}$. In like manner, pulse radiolysis of Ag_2SO_4 in sodium laurylsulphate micelles gives Ag_2, and the various decay processes may be analysed.[239] Electron-transfer between quinone radical anions and quinones (the former generated from e_{aq}^- reduction) may occur at the micelle surface in sodium laurylsulphate solution. The encounter rate constant for radical anion plus anionic micelle is an order of magnitude lower than for neutral molecule–micelle interactions because of charge–charge repulsions.[240]

The reactivity of positrons with nitrobenzene or cupric chloride has been studied[241] in the presence[242] and absence of additives (see p. 183, ref. 21). The quenching rate constant is lower in micelles than in water or hexadecane.

Quenching of Excited States.—Much research has been adapted from early observations that micelles could drastically alter the fluorescence lifetime of pyrene and the efficiency of excimer formation. The various mechanisms of excited-state transformation have been reviewed.[243] The polarity of the micellar surface may be estimated from the fluorescence spectrum of pyrene 2-carboxaldehyde[244] giving rise to the conclusion that sodium laurylsulphate ($\varepsilon = 45$) has a more polar surface than cetyltrimethylammonium bromide ($\varepsilon = 18$). The fluidity of the site of solubilization may be estimated from fluorescence polarization[245] in surfactants of general structure (104). The S_1

[236] A. Henglein and T. Proske *J. Am. Chem. Soc.*, 1978, **100**, 3706.
[237] T. Mitaya, A. Sakumoto, and M. Washino, *Bull. Chem. Soc. Jpn.*, 1977, **50**, 2951.
[238] T. Proske and A. Henglein, *Ber. Bunsenges. Phys. Chem.*, 1978, **82**, 711.
[239] A. Henglein and T. Proske, *Ber. Bunsenges. Phys. Chem.*, 1978, **82**, 471.
[240] M. Almgren, F. Greiser, and J. K. Thomas, *J. Phys. Chem.*, 1979, **83**, 3232.
[241] Y. C. Jean and H. J. Ache, *J. Am. Chem. Soc.*, 1977, **99**, 7504.
[242] Y. C. Jean and H. J. Ache, *J. Phys. Chem.*, 1976, **80**, 1693.
[243] J. K. Thomas, *Acc. Chem. Res.*, 1977, **10**, 133.
[244] K. Kalyanasundaram and J. K. Thomas, *J. Phys. Chem.*, 1979, **81**, 2176.
[245] M. Aoudia and M. A. J. Rodgers, *J. Am. Chem. Soc.*, 1979, **101**, 6777.

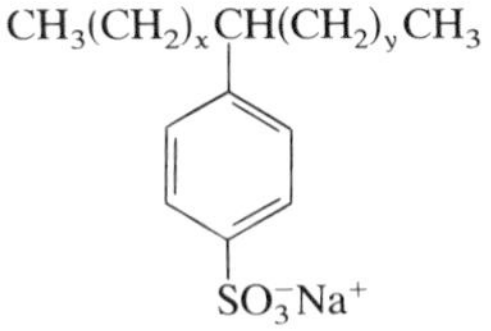

$CH_3(CH_2)_x CH(CH_2)_y CH_3$

(104) $x + y = 9$

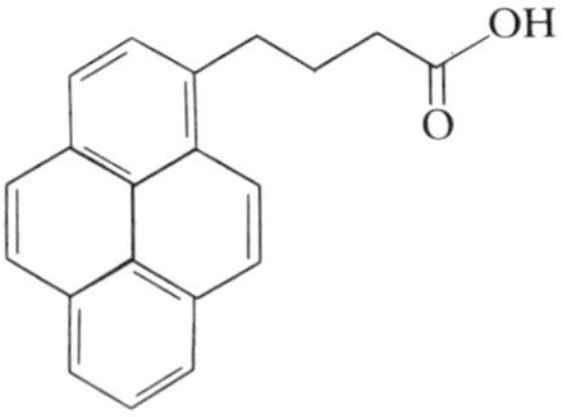

(105)

excited-state of pyrenes may be quenched within the micelle, or in aqueous solution, and when I$^-$ is the quenching agent, and (105) the reactant in sodium lauryl sulphate micelles, exit into aqueous solution precedes quenching.[246] The vibrational fine-structure of pyrene monomer fluorescence has been examined, and is solvent dependent; its appearance in micelles of sodium lauryl sulphate was taken to indicate considerable water penetration.[247]

Alternatively the pyrene S_1 state may be quenched within the micelle and nitroxides are efficient in this regard. For tetrasulphonate (106) solubilized in cetyltrimethylammonium chloride micelles (107) is by far the most efficient quencher, but in sodium lauryl sulphate micelles (108) is the most effective member of a series of nitroxides. Thus the binding efficiency of the probe is paramount.[248] A series of cationic pyreneammonium salts (109) was co-micellated with the nitroxide (110), and static (non-diffusional) quenching

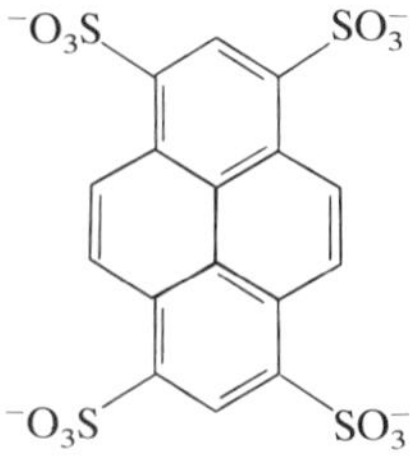

(106)

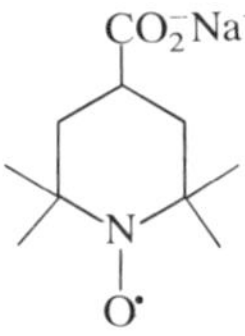

(107)

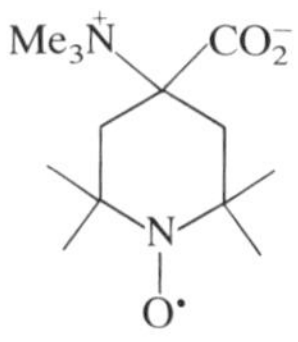

(108)

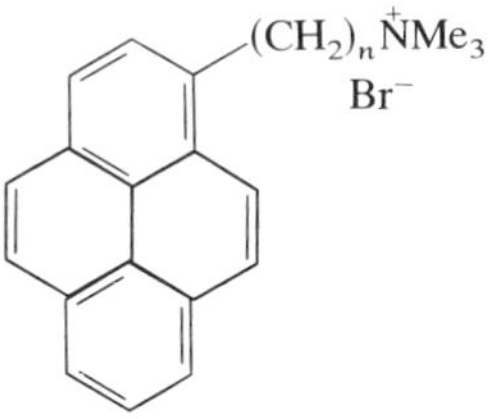

(109)

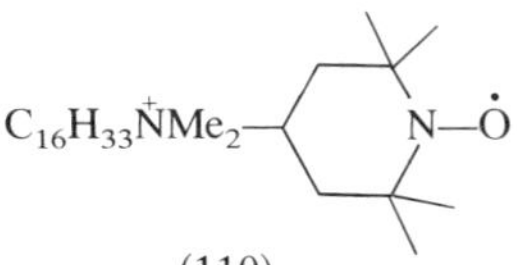

(110)

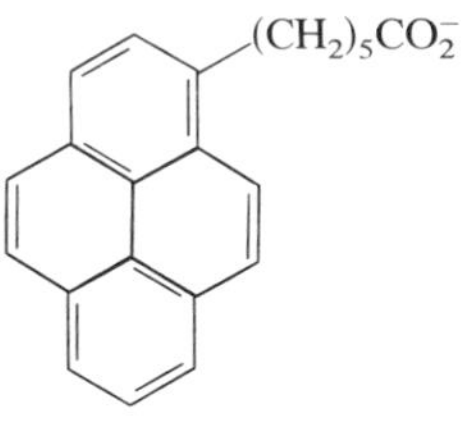

(111)

[246] F. H. Quina and V. G. Toscano, *J. Phys. Chem.*, 1977, **81**, 1750.
[247] K. Kalayanasundaraman and J. K. Thomas, *J. Am. Chem. Soc.*, 1977, **99**, 2039.
[248] S. S. Atik and L. A. Singer, *J. Am. Chem. Soc.*, 1978, **100**, 3234.

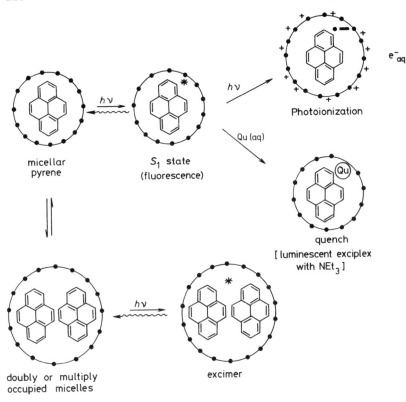

Figure 10 *Photophysical behaviour of pyrene in micelles*

observed above the c.m.c.[249] When both components are hydrophobic, efficient quenching may occur below the c.m.c. because sub-micellar aggregates are formed; this may also be observed in dilute solutions of (111) and cetyltrimethylammonium chloride.[250]

At high concentrations of pyrene (112), where there is a substantial probability of locating more than one pyrene molecule in the same micelle, excimer emission is the dominant photophysical process to follow excitation (Figure 10). The system may be treated quantitatively[251] by Poisson distribution statistics and applied successfully to S_1 decay in surfactant (112). A similar treatment of naphthalene fluorescence in micelles has been presented.[252]

[249] S. S. Atik, C. L. Kwan, and L. A. Singer, *J. Am. Chem. Soc.*, 1979, **101,** 5696.

[250] S. S. Atik and L. A. Singer, *J. Am. Chem. Soc.*, 1979, **101,** 6761.

[251] P. P. Infelta and M. Grätzel, *J. Chem. Phys.*, 1979, **70,** 179.

[252] A. Yekta, M. Aikawa, and N. J. Turro, *Chem. Phys. Lett.*, 1979, **63,** 543; P. P. Infelta, *ibid.*, 1979, **61,** 88; B. K. Selinger and A. R. Watkins, *ibid.*, 1979, **56,** 99.

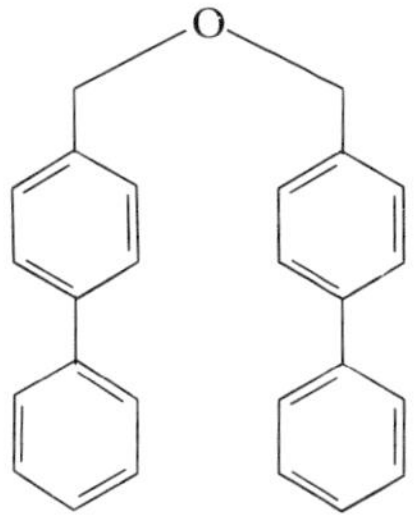

(113)

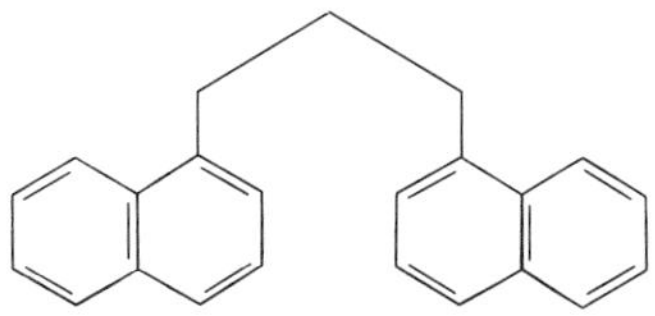

(114)

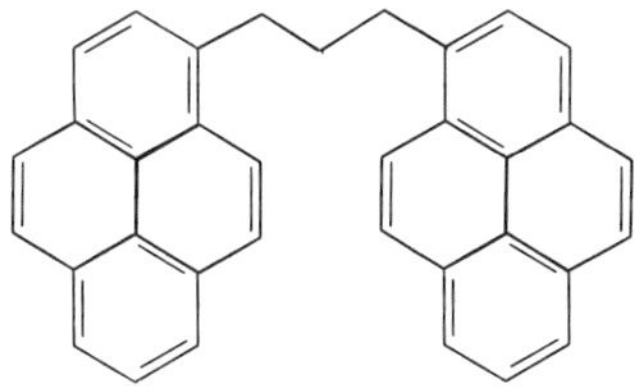

(115)

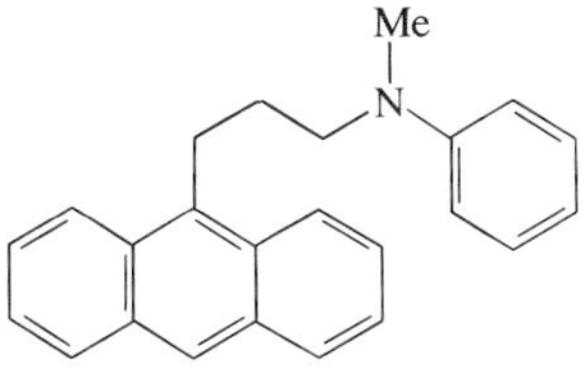

(116)

The concentration of excimers may be artificially enhanced by employing a bridged bis-arene and (113),[253] (114),[254] and (115)[255] were all synthesized at a similar time by different research groups. They give a useful method for measurement of micellar microviscosities since the ratio of excimer to normal fluorescence is viscosity dependent. By this technique cetyltrimethylammonium chloride is more viscous than ethylene glycol but sodium lauryl sulphate is a less viscous medium. A prior, and conceptually similar, publication deals with intramolecular fluorescence quenching of anthracene S_1 by tertiary aromatic amines such as (116). Since this process is only effective for the neutral amine it affords a method of determining the pK_a of S_1.[256] There is an interesting constrast between the excited-state behaviour of 1-naphthol and 2-naphthol in micelles.[257] The former fluoresces in sodium lauryl sulphate micelles but not in water, the latter fluorescing in both environments. The phenol and phenoxide ion chromophores are distinct. Micellar fluorescence of 1,5-dimethyl-naphthalene may be quenched by oxygen, which is preferentially solvated in the micelle, with high efficiency.[258] The system is well adapted to quantitative analysis leading to a quenching rate-constant $5 \times 10^7 \, \text{s}^{-1}$.

The photochemistry of cyanine (117) has been investigated. In water, slow

[253] J. Emert, C. Behrens, and M. Goldenberg, *J. Am. Chem. Soc.*, 1979, **101**, 771.
[254] N. J. Turro, M. Aikawa, and A. Yekta, *ibid.*, 1979, **101**, 773.
[255] K. Zachariasse, *Chem. Phys. Lett.*, 1978, **57**, 429.
[256] B. K. Selinger, *Aust. J. Chem.*, 1977, **30**, 2087.
[257] B. K. Selinger and A. Weller, *Aust. J. Chem.*, 1977, **30**, 2377.
[258] N. J. Turro, M. Aikawa, and A. Yekta, *Chem. Phys. Lett.*, 1979, **64**, 473.

Et
|
N   Ph

O ⟍   ⟍   ⟍   N
Ph ⟍   O

N⁺
|
Et   NO₃⁻

(117)

aggregation occurs which decreases both absorbance and fluorescence intensity. In sodium lauryl sulphate, fluorescence emission is strongly enhanced, and self-aggregation does not occur; the dye is then stable to both xenon and 347 nm ruby laser irradiation.[259]

The triplet state has received rather less attention than the singlet in micellar photochemistry. Phosphorescent decay of solubilized polynuclear aromatic hydrocarbons from their T_1 state may be observed in heavy-metal ion lauryl sulphate micelles.[260] This involves a conventional intersystem crossing from $S_1 \leadsto T_1$ promoted by spin–orbit coupling with the heavy atom. 1-Bromonaphthalene readily forms a triplet excited-state in micelles, which may be quenched by added sodium nitrite in water, the lifetime then being reduced from 2.8×10^{-4} s to 5×10^{-9} s.[261] There is efficient triplet energy transfer from N-methylphenothiazine (T_1) to *trans*-stilbene (S_0), which is irreversible but reversible energy-transfer to naphthalene (S_0) occurs.[262]

Excited states of metal-ion-containing surfactants are of very considerable current interest. Photoexcitation of ruthenium(II) tris-bipyridyl²⁺ has been studied and in sodium lauryl sulphate micelles efficient triplet energy-transfer to 9-methylanthracene occurs within the micelle. It is governed by the equation:

$$\ln\left(\frac{I}{I_0}\right) = \frac{[\text{9-Me-anthracene}]_{\bar{n}}}{[\text{NaLS}]_{\text{tot}} - [\text{NaLS}]_{\text{free}}}, \qquad (7)$$

and therefore provides a convenient method for measuring the surfactant aggregation number when the concentration of Ru(bipy)₃²⁺ is sufficiently low. A value of 60 ± 2 was obtained at 25 °C.[263] The bis-lauryl sulphate salt of this cation may be isolated, and is only water soluble in the presence of excess sodium lauryl sulphate.[264] Fluorescence of the charge-transfer excited state is quenched efficiently by aqueous Cu^{2+} and inefficiently by $Fe(CN)_6^{3-}$, an electron-transfer mechanism being involved. In the absence of addends, a fast excited-state decay occurs by intramicellar triplet–triplet quenching.[265] This

[259] R. Humphry-Baker, M. Grätzel, and R. Steiger, *J. Am. Chem. Soc.*, 1980, **102**, 847.
[260] R. Humphry-Baker, Y. Moroi, and M. Grätzel, *Chem. Phys. Lett.*, 1978, **58**, 207; K. Kalyanasundaram, F. Grieser, and J. K. Thomas, *ibid.*, 1977, **51**, 507.
[261] N. J. Turro, K. C. Liu, M. F. Chow, and P. Lee, *Photochem. Photobiol.* 1978, **27**, 523; J. K. Thomas, F. Grieser, and M. Wong, *Ber. Bunsenges. Phys. Chem.*, 1978, **82**, 937; M. Almgren, F. Grieser, and J. K. Thomas, *J. Am. Chem. Soc.*, 1979, **101**, 279.
[262] G. Rothenberger, P. P. Infelta, and M. Grätzel, *J. Phys. Chem.*, 1979, **83**, 1871.
[263] N. J. Turro and A. Yetka, *J. Am. Chem. Soc.*, 1978, **100**, 5951.
[264] D. Meisel, M. S. Matheson, and J. Rabani, *J. Am. Chem. Soc.*, 1978, **100**, 117.
[265] U. Lachish, M. Ottolenghi, and J. Rabani, *J. Am. Chem. Soc.*, 1977, **99**, 8062.

mechanism occurs several orders of magnitude faster in anionic micelles than in free solution. The triplet state of solubilized naphthalene transfers energy to aqueous terbium chloride (much used as a phosphorescence quench in biochemical studies), which is presumed to be strongly bound to the aqueous interface of the anionic micelle.[266] Similar reactions of terbium and europium chlorides with 1-bromonaphthalene or biphenyl T_1 states have been studied quantitatively.[267] The second-order corrected rate constants are quite small ($\leqslant 10^5$ M^{-1} s^{-1}), and reaction is facilitated by approximation and localization of the hydrocarbon T_1 state within the anionic micelle.

Photoionization and Electron Transfer from Excited States.—Realistic models for solar energy conversion require that two separate photocatalysed reactions occur simultaneously *e.g.* the photodissociation of water (Scheme 5). A difficulty, which has to be overcome before any practical application is conceivable, is the possibility of back reactions, so that some barrier to recombination must be enforced. The applicability of micelles and membranes is obvious, and this has stimulated an intense effort involving competition and collaboration by several groups.

One promising lead was the discovery (Colloid Science, Vol. 3, p. 268) that photoionization processes are facilitated in micelles. In the case of tetramethylbenzidine (118) photolysis in anionic micelles leads to a radical cation that is stable for days. Concomitant with this is expulsion of an electron from the micelle, which reacts stoicheiometrically to produce 0.5 moles of hydrogen and hydroxide ion. Many further examples are now available. Flash photolysis of chloropromazine (119) and its analogues in oxygenated anionic micelles gives efficient photoionization, and a radical cation whose lifetime is $>10^2$ longer in the micelle than in water.[268] Laser-induced photoionization of pyrene may be

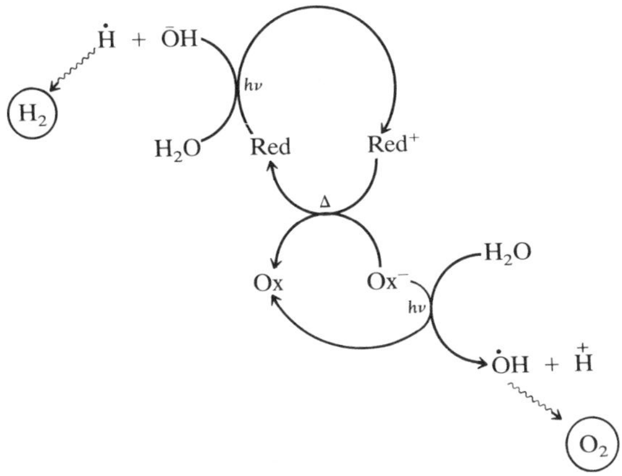

Scheme 5

[266] J. R. Escabi-Perez, F. Nome, and J. H. Fendler, *J. Am. Chem. Soc.*, 1977, **99**, 7749.
[267] M. Almgren, F. Grieser, and J. K. Thomas, *J. Am. Chem. Soc.*, 1979, **101**, 2021.
[268] T. Iwaoka and M. Kondo, *Chem. Lett.*, 1978, 731.

Me$_2$N—⟨⟩—⟨⟩—NMe$_2$

(118)

CH$_2$CH$_2$CH$_2$NMe$_2$
(119)

effected indirectly, since the S_1 state of the hydrocarbon undergoes electron transfer from dimethylaniline. In cationic micelles this results in loss of the cation radical and retention of pyrene$^{\cdot-}$ for a mean lifetime of 5×10^{-4} s. Substituted pyrenes form radical anions in this manner with a wide range of stabilities.[269]

One difficulty in photochemical energy conversion is the need to consider the efficient range of incident sunlight, say 400—700 nm. Using a neodymium laser at 530 nm, Thomas and co-workers demonstrated[270] that 3-aminoperylene in anionic micelles undergoes one-photon photoionization to give a solubilized radical cation and solvated electrons, which may be scavenged by oxygen. The result has survived criticism[271] that the observation of a one-photon process may reflect a micellar inhibition of ion recombination leading to an apparent change from bi- to mono-photonic ionization.[272] Microemulsions may offer some advantages over micelles in electron-transfer processes of this type.[273]

A second, related development has been in the application of transition-metal ions and complexes in electron transfer. Since excited Ru(bipy)$_3^{2+}$ is a powerful electron donor, this has been a subject of considerable study. Whitten and co-workers made a surfactant ruthenium derivative (120), which in preliminary work appeared to photocatalyse the decomposition of water at monolayers.[274] This was subsequently contested,[275] and it now seems that the *pure* material does not give rise to hydrogen and oxygen in appreciable quantities on photolysis. The ruthenium complex (120) may form part of a redox couple[276] in sodium lauryl sulphate micelles, which is a model for some later developments in which this type of system is used for photochemical energy storage. (Scheme 6). Hydrogen is not evolved in that system unless the enzyme hydrogenase is added to reoxidize methylviologen$^{\cdot+}$ (121). A conceptually similar system[277] employs Triton X-100-solubilized carbonyl-ruthenium(ii) tetraphenylporphyrin, 2-mercaptoethanol, and (121) and gave 12 turnovers (2.9 μmol) of hydrogen after eight hours irradiation.

[269] B. Katusin-Razem, M. Wong, and J. K. Thomas, *J. Am. Chem. Soc.*, 1978, **100,** 1679.
[270] J. K. Thomas and P. Piciulo, *J. Am. Chem. Soc.*, 1978, **100,** 3239.
[271] G. E. Hall, *J. Am. Chem. Soc.*, 1978, **100,** 8262.
[272] J. K. Thomas and P. Piciulo, *J. Am. Chem. Soc.*, 1979, **101,** 2502.
[273] J. Kiwi and M. Grätzel, *J. Am. Chem. Soc.*, 1978, **100,** 6314.
[274] G. Sprintschnik, H. W. Sprintschnik, P. P. Kirsch, and D. G. Whitten, *J. Am. Chem. Soc.*, 1976, **98,** 2237; *ibid.*, 1977, **99,** 4947; P. J. DeLaive, J. T. Lee, H. W. Sprintschnik, H. Abrüna, T. J. Meyer, and D. G. Whitten, *ibid.*, 1977, **99,** 7094.
[275] A Harriman, *J. Chem. Soc., Chem. Commun.*, 1977, 777.
[276] K. Kalyanasundaraman, *J. Chem. Soc., Chem. Commun.*, 1978, 628.
[277] I. Okura and N. Kim Thuan, *J. Chem. Soc., Chem. Commun.*, 1980, 84.

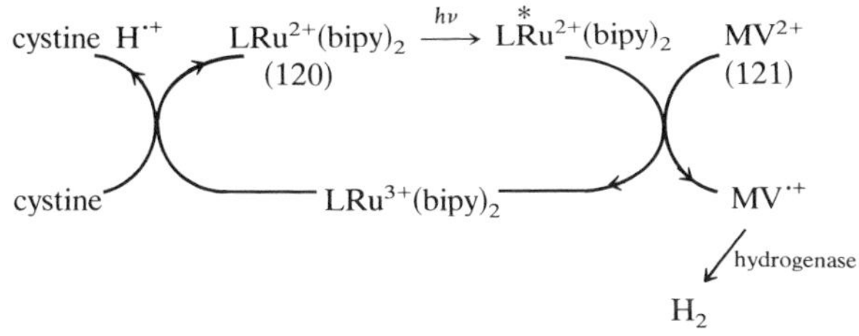

Scheme 6

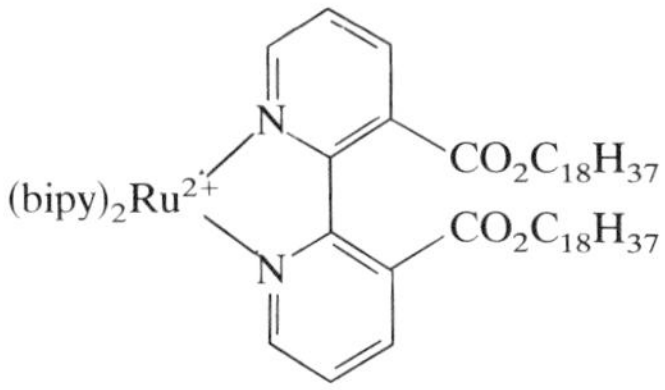

(120)

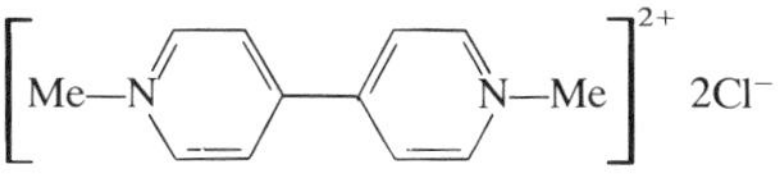

(121)

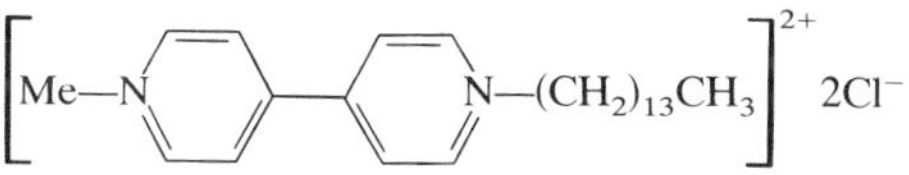

(122)

A considerable advance is the application of surfactant analogues of methyl-viologen as the electron-acceptor.[278] The dication $(122)^{2+}$ forms micelles above 7×10^{-3} M and does not have a particularly high affinity for cetyltrimethyl-ammonium chloride micelles. The monocationic reduced form is much less hydrophilic, however. A direct consequence is that the forward electron transfer is unaffected by cationic micelles

$$\text{Ru(bipy)}_3^{2+} \xrightarrow{530\,\text{nm}} \overset{*}{\text{Ru}}\text{(bipy)}_3^{2+} + (122)^{2+} \underset{k_r}{\overset{k_f}{\rightleftharpoons}} \text{Ru(bipy)}_3^{3+} + (122)^{\cdot+}, \qquad (8)$$

but the reverse reaction is strongly retarded. This couple has not yet been applied to photochemical water fission, but should (in principle) be effective. Grätzel and co-workers[279] have also developed a series of related redox catalysts containing colloidal metals and edta, the most effective being based on platinum, and capable of producing 12 litres of hydrogen per day per litre of water on irradiation with a 450 watt lamp in the visible region.

Four further papers from the same laboratory illustrate the potential of metal-ion photochemistry in micelles. The annelide (18) (see ref. 67, p. 189) forms a stable silver ion adduct (123) which reacts by electron-transfer from (124) on photochemical excitation.[280] The resulting Ag(0) complex is stable to

[278] P. A. Brugger and M. Grätzel, *J. Am. Chem. Soc.*, 1980, **102**, 2461.
[279] J. Kiwi and M. Grätzel, *Angew. Chem., Int. Ed. Engl.*, 1979, **18**, 624.
[280] R. Humphry-Baker, M. Grätzel, P. Tundo, and E. Pelizzetti, *Angew. Chem., Int. Ed. Engl.*, 1979, **18**, 630.

M

$C_{14}H_{29}$

(123) M = Ag$^+$
(124) M = Ag

$C_{14}H_{29}$
Cu^{2+}

(125)

aggregation and may be isolated by extraction into chloroform. The copper(II) complex (125) may be correspondingly photoreduced in the presence of N-methylphenothiazine.[281]

Photochemically-induced electron transfer from N-methylphenothiazine in anionic micelles to cupric ions occurs in the Stern layer. This process is complete within the 15 ns duration of the 347 nm ruby laser excitation. The primary photochemical event is formation of the T_1 state of N-methylphenothiazine and both this and the subsequent electron transfer occur with unit quantum yield completely suppressing competitive photochemical pathways. Added Co^{2+} and Ni^{2+} deactivate the triplet state.[282] A related system in which Eu^{3+} replaced Cu^{2+} was examined, and electron transfer was again observed. In this case back reaction occurs in the ground state and intermicellar events may be kinetically distinguished from the much faster intramicellar events.[283]

Organic Photoreactions in Micelles.—The reactivity of excited states may be strongly perturbed by micelles, for a variety of reasons. Enforced approximation may make intermolecular reactions more efficient in some cases. In others, the cage effect of the micelle can entrap transients, and especially inhibit radical-pair diffusion processes. Micelle-induced conformational changes may affect the product distribution in intramolecular hydrogen-atom abstraction, and the reactivity of external reagents towards excited states may be enhanced by ionic micelles.

An example of the last effect is the hydroxide-ion catalysed photochemical Nef reaction of (4-nitrophenyl)nitromethane. This requires a prior deprotonation to give (126), which reacts to form 4-nitrobenzaldehyde presumably *via* (127). In this case it is a favourable ground-state pre-equilibrium which leads to catalysis by cationic micelles.[284] By contrast, the quantum yield for alkaline hydrolysis of 3,4-dinitroanisole is decreased in tetradecyltrimethylammonium chloride micelles, simply because the lifetime of the triplet excited state is greatly reduced in micelles,[285] and this is the species which is reactive towards hydroxide ion (Scheme 7).

[281] R. Humphry-Baker, Y. Moroi, M. Grätzel, E. Pelizzetti, and P. Tundo, *J. Am. Chem. Soc.*, 1980, **102**, 3689.
[282] Y. Moroi, A. M. Braun, and M. Grätzel, *J. Am. Chem. Soc.*, 1979, **101**, 567.
[283] Y. Moroi, P. P. Infelta, and M. Grätzel, *J. Am. Chem. Soc.*, 1979, **101**, 573.
[284] K. Yamada, K. Shigehiro, T. Kujozuka, and H. Iida, *Bull. Chem. Soc. Jpn.*, 1978, **51**, 2447.
[285] J. B. S. Bonilha, H. Chaimovich, V. G. Toscano, and F. Quina, *J. Phys. Chem.*, 1979, **83**, 2463.

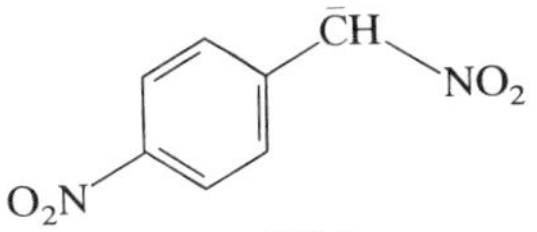

(126)

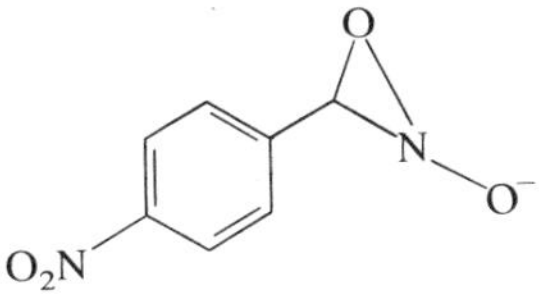

(127)

Dimerization of photoreactive olefins is generally enhanced in micelles. In iso-phorone (128) dimerization, the ratio of possible photodimers is altered on micellization,[286] and acenaphthene dimerization to (129) gives 95% of product in non-ionic or anionic surfactant solution, but 10% on irradiation in benzene solution under otherwise identical conditions. The *cis*:*trans* ratio varies from 22.3:1 in sodium lauryl sulphate micelles, but 0.06:1 at low substrate concentration in non-ionic micelles.[287] The crossed photodimer (130) is a significant product on co-irradiation of acenaphthene and acrylonitrile in micelles, but not in benzene.[288] Irradiation of 3-alkylcyclopentenones in micelles of potassium dodecanoate gives 98% of head-to-head dimer (131), which is never the dominant product in organic solvents. This result implies ordering of the

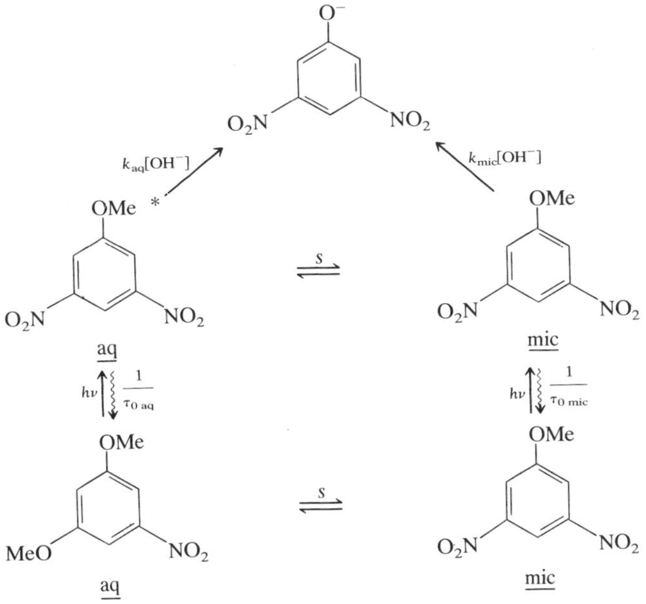

Scheme 7

[286] I. Rico, M. T. Maurette, E. Oliveros, M. Riviere, and A. Lattes, *Tetrahedron Lett.*, 1978, 4795.
[287] Y. Nakamura, Y. Ikamura, T. Kato, and Y. Morita, *J. Chem. Soc., Chem. Commun.*, 1977, 887.
[288] Y. Nakamura, Y. Ikamura, and Y. Morita, *Chem. Lett.*, 1978, 965.

(128)   (129)   (130)

reactant in micelles.[289] The photochemical behaviour of (132) is phase-dependent.[290] In the solid state, excimer fluorescence and olefin dimerization are predominant, these being coupled with *cis-trans* olefin isomerization in aqueous monolayers. In the softer and less organized environment of cetyltrimethylammonium bromide micelles only the single-molecule processes of monomer fluorescence and *cis–trans*-isomerization are observed.

(131)   (132)   (133)

Singlet $^1\Delta_g$ oxygen may react with micelle-bound substrates such as (133). The state lifetime is the same in water as in micelles[291] but substantially longer in D_2O (in the presence or absence of micelles) because the intersystem-crossing mechanism requires vibrational coupling with solvent, and the frequency matching of O–H is much better than that of O–D. Reactions of micelle-bound substrates are probably favoured by the preferential solubilization of oxygen in micelles, but the trapping efficiency is insensitive to the site of photoexcitation.[292] Similar conclusions are reached in a study involving several acceptors in dodecyltrimethylammonium chloride solution.[293] At high pressures of oxygen, the transfer of $^1\Delta_g$-excited molecules into the micelle is faster because of collisional-energy transfer, which is then faster than intersystem crossing.[294]

The photo-oxidation of protoporphyrin IX (134) produces two products, namely (135) and (136).[295] In organic solvents the former is dominant but in micelles or monolayers the product formally derived from 1,2-addition of singlet oxygen, (136) is preferred.

Photochemical intramolecular hydrogen-atom abstraction affords a method for determining the microviscosity of micelles, as the Barton reaction of (137)

[289] K. K. Lee and P. de Mayo, *J. Chem. Soc., Chem. Commun.*, 1979, 493.
[290] F. H. Quina and D. G. Whitten, *J. Am. Chem. Soc.*, 1977, **93**, 877.
[291] A. A. Gorman and M. A. J. Rodgers, *Chem. Phys. Lett.*, 1978, **55**, 52.
[292] B. A. Lindig and M. A. J. Rodgers, *J. Phys. Chem.*, 1979, **83**, 1683.
[293] Y. Usui, M. Tsukada, and H. Nakamura, *Bull. Chem. Soc. Jpn.*, 1978, **51**, 379.
[294] I. B. C. Matheson and R. Massoudi, *J. Am. Chem. Soc.*, 1980, **102**, 1942.
[295] B. E. Horsey and D. G. Whitten, *J. Am. Chem. Soc.*, 1978, **100**, 1293.

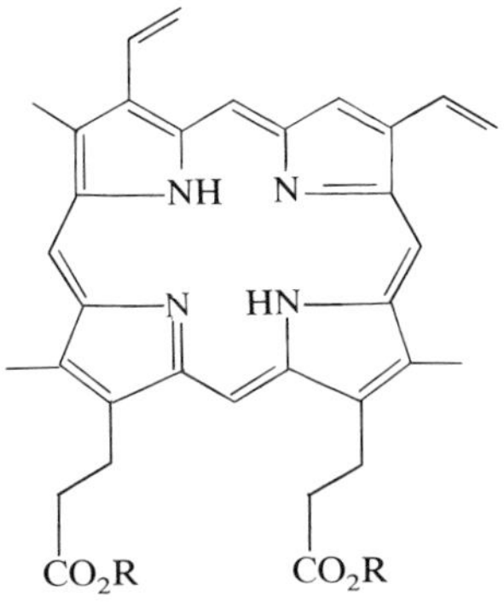

(134)

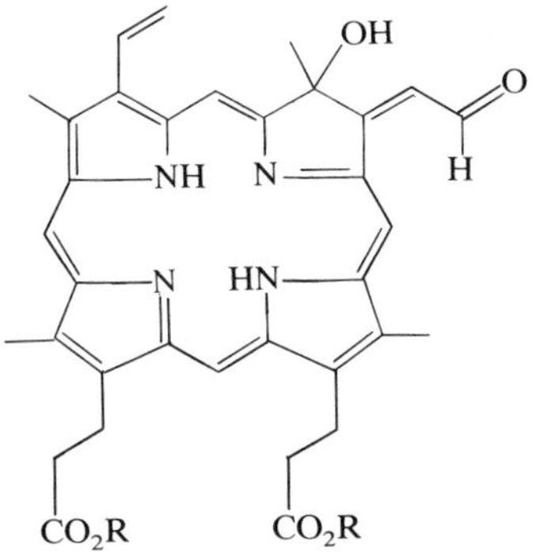

(135)

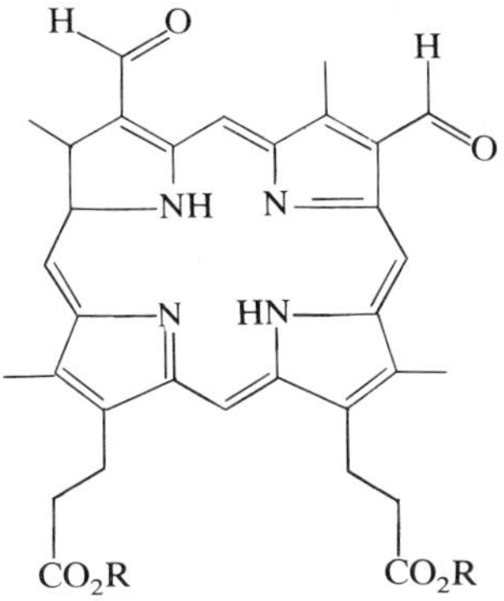

(136)

is dependent on local correlation times.[296] Formation of (138) is suppressed considerably in potassium dodecanoate micelles relative to non-viscous hydrocarbon solvent, a remarkably good relationship between yield and reciprocal viscosity being found. The microviscosity of the micelle is found to be 19 cP on this basis. Intermolecular hydrogen-atom abstraction in mixed micelles of benzophenone probes has been studied[297] (see also ref. 224). A related paper[298] suggests some degree of selectivity variation between (139) and (140) suggesting that the carbonyl-group of the latter is more deeply buried in sodium myristate micelles. Compound (141) has been photolysed in solution anionic micelles, and assembled multilayers[299] containing arachidic acid. The Norrish Type 2 process was very favourable in micelles giving *p*-methyl-acetophenone as the main photoproduct, but strongly disfavoured in

[296] K.-Y. Law and P. de Mayo, *J. Chem. Soc., Chem. Commun.*, 1978, 110.
[297] R. Breslow, S. Kotabatake, and J. Rothbard, *J. Am. Chem. Soc.*, 1978, **100**, 8156; R. Breslow, J. Rothbard, F. Herman, and M. L. Rodrigues, *ibid.*, 1978, **100**, 1213.
[298] M. Mitani, T. Suzuki, H. Takeuchi, and K. Koyama, *Tetrahedron Lett.*, 1979, 803.
[299] P. R. Worsham, D. W. Eaker, and D. G. Whitten, *J. Am. Chem. Soc.*, 1978, **100**, 7091.

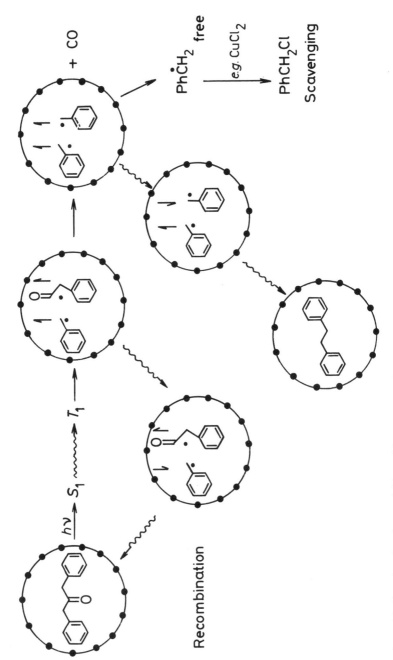

Figure 11 *Photodecarbonylation of dibenzyl ketone in micelles*

(137)

(139) $n = 0$
(140) $n = 2$

(138)

(141)

monoloayers. Whitten[300] has reviewed micellar and monolayer photochemistry.

A series of ingenious papers[301—5] by Turro and co-workers at Columbia complete this review. It is known that dibenzyl ketones undergo photodecarbonylation, and the initial products can recombine within the solvent cage or diffuse away and recombine randomly. The latter process is preferred in homogeneous solution In a micelle, where the ketone is present at high dilution, the initial radical pair are in enforced contact since their exit is disfavoured. The result is a dramatic increase in the efficiency of cage recombination[301] (Figure 11). The initial radical pair prior to decarbonylation is in its triplet state, which undergoes intersystem crossing and return to starting material in competition with CO expulsion. Intersystem crossing is easier for the ^{13}CO isomer of starting material, since it involves nuclear-electron hyperfine coupling, and only the carbon-13 isotope possesses a magnetic moment. A consequence of this is that starting material becomes progressively enriched in ^{13}C.[302] The effect is strikingly large in micelles and the quantum yield for photolysis of the ^{12}CO isomer is 30% higher than that for photolysis of the ^{13}C isomer. Since intersystem crossing may be perturbed by a magnetic nucleus, it should also be magnetic field dependent and in a laboratory magnet, geminate recombination in the micelle becomes much less efficient.[303] The mechanisms involved in photochemical transformations of dibenzyl ketones, (summarized in Figure 11) are fully corroborated by studies on phenyl adamantyl ketone[304] and on phenyl 1,1-dimethylbenzyl ketone.[305] In the last case, cage products are α-methylstyrene and benzaldehyde formed by intramolecular H-atom transfer in the initial cleavage product. Their production is enhanced in cationic, neutral, or anionic micelles compared to benzene solution, and unaffected by addition of Cu^{2+} salts to the surfactant solution.

[300] D. G. Whitten, *Angew. Chem., Int. Ed. Engl.*, 1978, **18**, 440.
[301] N. J. Turro and W. R. Cherry, *J. Am. Chem. Soc.*, 1978, **100**, 7431.
[302] N. J. Turro and B. Kraeutler, *J. Am. Chem. Soc.*, 1978, **100**, 7432.
[303] N. J. Turro, B. Kraeutler, and D. R. Anderson, *J. Amer. Chem. Soc.*, 1979, **101**, 7435.
[304] N. J. Turro, D. R. Anderson, and B. Kraeutler, *Tetrahedron Lett.*, 1980, **21**, 3.
[305] N. J. Turro and J. Mattey, *Tetrahedron Lett.*, 1980, **21**, 1799.

6

A Bibliography of Gas–Liquid Surface Tensions for Binary Liquid Mixtures

BY I. A. McLURE, I. L. PEGG, AND V. A. M. SOARES

1 Introduction

The second section of this bibliography contains entries for mixtures in the alphabetical order of the first component according to the Hill system as used in Chemical Abstracts. Each entry consists of the name of the mixture followed by the temperatures for which the results were reported with in parentheses the number of these temperatures if more than three, the deviation of the surface tensions from the mole fraction-weighted average of the surface tensions of the pure components that is taken, *faute de mieux*, as an operationally useful if theoretically unsupported measure of ideality, the method used for the surface-tension determinations, and a reference to the original publication. The gaps in some entries reflect the paucity of information given either in the abstract of the article or in the source article itself. The third section is a list of the references in numerical order. The fourth section is a compound index whereby all the binary mixtures in which a particular substance is involved can be located. Numbers in brackets in the index denote the number of mixtures that correspond to a particular formula pair. The mixtures with very few exceptions contain only non-electrolyte substances. The exceptions, such as Pb+Sn, are chosen as having some particular significance in the treatment of the surface tension of mixtures. The literature has been covered to the end of September 1981.

The Hill empirical formula is obtained for a particular substance by placing the elements in their alphabetical order except that if C is present it is placed first, followed by H, and then the other elements present in alphabetical order.

2 Compilation of Data on Binary Liquid Mixtures

Table 1

SYSTEM

Temperature	Deviation	Method	Ref.
$Ar(Argon)+CH_4(Methane)$			
90.67 K	Negative	Cap.Rise	[22]
			[126]
$Ar(Argon)+D_2(Deuterium)$			
			[153]
$Ar(Argon)+H_2(Hydrogen)$			
60,87,104 K		Dif.Cap.Rise	[175]
			[153]
$Ar(Argon)+He(Helium)$			
			[153]
$Ar(Argon)+NH_3(Ammonia)$			
			[123]
$Ar(Argon)+N_2(Nitrogen)$			
83.82 K	Negative	Cap.Rise	[22]
83.85 K	Negative		[126]
$Ar(Argon)+Ne(Neon)$			
			[153]
$Ar(Argon)+O_2(Oxygen)$			
			[76]
83.85 K	Negative		[142]
$Br_2(Bromine)+CCl_4(Carbon\ tetrachloride)$			
293.15 to 343.15 K	Negative	Cap.Rise&Bub.Pres.	[150]
$Br_2(Bromine)+C_6H_6(Benzene)$			
293.15 to 343.15 K		Cap.Rise&Bub.Pres.	[150]
$CBr_4(Tetrabromomethane)+CCl_4(Carbon\ tetrachloride)$			
298.15 K			[232]

Table 1 (*cont.*)
SYSTEM

	Temperature	Deviation	Method	Ref.

CCl_4(Carbon tetrachloride)+$CHCl_3$(Chloroform)

	298.15 K	Negative	Cap.Rise	[20]
	293.15 K			[103]
				[180]
	291.35 K	Neg.Min.	Cap.Rise	[188]

CCl_4(Carbon tetrachloride)+CH_3NO_2(Nitromethane)

| | 303.15,308.15,318.15 K | | Ring | [177] |

CCl_4(Carbon tetrachloride)+CS_2(Carbon disulphide)

| | 395.05 K | Negative | Cap.Rise | [3] |
| | 298.15 K | Negative | | [72] |

CCl_4(Carbon tetrachloride)+$C_2H_4Cl_2$(1,2-Dichloroethane)

| | 293.15,313.15 K | | | [79] |

CCl_4(Carbon tetrachloride)+$C_2H_4O_2$(Acetic acid)

| | 291.35 K | Neg.Min. | Cap.Rise | [188] |

CCl_4(Carbon tetrachloride)+$C_2H_5NO_2$(Nitroethane)

| | 303.15,308.15,318.15 K | | Ring | [177] |

CCl_4(Carbon tetrachloride)+C_2H_6O(Ethanol)

| | 278.15 to 333.15 K | | | [145] |

CCl_4(Carbon tetrachloride)+C_2H_6OS(Dimethyl sulphoxide)

| | | | | [94] |
| | 303.15,313.15,323.15 K | Negative | Cap.Rise | [274] |

CCl_4(Carbon tetrachloride)+C_3H_6O(Acetone)

| | 278.15 to 333.15 K | | | [145] |

CCl_4(Carbon tetrachloride)+$C_4H_8O_2$(Ethyl acetate)

	298.15 K	Negative	Max.Bub.Pres.	[16]
	298.15 K	Negative		[72]
	293.15 K	Positive		[242]
	288.15,308.15,323.15 K	Negative		[231]

CCl_4(Carbon tetrachloride)+$C_4H_8O_2$(Propyl formate)

| | 293.15 K | Negative | | [242] |

CCl_4(Carbon tetrachloride)+C_5H_{10}(Cyclopentane)

| | 298.15 K | Negative | Max.Bub.Pres. | [33] |

Table 1 (*cont.*)
SYSTEM

Temperature	Deviation	Method	Ref.

CCl_4(Carbon tetrachloride)+C_6H_5Cl(Chlorobenzene)

293.15 K	Positive	Drop Weight	[23]
293.15 K	Negative	Cap.Rise	[23]
323.15 to 443.15 K	Negative	Cap.Rise	[127]
			[176]

CCl_4(Carbon tetrachloride)+$C_6H_5NO_2$(Nitrobenzene)

298.15 K	Negative	Max.Bub.Pres.	[16]
288.15,293.15,317.15 K			[93]

CCl_4(Carbon tetrachloride)+C_6F_6(Hexafluorobenzene)

293.15,313.15 K	Negative		[154]

CCl_4(Carbon tetrachloride)+C_6H_6(Benzene)

313.15 K	Negative	Cap.Rise	[3]
283.15 to 351.15 K (3)			[229]
323.15 K	Negative	Max.Bub.Pres.	[17]
323.15 to 443.15 K	Negative	Cap.Rise	[127]
298.15 K	Positive		[72]
293.15 K	Negative		[195]
295.15 K	Positive		[194]
298.15 K	Negative	Max.Bub.Pres.	[16]
			[118]

CCl_4(Carbon tetrachloride)+C_6H_{12}(Cyclohexane)

280.73 to 313.15 K (5)	Negative	Dif.Cap.Rise	[44]
298.15 K	Negative	Cap.Rise	[23]

CCl_4(Carbon tetrachloride)+C_6H_{14}(Hexane)

293.15 to 343.15 K	Negative	Cap.Rise&Bub.Pres.	[150]

CCl_4(Carbon tetrachloride)+C_7H_8(Toluene)

313.15 K	Positive	Cap.Rise	[3]
293.15 K	Positive	Cap.Rise&Drop Wt.	[23]
323.15 to 443.15 K	Positive	Cap.Rise	[127]
278.15 to 333.15 K			[145]

$CHCl_3$(Chloroform)+CH_4O(Methanol)

298.15 K	Negative		[252]

$CHCl_3$(Chloroform)+CS_2(Carbon disulphide)

			[180]
291.35 K	Negative	Cap.Rise	[188]
286.15,319.55,334.35 K	Negative		[229]
288.15 K			[234]

Table 1 (*cont.*)

SYSTEM

	Temperature	Deviation	Method	Ref.

$CHCl_3$(Chloroform)+$C_2H_4Cl_2$(1,1-Dichloroethane)

| | 293.15 K | Positive | Cap.Rise | [8] |

$CHCl_3$(Chloroform)+$C_2H_4Cl_2$(1,2-Dichloroethane)

| | 293.15 K | Positive | Cap.Rise | [8] |

$CHCl_3$(Chloroform)+$C_2H_4O_2$(Acetic acid)

| | 291.35 K | Neg.Min. | Cap.Rise | [188] |

$CHCl_3$(Chloroform)+C_2H_6O(Ethanol)

| | 273.15 60 298.15 K (6) | Positive | | [71] |

$CHCl_3$(Chloroform)+C_2H_6OS(Dimethyl sulphoxide)

| | 303.15,313.15,323.15 K | Negative | Cap.Rise | [274] |

$CHCl_3$(Chloroform)+C_3H_6O(Acetone)

	291.16 K	Positive	Cap.Rise	[188]
	279.15 K	Positive		[194]
	288.15,298.15,305.15 K	Positive		[231]

$CHCl_3$(Chloroform)+$C_3H_6O_3$(Methyl carbonate)

| | 298.15 K | Negative | Max.Bub.Pres. | [25] |

$CHCl_3$(Chloroform)+$C_4H_8O_2$(Butanoic acid)

| | 273.15,306.15 K | Negative | | [227] |

$CHCl_3$(Chloroform)+$C_4H_{10}O$(Diethyl ether)

	291.35 K	Negative	Cap.Rise	[188]
	288.15 K	Negative		[234]
	290.15 K	Negative		[241]

$CHCl_3$(Chloroform)+$C_6H_5NO_2$(Nitrobenzene)

| | 293.15 K | | | [10] |

$CHCl_3$(Chloroform)+C_6H_6(Benzene)

	313.15 K	Pos.Max.	Cap.Rise	[3]
	298.15 K	Positive	Max.Bub.Pres.	[16]
	291.35 K	Negative	Cap.Rise	[188]

$CHCl_3$(Chloroform)+C_6H_7N(Aniline)

| | 273.15 to 298.15 K (6) | Negative | | [71] |
| | 288.15 K | Negative | | [234] |

Table 1 (*cont.*)
SYSTEM

Temperature	Deviation	Method	Ref.
$CHCl_3$(Chloroform)+$C_6H_{10}O$(Cyclohexanone)			
291.15 K	Positive		[191]
$CHCl_3$(Chloroform)+$C_6H_{10}O_3$(Ethyl acetoacetate)			
288.15 K	Positive		[234]
$CHCl_3$(Chloroform)+$C_6H_{12}O$(Cyclohexanol)			
293.15 K	Negative		[192]
$CHCl_3$(Chloroform)+C_7H_8O(o-Cresol)			
288.15 K			[251]
$CHCl_3$(Chloroform)+C_7H_8O(m-Cresol)			
288.15 K			[251]
$CHCl_3$(Chloroform)+C_7H_8O(p-Cresol)			
288.15 K			[251]
$CHCl_3$(Chloroform)+$C_7H_{14}O$(o-Methylcyclohexanol)			
288.15 K			[251]
$CHCl_3$(Chloroform)+$C_7H_{14}O$(m-Methylcyclohexanol)			
288.15 K			[251]
$CHCl_3$(Chloroform)+$C_7H_{14}O$(p-Methylcyclohexanol)			
288.15 K			[251]
$CHCl_3$(Chloroform)+$C_{10}H_{12}$(Tetralin)			
291.15 K			[191]
CH_2O_2(Formic acid)+C_2HCl_3O(Trichloroacetaldehyde)			
283.15 to 363.15 K			[201]
293.15 to 353.15 K			[270]
CH_2O_2(Formic acid)+$C_2H_4O_2$(Acetic acid)			
288.15 to 348.15 K (7)	Negative		[225]
CH_2O_2(Formic acid)+C_3H_6O(Acetone)			
298.15, 308.15, 313.15 K			[95]

Table 1 (*cont.*)
SYSTEM

Temperature	Deviation	Method	Ref.

CH_2O_2(Formic acid)+C_5H_5N(Pyridine)

298.15 K	Pos.Max.	Cap.Rise	[6]
287.15 K	Pos.Max.Infl.	Cap.Rise	[193]

CH_2O_2(Formic acid)+$C_5H_{10}O$(Pentan-2-one)

298.15,308.15,313.15 K			[95]

CH_2O_2(Formic acid)+C_6H_6(Benzene)

348.15,352.15 K	Neg.Min.	Cap.Rise	[58]

CH_2O_2(Formic acid)+C_7H_9N(o-Toluidine)

298.15 K	Positive		[256]

CH_2O_2(Formic acid)+C_7H_9N(m-Toluidine)

298.15 K	Pos.Max.		[256]

CH_2O_2(Formic acid)+H_2O(Water)

298.15,303.15,308.15 K	Negative		[210]
298.15,308.15 K	Negative		[220]
283.15,303.15 K		Cap.Rise	[275]
			[104]

CH_3I(Iodomethane)+CS_2(Carbon disulphide)

288.15 K	Neg.Min.		[234]

CH_3I(Iodomethane)+$C_2H_4O_2$(Acetic acid)

293.14 to 323.15 K			[158]

CH_3I(Iodomethane)+$C_3H_6O_2$(Methyl acetate)

293.15 to 323.15 K			[158]

CH_3I(Iodomethane)+C_3H_8(Propane)

293.15 to 347.15 K (5)			[119]

CH_3I(Iodomethane)+$C_4H_6O_3$(Acetic anhydride)

293.15 to 323.15 K			[158]

CH_3I(Iodomethane)+$C_4H_{10}O$(Diethyl ether)

288.15 K	Negative		[234]

CH_3NO(Formamide)+$C_3H_8O_3$(Propan-1,2,3-triol)

293.15 K	Positive	Max.Bub.Pres.	[277]

Table 1 (*cont.*)
SYSTEM

Temperature	Deviation	Method	Ref.

CH_3NO(Formamide)+$C_5H_{11}NO$(Diethyl formamide)

| | | | [113] |

CH_3NO(Formamide)+H_2O(Water)

| 293.15 K | Negative | Max.Bub.Pres. | [277] |
| | | | [113] |

CH_3NO_2(Nitromethane)+$C_2H_4O_2$(Acetic acid)

| 293.15 K | Negative | Cap.Rise | [96] |

CH_3NO_2(Nitromethane)+$C_3H_6O_2$(Propanoic acid)

| 293.15 K | Negative | Cap.Rise | [96] |

CH_3NO_2(Nitromethane)+$C_4H_8O_2$(Butanoic acid)

| 293.15 K | Negative | Cap.Rise | [96] |

CH_3NO_2(Nitromethane)+$C_4H_8O_2$(Dioxan)

| 293.15,303.15 K | Negative | Max.Bub.Pres. | [46] |

CH_3NO_2(Nitromethane)+$C_5H_{10}O_2$(Pentanoic acid)

| 293.15 K | Negative | Cap.Rise | [96] |

CH_3NO_2(Nitromethane)+C_6F_6(Hexafluorobenzene)

| 293.15 to 313.15 K | Negative | | [154] |

CH_3NO_2(Nitromethane)+C_6H_6(Benzene)

| 298.15 K | Negative | Max.Bub.Pres. | [16] |
| 293.15,303.15 K | Negative | Max.Bub.Pres. | [46] |

CH_3NO_2(Nitromethane)+$C_6H_{12}O_2$(Hexanoic acid)

| 293.15 K | Negative | Max.Bub.Pres. | [96] |

CH_3NO_2(Nitromethane)+$C_7H_{14}O_2$(Heptanoic acid)

| 293.15 K | Negative | Max.Bub.Pres. | [96] |

CH_3NO_2(Nitromethane)+$C_8H_{16}O_2$(Octanoic acid)

| 293.15 K | Negative | Max.Bub.Pres. | [96] |

CH_4(Methane)+CO(Carbon monoxide)

| 363.82 K | Negative | Cap.Rise | [22] |

Table 1 (*cont.*)
SYSTEM

Temperature	*Deviation*	*Method*	*Ref.*
CH_4(Methane)+C_3H_8(Propane)			
278.15 to 467.15 K		Cap.Rise&Drop Vol.	[130]
CH_4(Methane)+C_9H_{20}(Nonane)			
238.15 to 298.15 K		Pendant Drop	[183]
CH_4(Methane)+Kr(Krypton)			
109.90 to 118.40 K	Negative	Cap.Rise	[21]
CH_4(Methane)+N_2(Nitrogen)			
90.67 K	Negative	Cap.Rise	[22] [126]
CH_4(Methane)+CS_2(Carbon disulphide)			
298.15 K	Negative	Dif.Cap.Rise	[51]
CH_4N_2O(Urea)+C_6H_6O(Phenol)			
353.15,363.15,368.15 K		Max.Bub.Pres.	[114]
CH_4O(Methanol)+C_2HCl_3O(Trichloroacetaldehyde)			
293.15 to 353.15 K			[270]
CH_4O(Methanol)+C_2H_5I(Iodoethane)			
298.35,310.65,318.15 K	Positive		[231]
CH_4O(Methanol)+C_2H_6O(Ethanol)			
273.15,303.15 K	Positive	Drop Weight	[1]
278.15 to 333.15 K			[145]
298.15 K			[152]
323.15 K	Negative	Drop Weight	[1]
CH_4O(Methanol)+C_2H_6OS(Dimethyl sulphoxide)			
			[144]
CH_4O(Methanol)+C_2H_7NO(2-Aminoethanol)			
			[108]
CH_4O(Methanol)+C_3H_6O(Acetone)			
273.15,303.15 K	Positive	Drop Weight	[1]
298.15 K	Pos.Max.	Dif.Cap.Rise	[51]
293.15,308.15 K			[103]
			[276]

Table 1 (*cont.*)
SYSTEM

Temperature	Deviation	Method	Ref.

CH_4O(Methanol)+C_3H_8O(Propanol)

| 288.15 to 333.15 K (6) | | | [225] |

CH_4O(Methanol)+C_3H_8O(Propan-2-ol)

| | | | [276] |

CH_4O(Methanol)+$C_4H_8O_2$(Dioxan)

| 293.15,313.15 K | Negative | | [173] |

CH_4O(Methanol)+$C_4H_8O_2$(Ethyl acetate)

| 293.15 to 323.15 K (4) | Negative | | [253] |

CH_4O(Methanol)+$C_4H_8O_2$S(Tetramethylene sulphone)

| 293.15 to 323.15 K | | | [120] |

CH_4O(Methanol)+$C_4H_{10}O$(Butanol)

| 298.15 K | Positive | Max.Bub.Pres. | [35] |

CH_4O(Methanol)+$C_4H_{10}O$(2-Methylpropanol)

| 298.15 K | Pos.Neg. | Max.Bub.Pres. | [35] |

CH_4O(Methanol)+$C_4H_{10}O$(Butan-2-ol)

| 298.15 K | Pos.Neg. | Max.Bub.Pres. | [35] |

CH_4O(Methanol)+$C_4H_{10}O$(2-Methylpropan-2-ol)

| 298.15 K | Pos.Neg. | Max.Bub.Pres. | [35] |

CH_4O(Methanol)+$C_5H_{11}N$(N-Methylpyrrolidine)

| | | | [186] |

CH_4O(Methanol)+C_6H_6(Benzene)

273.15,303.15 K	Negative	Drop Weight	[1]
298.15 K			[146]
300.15 K			[160]
293.15,303.15,313.15 K	Pos.Max.		[284]

CH_4O(Methanol)+C_6H_{12}(Cyclohexane)

| 319.15,321.15 K | Neg.Min. | Dif.Cap.Rise | [43] |

CH_4O(Methanol)+$C_6H_{12}O$(Cyclohexanol)

| 293.14 K | Positive | | [192] |

Table 1 (*cont.*)

SYSTEM

	Temperature	Deviation	Method	Ref.
CH_4O(Methanol)+$C_6H_{12}O$(1-Methylpentan-2-one)				
				[276]
CH_4O(Methanol)+C_6H_{14}(Hexane)				
	303.45,318.15 K	Negative	Max.Bub.Pres.	[27]
	318.15 K			[149]
	293.15 K	Negative		[195]
CH_4O(Methanol)+$C_6H_{14}O$(Di-isopropyl ether)				
				[276]
CH_4O(Methanol)+C_7H_8O(o-Cresol)				
	288.15 K			[251]
CH_4O(Methanol)+C_7H_8O(m-Cresol)				
	288.15 K			[251]
CH_4O(Methanol)+C_7H_8O(p-Cresol)				
	288.15 K			[251]
CH_4O(Methanol)+$C_7H_{14}O$(o-Methylcyclohexanol)				
	293.15 K			[243]
CH_4O(Methanol)+$C_7H_{14}O$(m-Methylcyclohexanol)				
	293.15 K			[243]
CH_4O(Methanol)+$C_7H_{14}O$(p-Methylcyclohexanol)				
	293.15 K			[243]
CH_4O(Methanol)+C_9H_{12}(Mesitylene)				
	273.15 to 313.15 K (9)	Neg.Infl.	Max.Bub.Pres.	[88]
CH_4O(Methanol)+$C_{10}H_{14}N_2$(Nicotine)				
				[181]
CH_4O(Methanol)+$C_{10}H_{22}O$(Decanol)				
	298.15 K	Positive	Max.Bub.Pres.	[35]

Table 1 (*cont.*)
SYSTEM

Temperature	Deviation	Method	Ref.

CH_4O(Methanol)+H_2O(Water)

303.15 K to B.pt.			[198]
298.15 K	Negative	Cap.Rise	[66]
303.15 K	Negative		[210]
293.15 to 323.15 K (4)	Negative	Torsion Balance	[185]
288.15 K	Negative		[209]
278.15 to 333.15 K			[145]
298.15 K		Cap.Rise	[165]
263.15 to 323.15 K	Negative	Max.Bub.Pres.	[189]
293.15 K	Negative		[202]
283.15,303.15 K		Cap.Rise	[275]
			[269]

CO(Carbon monoxide)+N_2(Nitrogen)

83.82 K	Negative	Cap.Rise	[22]

CS_2(Carbon disulphide)+$C_2H_2Cl_2$(1,2-Dichloroethane)

291.35 K	Neg.Min.	Cap.Rise	[188]

CS_2(Carbon disulphide)+C_3H_6O(Acetone)

298.15 K	Negative	Dif.Cap.Rise	[42]
279.15 K	Negative		[194]
298.15 K	Negative	Dif.Cap.Rise	[51]
288.15 K	Negative		[234]
283.15,308.15 K	Negative		[230]
288.15,298.15,308.15 K	Negative		[231]

CS_2(Carbon disulphide)+$C_4H_6O_3$(Acetic anhydride)

298.15 K	Negative	Dif.Cap.Rise	[42]

CS_2(Carbon disulphide)+$C_4H_8O_2$(Ethyl acetate)

288.15 K	Negative		[234]

CS_2(Carbon disulphide)+$C_4H_{10}O$(Diethyl ether)

291.35 K	Negative	Cap.Rise	[188]
288.15 K	Negative		[234]
298.15 K	Negative		[234]
293.15 to 373.15 K (3)	Negative		[235]

CS_2(Carbon disulphide)+$C_6H_5NO_2$(Nitrobenzene)

288.15 K	Pos.Max.		[234]

Table 1 (*cont.*)
SYSTEM

	Temperature	Deviation	Method	Ref.

CS_2(Carbon disulphide)+C_6H_6(Benzene)

	Temperature	Deviation	Method	Ref.
	293.15, 303.15 K	Negative	Max.Bub.Pres.	[17]
	298.15 K	Negative		[72]
	288.15 K	Positive		[234]
	291.35 K	Negative	Cap.Rise	[188]

CS_2(Carbon disulphide)+C_6H_7N(Aniline)

	288.15 K	Negative		[234]

CS_2(Carbon disulphide)+$C_6H_{10}O_3$(Ethyl acetoacetate)

	288.15 K	Negative		[234]

CS_2(Carbon disulphide)+C_7H_8(Toluene)

	298.15 K	Negative		[224]

CS_2(Carbon disulphide)+C_7H_8O(Methylphenyl ether)

	288.15 K			[244]

CS_2(Carbon disulphide)+C_7H_8O(o-Cresol)

	288.15 K			[251]

CS_2(Carbon disulphide)+C_7H_8O(m-Cresol)

	288.15 K			[251]

CS_2(Carbon disulphide)+C_7H_8O(p-Cresol)

	288.15 K			[251]

CS_2(Carbon disulphide)+$C_7H_{14}O_2$(Pentyl acetate)

	288.15 K	Negative		[234]

CS_2(Carbon disulphide)+$C_8H_{11}N$(N,N-Dimethylaniline)

	288.15 K	Positive		[234]

C_2Cl_4(Tetrachloroethylene)+C_5H_{10}(Cyclopentane)

	298.15 K	Negative	Max.Bub.Pres.	[33]

C_2Cl_4(Tetrachloroethylene)+C_6H_{12}(Cyclohexane)

	298.15 K	Negative	Max.Bub.Pres.	[33]

C_2HCl_3(Trichloroethylene)+$C_6H_{10}O$(Cyclohexanone)

	293.15 K			[243]

Table 1 (*cont.*)
SYSTEM

	Temperature	Deviation	Method	Ref.

C_2HCl_3O(Trichloroacetaldehyde)+$C_2H_4O_2$(Acetic acid)

| | 283.15 to 363.15 K | | | [201] |
| | 293.15 to 353.15 K | | | [270] |

C_2HCl_3O(Trichloroacetaldehyde)+C_2H_6O(Ethanol)

| | 293.15 to 353.15 K | | | [270] |

C_2HCl_3O(Trichloroacetaldehyde)+C_3H_6O(Acetone)

| | | | | [112] |

C_2HCl_3O(Trichloroacetaldehyde)+$C_4H_8O_2$(Butanoic acid)

| | 283.15 to 363.15 K | | | [201] |
| | 293.15 to 363.15 K | | | [270] |

C_2HCl_3O(Trichloroacetaldehyde)+$C_4H_{10}O$(Diethyl ether)

| | | | | [112] |

C_2HCl_3O(Trichloroacetaldehyde)+$C_5H_8O_2$(Pentan-2,4-dione)

| | 283.15 to 363.15 K | | | [201] |
| | 293.15 to 353.15 K | | | [270] |

C_2HCl_3O(Trichloroacetaldehyde)+$C_5H_{12}O$(3-Methylbutanol)

| | 293.15 to 353.15 K | | | [270] |

C_2HCl_3O(Trichloroacetaldehyde)+C_6H_6(Benzene)

| | | | | [112] |

C_2HCl_3O(Trichloroacetaldehyde)+C_6H_6O(Phenol)

| | 283.15 to 363.15 K | | | [201] |
| | 293.15 to 353.15 K | | | [270] |

C_2HCl_3O(Trichloroacetaldehyde)+$C_6H_{12}O$(Cyclohexanol)

| | 293.15 to 353.15 K | | | [270] |

C_2HCl_3O(Trichloroacetaldehyde)+$C_6H_{12}O_2$(Isopentyl formate)

| | | | | [171] |

C_2HCl_3O(Trichloroacetaldehyde)+C_7H_8O(Benzyl alcohol)

| | 283.15 to 363.15 K | | | [201] |
| | 293.15 to 353.15 K | | | [270] |

Table 1 (*cont.*)
SYSTEM

Temperature	Deviation	Method	Ref.

C_2HCl_3O(Trichloroacetaldehyde)+C_7H_8O(o-Cresol)

283.15 to 363.15 K			[201]
293.15 to 353.15 K			[270]

C_2HCl_3O(Trichloroacetaldehyde)+$C_7H_{12}O_4$(Ethyl malonate)

			[171]

C_2HCl_3O(Trichloroacetaldehyde)+H_2O(Water)

273.15 to 348.15 K (16)	Negative	Max.Bub.Pres.	[189]
293.15 to 353.15 K			[270]

$C_2HCl_3O_2$(Trichloroacetic acid)+$C_2H_4O_2$(Acetic acid)

			[151]

$C_2HCl_3O_2$(Trichloroacetic acid)+C_2H_5NO(Acetamide)

323.15,333.15,343.15 K			[260]

$C_2HCl_3O_2$(Trichloroacetic acid)+C_3H_6O(Acetone)

303.15,313.15,323.15 K	Positive		[98]

$C_2HCl_3O_2$(Trichloroacetic acid)+$C_5H_4O_2$(2-Furaldehyde)

			[117]

$C_2HCl_3O_2$(Trichloroacetic acid)+H_2O(Water)

298.15,308.15 K			[220]

C_2H_2(Ethyne)+$C_5H_{11}NO_2$(N-Ethyl urethane)

			[182]

$C_2H_2Br_4$(1,1,2,2-Tetrabromoethane)+$C_4H_{10}O$(Diethyl ether)

293.15 K	Negative	Max.Bub.Pres.	[53]

$C_2H_2Br_4$(1,1,2,2-Tetrabromoethane)+C_6H_6(Benzene)

293.15,313.15,333.15 K	Negative	Max.Bub.Pres.	[53]

$C_2H_2Br_4$(1,1,2,2-Tetrabromoethane)+C_8H_{18}(2,2,4-Trimethylpentane)

333.15 K	Negative	Max.Bub.Pres.	[53]

$C_2H_2Cl_2$(1,2-Dichloroethylene)+C_6H_6(Benzene)

313.15 K	Negative	Cap.Rise	[3]

Table 1 (*cont.*)
SYSTEM

Temperature	Deviation	Method	Ref.

$C_2H_2Cl_2O_2$(Dichloroacetic acid)+$C_2H_4O_2$(Acetic acid)

[151]

$C_2H_2Cl_2O_2$(Dichloroacetic acid)+H_2O(Water)

| 298.15,308.15 K | Negative | | [220] |

$C_2H_3ClO_2$(Chloroacetic acid)+$C_2H_4O_2$(Acetic acid)

[151]

$C_2H_3ClO_2$(Chloroacetic acid)+C_2H_5NO(Acetamide)

| 323.15,343.15,363.15 K | | | [261] |

$C_2H_3ClO_2$(Chloroacetic acid)+C_3H_6O(Acetone)

| 293.15,303.15,313.15 K | Near Zero | | [98] [99] |

$C_2H_3ClO_2$(Chloroacetic acid)+$C_5H_4O_2$(2-Furaldehyde)

[117]

$C_2H_3ClO_2$(Chloroacetic acid)+H_2O(Water)

| 298.15,308.15 K | | | [220] |

C_2H_3N(Acetonitrile)+C_2H_6O(Ethanol)

| 293.15 K | Negative | | [205] |

C_2H_3N(Acetonitrile)+$C_5H_{10}O_3$(Ethyl carbonate)

| 298.15 K | Negative | Max.Bub.Pres. | [25] |

C_2H_3N(Acetonitrile)+$C_{19}H_{24}$(4,4-Dicumylmethane)

| 293.15,313.15,333.15 K | Positive | Max.Bub.Pres. | [56] |

C_2H_3N(Acetonitrile)+H_2O(Water)

| 293.15 K | Negative | | [205] |

$C_2H_4Br_2$(1,2-Dibromoethane)+$C_2H_4O_2$(Acetic acid)

| 287.15 to 405.15 K (4) | Negative | | [233] |

$C_2H_4Br_2$(1,2-Dibromoethane)+$C_3H_6Br_2$(1,2-Dibromopropane)

| 290.15,313.15,358.15 K | Negative | | [231] |

Table 1 (*cont.*)
SYSTEM

Temperature	Deviation	Method	Ref.

$C_2H_4Br_2$(1,2-Dibromoethane)+C_4H_9I(Iodobutane)

283.15 to 404.15 K (4) Negative			[233]
286.15 to 405.15 K (4) Negative			[229]

$C_2H_4Br_2$(1,2-Dibromoethane)+C_6H_6(Benzene)

293.15,303.15,308.15 K			[64]

$C_2H_4Br_2$(1,2-Dibromoethane)+C_6H_{12}(Cyclohexane)

298.15,303.15,308.15 K			[64]

$C_2H_4Br_2$(1,2-Dibromoethane)+C_7H_8(Toluene)

298.15,308.15,308.15 K			[64]

$C_2H_4Br_2$(1,2-Dibromoethane)+C_8H_{10}(o-Xylene)

298.15,303.15,308.15 K			[64]

$C_2H_4Br_2$(1,2-Dibromoethane)+C_8H_{10}(m-Xylene)

298.15,303.15,308.15 K			[64]

$C_2H_4Br_2$(1,2-Dibromoethane)+C_8H_{10}(p-Xylene)

298.15,303.15,308.15 K			[64]

$C_2H_4Cl_2$(1,1-Dichloroethane)+$C_2H_4Cl_2$(1,2-Dichloroethane)

293.15 K	Negative	Cap.Rise	[8]

$C_2H_4Cl_2$(1,2-Dichloroethane)+C_6H_6(Benzene)

298.15 K			[162]
296.15 K	Negative		[188]
287.15 to 343.15 K (5)	Negative		[230]
290.15,323.15 K	Negative		[231]

C_2H_4O(Acetaldehyde)+$C_2H_4O_2$(Acetic acid)

293.15 K			[271]

C_2H_4O(Acetaldehyde)+C_3H_6O(Acetone)

293.15 K			[271]

C_2H_4O(Acetaldehyde)+C_4H_6O(Crotonaldehyde)

293.15 K			[271]

C_2H_4O(Acetaldehyde)+$C_4H_6O_2$(Vinyl acetate)

291.15 K			[271]

Table 1 (*cont.*)
SYSTEM

	Temperature	Deviation	Method	Ref.

$C_2H_4O_2$(Acetic acid)+C_2H_5I(Iodoethane)

291.35 K	Neg.Min.	Cap.Rise	[188]

$C_2H_4O_2$(Acetic acid)+C_2H_5NO(Acetamide)

358.15 K	Zero		[86]
293.15,333.15,353.15 K	Positive		[260]

$C_2H_4O_2$(Acetic acid)+C_2H_6O(Ethanol)

298.15 K	Positive	Max.Bub.Pres.	[16]

$C_2H_4O_2$(Acetic acid)+C_3H_6O(Acetone)

298.15 K	Positive	Max.Bub.Pres.	[16]
298.15,308.15,313.15 K			[95]
273.15,293.15,303.15 K	Positive		[98]
298.15 K	Pos.Max.		[125]
293.15 K			[271]

$C_2H_4O_2$(Acetic acid)+$C_3H_6O_2$(Methyl acetate)

293.15 to 353.15 K			[84]

$C_2H_4O_2$(Acetic acid)+C_4H_6O(Crotonaldehyde)

293.15 K			[271]

$C_2H_4O_2$(Acetic acid)+$C_4H_6O_2$(Vinyl acetate)

			[169]
293.15 K			[271]

$C_2H_4O_2$(Acetic acid)+$C_4H_6O_3$(Acetic anhydride)

293.15 to 353.15 K			[84]

$C_2H_4O_2$(Acetic acid)+C_4H_8O(Butan-2-one)

293.15,303.15,313.15 K			[95]

$C_2H_4O_2$(Acetic acid)+$C_4H_8O_2$(Dioxan)

298.15,313.15 K	Positive		[78]

$C_2H_4O_2$(Acetic acid)+$C_4H_8O_2$(Ethyl acetate)

298.15 K	Negative	Max.Bub.Pres.	[16]

$C_2H_4O_2$(Acetic acid)+$C_5H_4O_2$(2-Furaldehyde)

			[117]

Table 1 (*cont.*)
SYSTEM

Temperature	Deviation	Method.	Ref.

$C_2H_4O_2$(Acetic acid)+C_5H_5N(Pyridine)

287.15 K	Pos.Neg.Infl.	Cap.Rise	[193]
	Positive		[230]
293.15,313.15,353.15 K	Positive		[231]
295.15 K	Positive		[194]

$C_2H_4O_2$(Acetic acid)+$C_5H_{10}O$(Pentan-2-one)

298.15,308.15,313.15 K			[95]

C_2H_4O(Acetic acid)+C_6H_6(Benzene)

288.15,303.15 K	Neg.Min.	Drop Weight	[1]
298.15 K	Negative	Max.Bub.Pres.	[16]
293.15,308.15 K	Negative	Max.Bub.Pres.	[17]
291.35 K	Neg.Min.	Cap.Rise	[188]
			[161]

$C_2H_4O_2$(Acetic acid)+C_6H_7N(Aniline)

298.15 K	Positive		[256]
			[70]

$C_2H_4O_2$(Acetic acid)+C_7H_9N(o-Toluidine)

298.15 K	Positive		[256]

$C_2H_4O_2$(Acetic acid)+C_7H_9N(m-Toluidine)

298.15 K	Positive		[256]

$C_2H_4O_2$(Acetic acid)+C_9H_7N(Quinoline)

298.15 K	Pos.Neg.	Cap.Rise	[5]

$C_2H_4O_2$(Acetic acid)+HNO_3(Nitric acid)

273.15,293.15,313.15 K	Negative		[230]

$C_2H_4O_2$(Acetic acid)+H_4N_2(Hydrazine)

343.15,348.15,358.15 K	Positive	Max.Bub.Pres.	[85]

$C_2H_4O_2$(Acetic acid)+H_2O(Water)

298.15 K	Pos.Zero.Neg.	Drop Weight	[89]
293.15,303.15 K	Negative		[222]
			[104]
293.15 K	Negative		[221]
291.35 K	Negative	Cap.Rise	[188]
298.15,308.15 K	Negative		[220]
283.15 to 313.15 K (4)	Negative		[210]
293.15 K	Negative		[202]
283.15,303.15 K		Cap.Rise	[275]

Table 1 (*cont.*)
SYSTEM

Temperature	Deviation	Method	Ref.

C_2H_5I(Iodoethane)+$C_4H_8O_2$(Ethyl acetate)

291.15,308.15,323.15 K	Negative		[231]

C_2H_5I(Iodoethane)+$C_4H_{10}O$(Diethyl ether)

	Negative		[224]

C_2H_5I(Iodoethane)+C_6H_6(Benzene)

291.35 K	Neg.Min.	Cap.Rise	[188]

C_2H_5NO(Acetamide)+$C_5H_4O_2$(2-Furaldehyde)

			[118]

C_2H_5NO(Acetamide)+C_6H_6O(Phenol)

333.15,343.15,353.15 K		Max.Bub.Pres.	[114]

$C_2H_5NO_2$(Nitroethane)+C_6H_6(Benzene)

303.15 K		Max.Bub.Pres.	[12]

$C_2H_5NO_2$(Nitroethane)+C_6H_{12}(Cyclohexane)

303.15 K		Max.Bub.Pres.	[12]

$C_2H_5NO_2$(Nitroethane)+C_6H_{14}(Hexane)

303.15 K		Max.Bub.Pres.	[12]

$C_2H_5NO_2$(Nitroethane)+C_6H_{14}(3-Methylpentane)

290.60 to 307.16 K	Neg.Infl.	Dif.Cap.Rise	[267]

$C_2H_5NO_2$(Nitroethane)+C_7H_8(Toluene)

303.15 K		Max.Bub.Pres.	[12]

$C_2H_5NO_2$(Nitroethane)+C_7H_{14}(Methylcyclohexane)

303.15 K		Max.Bub.Pres.	[12]

C_2H_6O(Ethanol)+C_2H_6OS(Dimethyl sulphoxide)

			[144]

C_2H_6O(Ethanol)+C_3H_6O(Acetone)

273.15,293.15,318.15 K	Positive	Drop Weight	[1]
298.15 K	Positive	Drop Weight	[89]
298.15 K	Positive	Max.Bub.Pres.	[16]

Table 1 (*cont.*)
SYSTEM

Temperature	Deviation	Method	Ref.

C_2H_6O(Ethanol)+$C_3H_8O_3$(Glycerol)

| 298.15 K | Negative | | [217] |

C_2H_6O(Ethanol)+C_4H_5NS(Allyl isothiacyanate)

| 254.85,273.15,293.15 K | Negative | | [259] |

C_2H_6O(Ethanol)+$C_4H_6O_3$(Acetic anhydride)

| | | | [184] |
| | | | [141] |

C_2H_6O(Ethanol)+$C_4H_8O_2$(Dioxan)

| 293.15,313.5,333.15 K | Negative | | [173] |

C_2H_6O(Ethanol)+$C_4H_8O_2S$(Tetramethylene sulphone)

| 293.15 to 323.15 K | | | [120] |

C_2H_6O(Ethanol)+$C_5H_{12}O$(Pentanol)

| 273.15 to 373.15 K (6) | Positive | Max.Bub.Pres. | [69] |

C_2H_6O(Ethanol)+$C_6H_4Cl_2$(p-Dichlorobenzene)

| 328.15 K & F.pt. | Negative | | [134] |

C_2H_6O(Ethanol)+$C_6H_5NO_2$(Nitrobenzene)

| 278.65 K | Negative | Max.Bub.Pres. | [147] |

C_2H_6O(Ethanol)+C_6H_6(Benzene)

298.15,318.15 K	Negative	Drop Weight	[1]
295.15 K	Pos.Neg.	Max.Bub.Pres.	[197]
303.15 K	Negative	Max.Bub.Pres.	[36]
298.15 K	Pos.Neg.	Drop Weight	[89]
			[160]
298.15 K	Negative	Max.Bub.Pres.	[16]
278.15 to 333.15 K			[145]
298.15 K	Zero		[146]
298.15 K	Negative		[240]
283.15,319.35,351.35 K	Negative		[233]
293.15 K	Negative		[195]

C_2H_6O(Ethanol)+C_6H_6O(Phenol)

| 293.15,308.15 K | Negative | Drop Weight | [1] |

C_2H_6O(Ethanol)+C_6H_7N(Aniline)

| 293.15 to 298.15 K (6) | Negative | | [71] |
| | | | [115] |

Table 1 (*cont.*)
SYSTEM

	Temperature	Deviation	Method	Ref.

C_2H_6O(Ethanol)+$C_6H_{10}O$(Cyclohexanone)

| | 291.15 K | | | [191] |

C_2H_6O(Ethanol)+C_6H_{12}(Cyclohexane)

| | 303.15 K | Negative | Max.Bub.Pres. | [36] |
| | 395.15 K | Negative | Max.Bub.Pres. | [197] |

C_2H_6O(Ethanol)+C_6H_{14}(Hexane)

| | 295.15 K | Negative | Max.Bub.Pres. | [197] |
| | 293.15 K | Negative | | [195] |

C_2H_6O(Ethanol)+$C_7H_6O_2$(Salicylaldehyde)

| | 290.15 K | | | [250] |

C_2H_6O(Ethanol)+C_7H_8(Toluene)

| | 303.15 K | Pos.Neg. | Max.Bub.Pres. | [36] |

C_2H_6O(Ethanol)+C_7H_8O(o-Cresol)

| | 291.15 K | Positive | | [237] |

C_2H_6O(Ethanol)+C_7H_8O(m-Cresol)

| | 291.15 K | Positive | | [237] |

C_2H_6O(Ethanol)+C_7H_8O(p-Cresol)

| | 291.15 K | Positive | | [237] |

C_2H_6O(Ethanol)+$C_7H_8O_2$(o-Methoxyphenol)

| | 290.15 K | | | [250] |

C_2H_6O(Ethanol)+C_7H_{14}(Methylcyclohexane)

| | 303.15 K | Negative | Max.Bub.Pres. | [36] |

C_2H_6O(Ethanol)+$C_{10}H_{12}$(Tetralin)

| | 291.15 K | | | [191] |

C_2H_6O(Ethanol)+$C_{10}H_{22}O$(Decanol)

| | 298.15 K | Positive | Max.Bub.Pres. | [35] |

C_2H_6O(Ethanol)+$C_{12}H_{10}$(Biphenyl)

| | | | | [77] |

Table 1 (*cont.*)
SYSTEM

Temperature	Deviation	Method	Ref.

C_2H_6O(Ethanol)+$C_{12}H_{10}O$(Diphenyl ether)

			[77]

C_2H_6O(Ethanol)+H_2O(Water)

Temperature	Deviation	Method	Ref.
293.15 to 323.15 K (4)	Negative	Torsion Balance	[185]
288.15 to 363.15 K		Cap.Rise	[132]
263.15 to 333.15 K (15)		Max.Bub.Pres.	[189]
278.15 to 333.15 K			[145]
253.15 to 313.15 K		Several	[156]
298.15 K		Cap.Rise	[165]
288.15 K			[209]
288.15 K	Negative		[211]
289.15 to 423.15 K (6)			[212]
288.15 K	Negative		[213]
273.15 to 303.15 K (4)	Negative		[210]
298.15 K	Negative		[214]
298.15 K	Negative		[215]
298.15 K	Negative		[216]
298.15 K	Negative		[217]
283.15 to 303.15 K		Cap.Rise	[275]

C_2H_6OS(Dimethyl sulphoxide)+C_3H_6O(Acetone)

303.15 K	Negative	Max.Bub.Pres.	[37]
			[127]

C_2H_6OS(Dimethyl sulphoxide)+$C_4H_8O_2$(Dioxan)

			[121]

C_2H_6OS(Dimethyl sulphoxide)+C_6F_6(Hexafluorobenzene)

293.15,313.15 K			[154]

C_2H_6OS(Dimethyl sulphoxide)+C_6H_5Br(Bromobenzene)

303.15,313.15,323.15 K	Negative	Cap.Rise	[274]

C_2H_6OS(Dimethyl sulphoxide)+C_6H_5Cl(Chlorobenzene)

303.15,313.15,323.15 K	Negative	Cap.Rise	[274]

C_2H_6OS(Dimethyl sulphoxide)+$C_6H_5NO_2$(Nitrobenzene)

303.15,313.15,323.15 K	Positive	Cap.Rise	[274]

C_2H_6OS(Dimethyl sulphoxide)+C_6H_6(Benzene)

303.15,313.15,323.15 K	Negative	Cap.Rise	[274]
			[144]

Table 1 (*cont.*)
SYSTEM

Temperature	Deviation	Method	Ref.

C_2H_6OS(Dimethyl sulphoxide)+C_7H_8(Toluene)

| 303.15,313.15,323.15 K | Negative | Cap.Rise | [274] |

C_2H_6OSi(Poly(dimethylsiloxane))+$C_6H_{18}OSi_2$(Hexamethyldisiloxane)

| 297.15 K | Negative | Ring | [263] |

C_2H_6OSi(Poly(dimethylsiloxane))+C_7H_8(Toluene)

| 303.15 to 363.15 K (5) | Negative | Max.Bub.Pres. | [266] |

C_2H_6OSi(Poly(dimethylsiloxane))+$C_8H_{24}O_2Si_3$(Octamethyltrisiloxane)

| 297.15 K | Negative | Ring | [263] |

C_2H_6OSi(Poly(dimethylsiloxane))+$C_{12}H_{36}O_4Si_5$(Dodecamethylpentasiloxane)

| 297.15 K | Negative | Ring | [263] |

$C_2H_6O_2$(Ethane-1,2-diol)+$C_4H_8O_2$(Ethyl acetate)

| 298.15 K | | Pendant Drop | [167] |

$C_2H_6O_2$(Ethane-1,2-diol)+$C_6H_{12}O_2$(Butyl acetate)

| 298.15 K | | Pendant Drop | [167] |

$C_2H_6O_2$(Ethane-1,2-diol)+$C_7H_8O_2$(Phenyl acetate)

| 298.15 K | | Pendant Drop | [167] |

$C_2H_6O_2$(Ethane-1,2-diol)+$C_9H_{10}O_2$(Ethyl benzoate)

| 298.15 K | | Pendant Drop | [167] |

$C_2H_6O_2$(Ethane-1,2-diol)+$C_7H_{13}NO$(N-Methyl caprolactam)

| 293.15 to 333.15 K | | | [109] |

$C_2H_6O_2$(Ethane-1,2-diol)+H_2O(Water)

| 293.15 K | Negative | Max.Bub.Pres. | [277] |

C_2H_7N(Monoethylamine)+H_2O(Water)

| 313.15,333.15 K | Negative | Max.Bub.Pres. | [100] |

$C_2H_8N_2$(1,2-Ethanediamine)+H_2O(Water)

| | | | [137] |

C_3F_8(Octafluoropropane)+C_4H_{10}(Butane)

| 230.15 to 250.15 K | Negative | Dif.Cap.Rise | [294] |

Table 1 (*cont.*)
SYSTEM

Temperature	Deviation	Method	Ref.

C_3F_8(Octafluoropropane)+C_5H_{12}(Pentane)

| 250.15 to 258.15 K | Negative | Dif.Cap.Rise | [294] |

$C_3H_4O_3$(Pyruvic acid)+C_5H_5N(Pyridine)

| | | | [170] |

$C_3H_4O_3$(Pyruvic acid)+H_2O(Water)

| | | | [170] |

C_3H_5N(Propanenitrile)+C_5H_{12}(Pentane)

| 293.15 to 353.15 K (10) | Neg.Infl. | Dif.Cap.Rise | [291] |

C_3H_5N(Propanenitrile)+C_7H_{16}(Heptane)

| 293.15 to 353.15 K (10) | Negative | Dif.Cap.Rise | [291] |

$C_3H_5N_3O_9$(Nitroglycerine)+$C_6H_5NO_2$(Nitrobenzene)

| 291.15 to 294.15 K | | | [248] |

C_3H_6O(Acetone)+C_3H_8O(Propan-2-ol)

| 313.15 K | Negative | Cap.Rise | [50] |
| | | | [276] |

C_3H_6O(Acetone)+C_4H_6O(Crotonaldehyde)

| 293.15 K | | | [271] |

C_3H_6O(Acetone)+$C_4H_6O_2$(Vinyl acetate)

| 293.15 K | | | [271] |

C_3H_6O(Acetone)+$C_4H_6O_3$(Acetic anhydride)

| 298.15 K | Pos.Near.Zero | Dif.Cap.Rise | [42] |
| 293.15,298.15,313.15 K | Near.Zero | | [125] |

C_3H_6O(Acetone)+C_4H_8O(Butan-2-one)

| 293.15,313.15,333.15 K | | | [200] |

C_3H_6O(Acetone)+$C_4H_8O_2$(Butanoic acid)

| 298.15,308.15,313.15 K | | | [95] |

C_3H_6O(Acetone)+$C_4H_{10}O$(Butanol)

| 298.15 K | Positive | | [203] |

Table 1 (*cont.*)
SYSTEM

Temperature	Deviation	Method	Ref.
C_3H_6O(Acetone)+$C_4H_{10}O$(Diethyl ether)			
288.15,303.15 K	Negative		[172]
288.15 K	Negative		[234]
C_3H_6O(Acetone)+$C_5H_{10}O_2$(Propyl acetate)			
288.15 K	Negative		[226]
C_3H_6O(Acetone)+$C_5H_4O_2$(2-Furfururaldehyde)			
			[118]
C_3H_6O(Acetone)+C_6H_5Cl(Chlorobenzene)			
288.15 K	Negative		[226]
C_3H_6O(Acetone)+C_6H_6(Benzene)			
298.15 K	Positive		[72]
291.15 K	Negative	Cap.Rise	[188]
295.15 K			[199]
298.15 K			[16]
288.15 K	Positive		[226]
C_3H_6O(Acetone)+C_6H_6O(Phenol)			
273.15,308.15 K	Negative	Drop Weight	[1]
303.15,313.15,323.15 K	Negative	Max.Bub.Pres.	[114]
C_3H_6O(Acetone)+$C_6H_{10}O$(Cyclohexanone)			
291.15 K	Negative		[191]
C_3H_6O(Acetone)+$C_6H_{12}O$(1-Methylpentan-2-one)			
			[276]
C_3H_6O(Acetone)+$C_6H_{12}O$(Cyclohexanol)			
293.15 K	Negative		[192]
C_3H_6O(Acetone)+C_6H_{14}(Hexane)			
308.15 K			[103]
288.15 to 323.15 K	Neg.Min.	Max.Bub.Pres.	[179]
C_3H_6O(Acetone)+$C_6H_{14}O$(Di-isopropyl ether)			
			[276]
C_3H_6O(Acetone)+$C_7H_6O_2$(Salicylaldehyde)			
290.15 K			[250]

Table 1 (*cont.*)
SYSTEM

Temperature	Deviation	Method	Ref.

C_3H_6O(Acetone)+C_7H_8O(Methyl phenyl ether)

288.15 K			[244]

C_3H_6O(Acetone)+C_7H_8O(o-Cresol)

291.15 K	Positive		[237]

C_3H_6O(Acetone)+C_7H_8O(m-Cresol)

291.15 K	Pos.Max.		[237]

C_3H_6O(Acetone)+C_7H_8O(p-Cresol)

291.15 K			[237]

C_3H_6O(Acetone)+$C_7H_8O_2$(o-Methoxyphenol)

290.15 K			[250]

C_3H_6O(Acetone)+$C_{10}H_{12}$(Tetralin)

291.15 K	Negative		[191]

C_3H_6O(Acetone)+$C_{16}H_{34}$(Hexadecane)

313.15,323.15 K	Pos.Neg.Infl.	Dif.Cap.Rise	[264]
294.15 to 305.15 K	Neg.Infl.	Dif.Cap.Rise	[300]

C_3H_6O(Acetone)+H_2O(Water)

273.15,298.15,318.15 K	Negative	Drop Weight	[1]
273.15 to 313.15 K (9)	Negative		[204]
298.15 to 343.15 K (6)	Negative	Max.Bub.Pres.	[124]
293.15 K	Negative		[120]
278.15 to 333.15 K			[145]
298.15 K	Negative		[159]
293.15 K to B.pt.		Cap.Rise	[163]
298.15 K	Negative		[203]

$C_3H_6O_2$(Methyl acetate)+$C_4H_6O_3$(Acetic anhydride)

293.15 to 353.15 K			[84]

$C_3H_6O_2$(Methyl acetate)+$C_4H_8O_2$(Ethyl acetate)

284.15 to 333.15 K (6)			[225]

$C_3H_6O_2$(Propanoic acid)+$C_4H_{11}N$(Diethylamine)

298.15 K	Pos.Max.		[255]

$C_3H_6O_2$(Propanoic acid)+C_5H_5N(Pyridine)

303.15 K	Inflexion	Cap.Rise	[7]

Table 1 (*cont.*)
SYSTEM

Temperature	Deviation	Method	Ref.

$C_3H_6O_2$(Propanoic acid)+C_5H_{11}N(Piperidine)

| 298.15 K | Pos.Max. | | [258] |

$C_3H_6O_2$(Propanoic acid)+C_6H_6(Benzene)

| 300.15 K | | | [161] |

$C_3H_6O_2$(Ethyl formate)+C_7H_8(Toluene)

| 288.15,313.15 K | Negative | | [226] |

$C_3H_6O_2$(Propanoic acid)+C_7H_9N(o-Toluidine)

| 298.15 K | Positive | | [256] |

$C_3H_6O_2$(Propanoic acid)+C_7H_9N(m-Toluidine)

| 298.15 K | Negative | | [256] |

$C_3H_6O_2$(Propanoic acid)+$C_{12}H_{11}$N(Diphenylamine)

| 327.15 K | | | [257] |

$C_3H_6O_2$(Propanoic acid)+H_2O(Water)

298.15,308.15 K	Negative		[220]
283.15,303.15 K		Cap.Rise	[275]
			[104]

$C_3H_6O_3$(Methyl carbonate)+$C_6H_5NO_2$(Nitrobenzene)

| 298.15 K | Negative | Max.Bub.Pres. | [25] |

$C_3H_6O_3$(Methyl carbonate)+C_6H_6(Benzene)

| 298.15 K | Negative | Max.Bub.Pres. | [25] |

C_3H_7NO(Dimethylformamide)+C_6F_6(Hexafluorobenzene)

| 293.15,313.15 K | Negative | | [154] |

C_3H_7NO(Dimethylformamide)+H_2O(Water)

| 223.15,298.15 K | | | [148] |
| | | | [122] |

C_3H_8O(Methyl ethyl ether)+$C_4H_{10}O$(Diethyl ether)

| 278.15 K | Negative | Ring | [67] |

C_3H_8O(Propanol)+$C_5H_{12}O$(Pentanol)

| 273.15 to 373.15 K (6) | Positive | Max.Bub.Pres. | [69] |

Table 1 (*cont.*)
SYSTEM

Temperature	Deviation	Method	Ref.

C_3H_8O(Propanol)+$C_6H_4Cl_2$(p-Dichlorobenzene)

328.15 K & F.pt.	Negative		[135]

C_3H_8O(Propan-2-ol)+$C_6H_4Cl_2$(p-Dichlorobenzene)

328.15 K & F.pt.	Negative		[135]

C_3H_8O(Propan-2-ol)+$C_6H_5NO_2$(Nitrobenzene)

278.15 K	Negative	Max.Bub.Pres.	[147]

C_3H_8O(Propanol)+C_6H_6(Benzene)

293.15 K	Negative	Max.Bub.Pres.	[92]
298.15 K	Negative		[146]

C_3H_8O(Propan-2-ol)+C_6H_6(Benzene)

293.15,303.15 K			[140]
305.15 K	Negative		[197]
300.65 K			[160]
293.15 K	Negative		[195]
298.15 K	Negative		[146]

C_3H_8O(Propanol)+C_6H_7N(Aniline)

288.15 to 338.15 K (6)	Negative		[228]

C_3H_8O(Propan-2-ol)+$C_6H_{12}O$(1-Methylpentan-2-one)

			[276]

C_3H_8O(Propan-2-ol)+$C_6H_{14}O$(Di-isopropyl ether)

			[276]

C_3H_8O(Propanol)+C_6H_{14}(Hexane)

308.15 K			[149]

C_3H_8O(Propanol)+C_7H_{16}(Heptane)

293.15 to 333.15 K (6)	Negative	Dif.Cap.Rise	[289]

C_3H_8O(Propanol)+$C_{10}H_{22}O$(Decanol)

298.15 K	Positive	Max.Bub.Pres.	[35]

C_3H_8O(Propan-2-ol)+$C_{10}H_{22}O$(Decanol)

298.15 K	Positive	Max.Bub.Pres.	[35]

Table 1 (*cont.*)
SYSTEM

Temperature	Deviation	Method	Ref.

C_3H_8O(Propan-2-ol)+$C_{12}H_{11}N$(Diphenylamine)

333.15 K	Negative		[246]

C_3H_8O(Propanol)+H_2O(Water)

298.15 K	Negative	Max.Bub.Pres.	[66]
263.15 to 323.15 K (13)	Negative	Max.Bub.Pres.	[189]
283.15,303.15 K		Cap.Rise	[275]
			[269]

C_3H_8O(Propan-2-ol)+H_2O(Water)

293.15,303.15 K	Negative		[140]

$C_3H_8O_2$(Methylene dimethyl ether)+$C_6H_{12}O_2$(Isobutyl acetate)

291.35 K	Positive	Cap.Rise	[188]

$C_3H_8O_2$(Propan-1,3-diol)+H_2O(Water)

293.15 K	Negative	Max.Bub.Pres.	[277]

$C_3H_8O_3$(Propan-1,2,3-triol)+H_2O(Water)

291.15 K	Positive		[218]
298.15 K	Positive		[217]
293.15 K	Negative	Max.Bub.Pres.	[277]

C_4F_8(Octafluorocyclobutane)+C_4H_{10}(Butane)

233.70,244.00,254.40 K	Neg.Min.	Dif.Cap.Rise	[265]

C_4F_{10}(Decafluoro-2-methylpropane)+C_4H_{10}(Butane)

233.08 to 260.15 K	Negative	Dif.Cap.Rise	[292]

C_4F_{10}(Decafluorobutane)+C_4H_{10}(Butane)

235.15 to 253.15 K	Neg.Min.	Dif.Cap.Rise	[294]

$C_4H_5Cl_3O_2$(Ethyl trichloroacetate)+$C_4H_8O_2$(Ethyl acetate)

287.15 to 343.15 K (7)	Positive		[225]

C_4H_5NS(Allyl isothiocyanate)+C_7H_9N(m-Toluidine)

323.15,348.15 K	Pos.Max.		[246]

C_4H_5NS(Allyl isothiocyanate)+$C_8H_{11}N$(N-Ethylaniline)

298.15,323.15,348.15 K	Pos.Max.		[246]

Table 1 (*cont.*)
SYSTEM

	Temperature	Deviation	Method	Ref.

C_4H_6O(Crotonaldehyde)+$C_4H_6O_2$(Vinyl acetate)

293.15 K				[271]

$C_4H_6O_3$(Acetic anhydride)+$C_4H_8O_2$(Dioxan)

| 298.15,313.15 K | Negative | | | [78] |

$C_4H_6O_3$(Acetic anhydride)+H_2O(Water)

| | | | | [141] |

$C_4H_6O_3$(Acetic anhydride)+H_2O_4S(Sulphuric acid)

| | Pos.Neg.Max. | | | [231] |

$C_4H_7O_2$(Butan-1,4-diol)+H_2O(Water)

| 293.15 K | Negative | Max.Bub.Pres. | | [277] |

C_4H_8(Five Poly(isobutylene) Fluids)+C_7H_{16}(Heptane)

| 297.15 K | Negative | Ring | | [268] |

C_4H_8(Poly(isobutylene))+C_8H_{18}(2,2,4-Trimethylpentane)

| 297.15 K | Negative | Ring | | [263] |

C_4H_8(Five Poly(isobutylene) Fluids)+$C_{10}H_{12}$(Tetralin)

| 297.15 K | Negative | Ring | | [268] |

C_4H_8O(Butan-2-one)+C_6H_6(Benzene)

| 291.15 K | Negative | | | [48] |

C_4H_8O(Butan-2-one)+C_6H_{12}(Cyclohexane)

| 291.15 K | Negative | | | [164] |

C_4H_8O(Butan-2-one)+H_2O(Water)

| 273.15 to 323.15 K (11) | Negative | | | [204] |

C_4H_8O(Tetrahydrofuran)+H_2O(Water)

| 293.15 K | Negative | | | [195] |

$C_4H_8O_2$(Butanoic acid)+C_5H_5N(Pyridine)

| 273.15 to 313.15 K (5) | Positive | Max.Bub.Pres. | | [57] |

$C_4H_8O_2$(Butanoic acid)+$C_5H_{10}O$(Pentan-2-one)

| 298.15,308.15,313.15 K | | | | [95] |

Table 1 (*cont.*)
SYSTEM

	Temperature	*Deviation*	*Method*	*Ref.*

$C_4H_8O_2$(Butanoic acid)+$C_5H_{11}N$(Piperidine)

| | 298.15 K | Pos.Max. | | [258] |

$C_4H_8O_2$(Ethyl acetate)+$C_5H_{12}O$(3-Methylbutanol)

| | 291.35 K | Neg.Min. | Cap.Rise | [188] |
| | 273.15,298.45,304.15 K | Negative | | [227] |

$C_4H_8O_2$(Dioxan)+C_6H_5Cl(Chlorobenzene)

| | 303.15 K | Negative | Max.Bub.Pres. | [45] |

$C_4H_8O_2$(Butanoic acid)+$C_6H_5NO_2$(Nitrobenzene)

| | 278.65 K | Negative | Max.Bub.Pres. | [147] |

$C_4H_8O_2$(Dioxan)+$C_6H_5NO_2$(Nitrobenzene)

| | 293.15,303.15 K | Zero | Max.Bub.Pres. | [38] |

$C_4H_8O_2$(Ethyl acetate)+$C_6H_5NO_2$(Nitrobenzene)

| | 293.15 K | Negative | | [238] |

$C_4H_8O_2$(Butanoic acid)+C_6H_6(Benzene)

| | 300.15 K | | | [161] |

$C_4H_8O_2$(Dioxan)+C_6H_6(Benzene)

| | 293.15,303.15 K | Zero | Max.Bub.Pres. | [38] |
| | 293.15 to 318.15 K (6) | Zero.Ideal | Max.Bub.Pres. | [279] |

$C_4H_8O_2$(Ethyl acetate)+C_6H_6(Benzene)

| | 298.15 K | Negative | | [72] |

$C_4H_8O_2$(Dioxan)+C_6H_7N(Aniline)

| | 303.15 K | Positive | Max.Bub.Pres. | [45] |

$C_4H_8O_2$(Ethyl acetate)+C_6H_7N(Aniline)

| | 288.15 K | Negative | | [234] |

$C_4H_8O_2$(Ethyl acetate)+$C_6H_{10}O$(Cyclohexanone)

| | 293.15 K | | | [243] |

$C_4H_8O_2$(Dioxan)+C_6H_{12}(Cyclohexane)

| | 293.15,303.15 K | Negative | Max.Bub.Pres. | [38] |
| | 293.15 to 318.15 K (6) | Negative | Max.Bub.Pres. | [279] |

Table 1 (*cont.*)
SYSTEM

	Temperature	Deviation	Method	Ref.

$C_4H_8O_2$(Ethyl acetate)+$C_6H_{12}O$(Cyclohexanol)

| | 293.15 K | Negative | | [192] |

$C_4H_8O_2$(Dioxan)+C_6H_{14}(Hexane)

| | 293.15 to 343.15 K | Negative | Cap.Rise&Bub.Pres. | [150] |

$C_4H_8O_2$(Dioxan)+C_7H_8(Toluene)

| | 303.15 K | Negative | Max.Bub.Pres. | [45] |

$C_4H_8O_2$(Methyl propanoate)+C_7H_8(Toluene)

| | 288.15,313.15 K | Negative | | [226] |

$C_4H_8O_2$(Ethyl acetate)+C_7H_8O(o-Cresol)

| | 288.15 K | | | [251] |

$C_4H_8O_2$(Ethyl acetate)+C_7H_8O(m-Cresol)

| | 288.15 K | | | [251] |

$C_4H_8O_2$(Ethyl acetate)+C_7H_8O(p-Cresol)

| | 288.15 K | | | [251] |

$C_4H_8O_2$(Butanoic acid)+C_7H_9N(N-Methylaniline)

| | | | | [138] |

$C_4H_8O_2$(Butanoic acid)+C_7H_9N(o-Toluidine)

| | 298.15 K | Negative | | [256] |

$C_4H_8O_2$(Butanoic acid)+C_7H_9N(m-Toluidine)

| | 298.15 K | Negative | | [256] |

$C_4H_8O_2$(Ethyl acetate)+$C_7H_{14}O$(o-Methylcyclohexanol)

| | 288.15 K | | | [251] |

$C_4H_8O_2$(Ethyl acetate)+$C_7H_{14}O$(m-Methylcyclohexanol)

| | 288.15 K | | | [251] |

$C_4H_8O_2$(Ethyl acetate)+$C_7H_{14}O$(p-Methylcyclohexanol)

| | 288.15 K | | | [251] |

Table 1 *(cont.)*
SYSTEM

	Temperature	Deviation	Method	Ref.

$C_4H_8O_2$(Dioxan)+C_8H_{10}(m-Xylene)

	303.15 K	Negative	Max.Bub.Pres.	[45]

$C_4H_8O_2$(Butanoic acid)+C_8H_{11}N(N,N-Dimethylaniline)

				[138]

$C_4H_8O_2$(Butanoic acid)+C_8H_{11}N(N-Ethylaniline)

				[138]

$C_4H_8O_2$(Ethyl acetate)+$C_8H_{14}O_4$(Diethyl succinate)

	287.15 to 343.15 K (8)	Positive		[225]

$C_4H_8O_2$(Dioxan)+$C_9H_{10}O_2$(Benzyl acetate)

	287.15 to 343.15 K (6)	Negative		[225] [74]

$C_4H_8O_2$(Butanoic acid)+$C_{10}H_8$(Naphthalene)

			Max.Bub.Pres.	[139]

$C_4H_8O_2$(Butanoic acid)+$C_{10}H_{15}$N(N,N-Diethylaniline)

				[138]

$C_4H_8O_2$(Ethyl acetate)+$C_{10}H_{22}$O(Di-isopentyl ether)

	300.15 to 333.15 K (6)			[225]

$C_4H_8O_2$(Butanoic acid)+$C_{12}H_{11}$N(Diphenylamine)

	327.15 K			[257]

$C_4H_8O_2$(Methyl propanoate)+$C_{14}H_{20}O_2$(Amyl hydrocinnamate)

	288.15,313.15 K		Drop Weight	[226]

$C_4H_8O_2$(Butanoic acid)+H_2O(Water)

	298.15,308.15 K	Negative		[220]
	283.15,303.15 K		Cap.Rise	[275] [104]

$C_4H_8O_2$(Dioxan)+H_2O(Water)

	283.15 to 353.15 K (9)	Negative	Cap.Rise	[61]
	293.15 to 353.15 K (4)	Neg.Min.		[73]
	298.15 to 313.15 K	Negative		[78]
	293.15 K	Neg.Min.		[195]

Table 1 (*cont.*)

SYSTEM

	Temperature	*Deviation*	*Method*	*Ref.*

$C_4H_8O_2$(Ethyl acetate)+H_2O(Water)

| | 298.15 K | | Pendant Drop | [167] |

$C_4H_8O_2$(2-Methylpropanoic acid)+H_2O(Water)

	298.15,308.15 K	Negative		[220]
	299.35,295.15 K	Neg.Infl.		[223]
	293.15,313.15 K	Neg.Infl.		[287]
	299.46 to 307.16 K (5)	Neg.Infl.	Dif.Cap.Rise	[267]

$C_4H_8O_2S$(Tetramethylene sulphone)+H_2O(Water)

| | 293.15 to 323.15 K | | | [120] |

C_4H_9NO(Morpholine)+H_2O(Water)

| | 293.15 K | Negative | Cap.Rise | [65] |

$C_4H_{10}N_2$(Piperazine)+H_2O(Water)

| | | | | [137] |

$C_4H_{10}O$(Butanol)+$C_5H_{12}O$(Pentanol)

| | 273.15 to 373.15 K (6) | Positive | Max.Bub.Pres. | [69] |

$C_4H_{10}O$(2-Methylpropanol)+$C_6H_4Cl_2$(p-Dichlorobenzene)

| | 328.15 K & F.pt. | | | [135] |

$C_4H_{10}O$(Butanol)+$C_6H_5NO_2$(Nitrobenzene)

| | 278.65 K | Negative | Max.Bub.Pres. | [147] |

$C_4H_{10}O$(Diethyl ether)+$C_6H_5NO_2$(Nitrobenzene)

| | 288.15 K | Positive | | [234] |

$C_4H_{10}O$(2-Methylpropanol)+$C_6H_5NO_2$(Nitrobenzene)

| | 278.65 K | Negative | Max.Bub.Pres. | [147] |

$C_4H_{10}O$(Butanol)+C_6H_6(Benzene)

	298.15 K			[146]
				[160]
	293.15 K	Negative	Max.Bub.Pres.	[197]

Table 1 (*cont.*)

SYSTEM

Temperature	Deviation	Method	Ref.

$C_4H_{10}O$ (Diethyl ether)+C_6H_6 (Benzene)

298.15 K	Negative	Max.Bub.Pres.	[16]
291.55 K	Negative	Cap.Rise	[185]
298.15 K	Negative		[224]
288.15 K	Positive		[234]
293.15 to 373.05 K (3)			[235]

$C_4H_{10}O$ (2-Methylpropan-2-ol)+C_6H_6 (Benzene)

303.75 K	Negative	Max.Bub.Pres.	[36]

$C_4H_{10}O$ (Diethyl ether)+C_6H_7N (Aniline)

288.15 K	Negative		[234]

$C_4H_{10}O$ (Butanol)+C_6H_{12} (Cyclohexane)

295.15 K	Negative	Max.Bub.Pres.	[197]

$C_4H_{10}O$ (2-Methylpropan-2-ol)+C_6H_{12} (Cyclohexane)

303.15 K	Negative	Max.Bub.Pres.	[36]

$C_4H_{10}O$ (Butanol)+C_6H_{14} (Hexane)

293.15 K	Negative		[195]
295.15 K	Negative	Max.Bub.Pres.	[197]

$C_4H_{10}O$ (Diethyl ether)+$C_7H_6O_2$ (Salicylaldehyde)

300.75 K			[250]

$C_4H_{10}O$ (2-Methylpropan-2-ol)+C_7H_8 (Toluene)

303.15 K	Negative	Max.Bub.Pres.	[36]

$C_4H_{10}O$ (Diethyl ether)+C_7H_8O (Methylphenyl ether)

288.75 K			[244]

$C_4H_{10}O$ (Diethyl ether)+C_7H_8O (o-Cresol)

291.15 K	Positive		[237]

$C_4H_{10}O$ (Diethyl ether)+C_7H_8O (p-Cresol)

291.15 K	Positive		[237]

$C_4H_{10}O$ (Diethyl ether)+$C_7H_8O_2$ (o-Methoxyphenol)

291.95 K			[243]
291.15 K			[250]

Table 1 *(cont.)*
SYSTEM

	Temperature	Deviation	Method	Ref.
$C_4H_{10}O$(2-Methylpropan-2-ol)+C_7H_{14}(Methylcyclohexane)				
	303.15 K	Negative	Max.Bub.Pres.	[36]
$C_4H_{10}O$(Diethyl ether)+$C_7H_{14}O_2$(Pentyl acetate)				
	288.15 K	Negative		[234]
$C_4H_{10}O$(Diethyl ether)+$C_8H_{11}N$(N,N-Dimethylaniline)				
	288.15 K	Negative		[234]
$C_4H_{10}O$(Butanol)+$C_{10}H_8$(Naphthalene)				
			Max.Bub.Pres.	[139]
$C_4H_{10}O$(2-Methylpropanol)+$C_{10}H_8$(Naphthalene)				
			Max.Bub.Pres.	[139]
$C_4H_{10}O$(Butanol)+$C_{10}H_{22}O$(Decanol)				
	298.15 K	Positive	Max.Bub.Pres.	[35]
$C_4H_{10}O$(2-Methylpropanol)+$C_{10}H_{22}O$(Decanol)				
	298.15 K	Positive	Max.Bub.Pres.	[49]
$C_4H_{10}O$(Butan-2-ol)+$C_{10}H_{22}O$(Decanol)				
	298.15 K	Positive	Max.Bub.Pres.	[49]
$C_4H_{10}O$(2-Methylpropan-2-ol)+$C_{10}H_{22}O$(Decanol)				
	298.15 K	Positive	Max.Bub.Pres.	[49]
$C_4H_{10}O$(Butanol)+$C_{12}H_{11}N$(Diphenylamine)				
	333.15 K	Positive		[245]
$C_4H_{10}O$(2-Methylpropanol)+$C_{12}H_{11}N$(Diphenylamine)				
	333.15 K	Positive		[245]
$C_4H_{10}O$(Butan-2-ol)+$C_{13}H_{12}$(Diphenylmethane)				
	293.15 to 363.15 K (4)	Max.Min.		[81]
$C_4H_{10}O$(Butanol)+H_2O(Water)				
	273.15 to 323.15 K		Max.Bub.Pres.	[189]
	283.15 to 303.15 K		Cap.Rise	[275]

Table 1 (*cont.*)
SYSTEM

	Temperature	Deviation	Method	Ref.

$C_4H_{10}O$(2-Methylpropanol)+H_2O(Water)

	268.15 to 318.15 K		Max.Bub.Pres.	[189]
	288.15 to 353.15 K		Cap.Rise	[286]

$C_4H_{10}O$(Diethyl ether)+H_2O_4S(Sulphuric acid)

	290 K	Negative	Drop Weight	[59]

$C_4H_{10}O_3$(Diethyleneglycol)+$C_7H_{13}NO$(N-Methyl caprolactam)

	293.15 to 333.15 K			[109]

$C_4H_{11}N$(Diethylamine)+C_7H_5NS(Phenyl thiocyanate)

	303.15,313.15 K	Pos.Max.		[246]

$C_4H_{11}N$(Diethylamine)+H_2O(Water)

	313.15,333.15 K	Negative	Max.Bub.Pres.	[100]
	298.15 K			[155]

$C_4H_{11}NO_2$(bis(2-Hydroxyethyl)amine)+H_2O(Water)

	293 to 363 K			[168]

$C_4H_{12}Si$(Tetramethylsilane)+C_5H_{12}(2,2-Dimethylpropane)

	263.15,273.15,279.15 K	Negative	Dif.Cap.Rise	[298]

$C_4H_{12}Si$(Tetramethylsilane)+$C_6H_{18}OSi_2$(Hexamethyldisiloxane)

	233.15 to 233.15 K	Ideal	Dif.Cap.Rise	[296]

$C_4H_{12}Si$(Tetramethylsilane)+$C_{12}H_{36}O_4Si_5$(Dodecamethylpentasiloxane)

	233.15 to 273.15 K	Positive	Dif.Cap.Rise	[296]

$C_5H_4O_2$(2-Furaldehyde)+C_7H_8O(o-Cresol)

	303.15,313.15,323.15 K	Negative		[97]

$C_5H_4O_2$(2-Furaldehyde)+C_7H_8O(m-Cresol)

	303.15,313.15,323.15 K	Positive		[97]

$C_5H_4O_2$(2-Furaldehyde)+C_7H_8O(p-Cresol)

	303.15,313.15,323.15 K	Zero		[97]

C_5H_5N(Pyridine)+C_6H_6(Benzene)

	293.15,313.15,333.15 K			[200]

Table 1 (*cont.*)
SYSTEM

	Temperature	Deviation	Method	Ref.
C_5H_5N(Pyridine)+C_6H_6O(Phenol)				
	292.15,318.15 K	Pos.Max.		[254]
C_5H_5N(Pyridine)+C_6H_7N(2-Methylpyridine)				
	293.15 to 333.15 K			[200]
C_5H_5N(Pyridine)+C_7H_9N(p-Toluidine)				
				[39]
C_5H_5N(Pyridine)+$C_8H_{11}N$(N-Ethylaniline)				
	273.15,315.15,373.15 K	Positive		[227]
C_5H_5N(Pyridine)+H_2O(Water)				
	273.15 to 323.15 K (11)	Negative		[204]
	273.15,298.15 K	Negative		[207]
$C_5H_8N_2$(1,3-Dimethylpyrazole)+$C_5H_8N_2$(1,5-Dimethylpyrazole)				
				[166]
C_5H_{10}(Pent-1-ene)+C_6H_7N(Aniline)				
	292.15 K	Neg.Infl.		[223]
C_5H_{10}(Cyclopentane)+C_7H_8(Toluene)				
	298.15 K	Negative	Max.Bub.Pres.	[33]
$C_5H_{10}O_2$(Propyl acetate)+$C_5H_{10}O_3$(Ethyl acetate)				
	288.15,313.15 K	Negative		[226]
$C_5H_{10}O_2$(3-Methylbutanoic acid)+$C_5H_{11}N$(Piperidine)				
	298.15 K	Pos.Max.		[258]
$C_5H_{10}O_2$(Methyl butanoate)+C_6H_5Cl(Chlorobenzene)				
	283.15 K	Negative		[226]
$C_5H_{10}O_2$(Ethyl propanoate)+C_6H_5Cl(Chlorobenzene)				
	283.15 K	Negative		[226]
$C_5H_{10}O_2$(3-Methylbutanoic acid)+$C_6H_5NO_2$(Nitrobenzene)				
	278.65 K	Negative	Max.Bub.Pres.	[147]

Table 1 (*cont.*)
SYSTEM

	Temperature	*Deviation*	*Method*	*Ref.*

$C_5H_{10}O_2$(Ethyl propanoate)+C_6H_6(Benzene)

| | 283.15 K | | | [226] |

$C_5H_{10}O_2$(Methyl butanoate)+C_6H_6(Benzene)

| | 288.15,313.15 K | Pos.Neg. | | [226] |

$C_5H_{10}O_2$(Propyl acetate)+C_6H_6(Benzene)

| | 288.15,313.15 K | Positive | | [226] |

$C_5H_{10}O_2$(Propyl acetate)+$C_6H_{12}O_2$(Butyl acetate)

| | 293.15,313.15,333.15 K | | | [200] |

$C_5H_{10}O_2$(Propyl acetate)+$C_6H_{12}O_2$(Pentyl formate)

| | 293.15 to 333.15 K (4) | | | [225] |

$C_5H_{10}O_2$(Ethyl propanoate)+C_7H_8(Toluene)

| | 283.15 K | Positive | | [226] |

$C_5H_{10}O_2$(3-Methylbutanoic acid)+$C_{10}H_8$(Naphthalene)

| | | | Max.Bub.Pres. | [139] |
| | 298.15,308.15 K | Negative | | [220] |

$C_5H_{10}O_2$(3-Methylbutanoic acid)+$C_{12}H_{11}N$(Diphenylamine)

| | 327.15 K | | | [257] |

$C_5H_{10}O_2$(Pentanoic acid)+H_2O(Water)

| | | | | [104] |
| | 298.15,308.15 K | Negative | | [220] |

$C_5H_{10}O_3$(Ethyl carbonate)+$C_6H_5NO_2$(Nitrobenzene)

| | 298.15 K | Negative | Max.Bub.Pres. | [25] |

$C_5H_{10}O_3$(Ethyl carbonate)+C_6H_6(Benzene)

| | 298.15 K | Negative | Max.Bub.Pres. | [25] |

$C_5H_{11}N$(Piperidine)+$C_6H_{12}O_2$(4-Methylpentanoic acid)

| | 298.15 K | Pos.Max. | | [258] |

$C_5H_{11}N$(Piperidine)+C_7H_8(Toluene)

| | 288.15 to 405.15 K (4) | | | [229] |

Table 1 (*cont.*)
SYSTEM

	Temperature	Deviation	Method	Ref.
$C_5H_{11}N$(Piperidine)+$C_7H_{14}O_2$(Heptanoic acid)				
	298.15 K	Pos.Max.		[258]
$C_5H_{11}N$(Piperidine)+$C_8H_{16}O_2$(Octanoic acid)				
	298.15 K	Pos.Max.		[258]
$C_5H_{11}NO$(Diethylformamide)+H_2O(Water)				
				[113]
C_5H_{12}(Pentane)+C_6F_{14}(Tetradecafluorohexane)				
	303.15 K	Neg.Min.	Dif.Cap.Rise	[295]
C_5H_{12}(Pentane)+C_6H_5Br(Bromobenzene)				
	288.15 K	Negative	Cap.Rise	[54]
C_5H_{12}(Pentane)+C_6H_5Cl(Chlorobenzene)				
	288.15 K	Negative	Cap.Rise	[54]
C_5H_{12}(Pentane)+C_6H_6(Benzene)				
	288.15 K	Negative	Cap.Rise	[54]
C_5H_{12}(Pentane)+C_6H_{12}(Cyclohexane)				
	288.15 K	Negative	Cap.Rise	[54]
C_5H_{12}(Pentane)+$C_6H_{18}OSi_2$(Hexamethyldisiloxane)				
	303.15 K	Positive	Dif.Cap.Rise	[290]
C_5H_{12}(Pentane)+C_7H_8(Toluene)				
	288.15 K	Negative	Cap.Rise	[54]
C_5H_{12}(Pentane)+C_8F_{18}(Octadecafluoro-octane)				
	303 K	Neg.Min.	Dif.Cap.Rise	[19]
C_5H_{12}(Pentane)+C_8H_{10}(m-Xylene)				
	288.15 K	Negative	Cap.Rise	[54]
C_5H_{12}(Pentane)+C_8H_{10}(p-Xylene)				
	288.15 K	Negative	Cap.Rise	[54]
C_5H_{12}(Pentane)+$C_8H_{24}O_2Si_3$(Octamethyltrisiloxane)				
	303.15 K	Positive	Dif.Cap.Rise	[290]

Table 1 (*cont.*)
SYSTEM

	Temperature	*Deviation*	*Method*	*Ref.*

C_5H_{12}(Pentane)+$C_{10}H_{30}O_3Si_4$(Decamethyltetrasiloxane)

| | 303.15 K | Positive | Dif.Cap.Rise | [290] |

C_5H_{12}(Pentane)+$C_{12}H_{36}O_4Si_5$(Dodecamethylpentasiloxane)

| | 303.15 K | Positive | Dif.Cap.Rise | [290] |

C_5H_{12}(Pentane)+$C_{12}F_{27}$N(Heptacosafluorotributylamine)

| | 313.15 K | Neg.Min. | Dif.Cap.Rise | [295] |

$C_5H_{12}N_2O$(N,N,N,N-Tetramethylurea)+C_6F_6(Hexafluorobenzene)

| | 293.15,313.15 K | Negative | | [154] |

$C_5H_{12}O$(3-Methylbutanol)+$C_6H_4Cl_2$(p-Dichlorobenzene)

| | 328.15 K & F.pt. | Negative | | [135] |

$C_5H_{12}O$(3-Methylbutanol)+$C_6H_5NO_2$(Nitrobenzene)

| | 278.65 K | Negative | Max.Bub.Pres. | [147] |

$C_5H_{12}O$(Pentanol)+C_6H_6(Benzene)

| | | | | [160] |

$C_5H_{12}O$(2-Methylbutan-2-ol)+C_6H_{14}(Hexane)

| | 293.15 K | Negative | | [195] |
| | 295.15 K | Negative | | [197] |

$C_5H_{12}O$(3-Methylbutanol)+$C_{10}H_8$(Naphthalene)

| | | | Max.Bub.Pres. | [139] |

$C_5H_{12}O$(3-Methylbutanol)+$C_{12}H_{11}$N(Diphenylamine)

| | 333.15 K | Positive | | [245] |

C_6F_6(Hexafluorobenzene)+$C_6H_5NO_2$(Nitrobenzene)

| | 293.15,313.15 K | Negative | | [154] |

C_6F_6(Hexafluorobenzene)+C_6H_6(Benzene)

| | 293.15,313.15 K | Negative | | [154] |
| | 303.15 K | Negative | Dif.Cap.Rise | [293] |

C_6F_6(Hexafluorobenzene)+C_6H_{10}(Cyclohexene)

| | 303.15 K | Negative | Dif.Cap.Rise | [293] |

Table 1 (*cont.*)
SYSTEM

Temperature	Deviation	Method	Ref.

C_6F_6(Hexafluorobenzene)+C_6H_{12}(Cyclohexane)

| 303.15 K | Neg.Min. | Dif.Cap.Rise | [293] |

C_6F_{12}(Dodecafluorocyclohexane)+C_6H_{12}(Cyclohexane)

| 338.7,343.2 | Positive | Dif.Cap.Rise | [297] |

C_6F_{14}(Tetradecafluorohexane)+C_6H_{14}(2-Methylpentane)

| 303.15 K | Neg.Min. | Dif.Cap.Rise | [295] |

C_6F_{14}(Tetradecafluorohexane)+C_6H_{14}(Hexane)

| 303 K | Max.Min. | Dif.Cap.Rise. | [19] |

C_6F_{14}(Tetradecafluorohexane)+$C_6H_{18}OSi_2$(Hexamethyldisiloxane)

| 303.15 K | Neg.Min. | Dif.Cap.Rise | [295] |

C_6F_{14}(Tetradecafluorohexane)+C_7H_{16}(Heptane)

| 328.15,338.15 K | Max.Min. | Dif.Cap.Rise | [295] |

C_6F_{14}(Tetradecafluorohexane)+C_8H_{18}(Octane)

| 338 K | Neg.Max. | Dif.Cap.Rise | [19] |

C_6F_{14}(Tetradecafluorohexane)+$C_8H_{24}O_2Si_3$(Octamethyltrisiloxane)

| 323.15 K | Negative | Dif.Cap.Rise | [295] |

C_6H_4BrCl(p-Bromochlorobenzene)+$C_6H_4Br_2$(p-Dibromobenzene)

| 383.15,383.15,393.15 K | Negative | Dif.Cap.Rise | [133] |

C_6H_4BrCl(p-Bromochlorobenzene)+$C_6H_4ClNO_2$(m-Chloronitrobenzene)

| 383.15,388.15,393.15 K | Negative | Dif.Cap.Rise | [133] |

$C_6H_4BrNO_2$(m-Bromonitrobenzene)+$C_6H_4ClNO_2$(m-Chloronitrobenzene)

| 323.15 to 363.15 K (4) | Positive | | [249] |

$C_6H_4BrNO_2$(o-Bromonitrobenzene)+$C_6H_4ClNO_2$(o-Chloronitrobenzene)

| 316.15 to 363.15 K (9) | | | [249] |

$C_6H_4BrNO_2$(p-Bromonitrobenzene)+$C_6H_4ClNO_2$(p-Chloronitrobenzene)

| 373.15 to 449.15 K (4) | | | [249] |

$C_6H_4Br_2$(p-Dibromobenzene)+$C_6H_4ClNO_2$(m-Chloronitrobenzene)

| 383.15,388.15,393.15 K | Negative | Dif.Cap.Rise | [133] |

Table 1 (*cont.*)
SYSTEM

	Temperature	Deviation	Method	Ref.

$C_6H_4Br_2$(p-Dibromobenzene)+C_6H_6ClN(p-Chloroaniline)

| 383.15,388.15,393.15 K | Pos.Neg. | Dif.Cap.Rise | [133] |

$C_6H_4Cl_2$(o-Dichlorobenzene)+C_6H_6(Benzene)

| 293.15,303.15 K | Negative | Cap.Rise | [55] |

$C_6H_4Cl_2$(o-Dichlorobenzene)+C_6H_{12}(Cyclohexane)

| 293.15,303.15 K | Negative | Cap.Rise | [55] |

$C_6H_4Cl_2$(p-Dichlorobenzene)+C_7H_{16}O(Heptanol)

| 328.15 K & F.pt. | Negative | | [135] |

$C_6H_4Cl_2$(o-Dichlorobenzene)+C_8H_{10}(o-Xylene)

| 293.15,303.15 K | Negative | Cap.Rise | [55] |

$C_6H_4Cl_2$(m-Dichlorobenzene)+C_8H_{10}(m-Xylene)

| 293.15,303.15 K | Negative | Cap.Rise | [55] |

$C_6H_4Cl_2$(o-Dichlorobenzene)+C_8H_{10}(p-Xylene)

| 293.15,303.15 K | Negative | Cap.Rise | [55] |

$C_6H_4Cl_2$(p-Dichlorobenzene)+C_8H_{18}O(Octadecanol)

| 328.15 K & F.pt. | Negative | | [135] |

C_6H_5Br(Bromobenzene)+C_6H_5Cl(Chlorobenzene)

| 293.15,296.15,303.15 K | Negative | Cap.Rise | [47] |
| 284.15 to 343.15 K (7) | | | [225] |

C_6H_5Br(Bromobenzene)+C_6H_6(Benzene)

293.15,298.15,303.15 K	Negative	Cap.Rise	[47]
293.15,313.15,323.15 K			[200]
293.15,303.15 K	Negative	Cap.Rise	[278]

C_6H_5Br(Bromobenzene)+C_6H_{12}(Cyclohexane)

| 293.15,303.15 K | Negative | Cap.Rise | [278] |

C_6H_5Br(Bromobenzene)+C_7H_8(Toluene)

293.15,298.15,303.15 K	Negative	Cap.Rise	[47]
293.15,313.15,333.15 K			[200]
293.15,308.15,323.15 K	Positive		[231]

Table 1 (*cont.*)
SYSTEM

Temperature	Deviation	Method	Ref.
C_6H_5Br(Bromobenzene)$+C_8H_{10}$(o-Xylene)			
293.15,303.15 K	Negative	Cap.Rise	[278]
C_6H_5Br(Bromobenzene)$+C_8H_{10}$(m-Xylene)			
293.15,303.15 K	Negative	Cap.Rise	[278]
C_6H_5Br(Bromobenzene)$+C_8H_{10}$(p-Xylene)			
293.15,303.15 K	Negative	Cap.Rise	[14]
C_6H_5Cl(Chlorobenzene)$+C_6H_6$(Benzene)			
293.15,298.15,303.15 K	Negative	Cap.Rise	[47]
293.15,303.15 K	Negative	Cap.Rise	[101]
			[118]
323.15 to 343.15 K	Negative	Cap.Rise	[127]
293.15,313.15,333.15 K			[200]
283.15,313.15 K	Negative		[226]
C_6H_5Cl(Chlorobenzene)$+C_6H_6ClN$(m-Chloroaniline)			
293.15 K		Max.Bub.Pres.	[276]
C_6H_5Cl(Chlorobenzene)$+C_6H_7N$(Aniline)			
293.15 K		Max.Bub.Pres.	[116]
C_6H_5Cl(Chlorobenzene)$+C_6H_{12}$(Cyclohexane)			
293.15,313.15 K	Negative	Cap.Rise&Drop Wt.	[23]
293.15,303.15 K	Negative	Cap.Rise	[101]
C_6H_5Cl(Chlorobenzene)$+C_7H_8$(Toluene)			
293.15,298.15,303.15 K	Negative	Cap.Rise	[47]
			[176]
323.15 to 443.15 K		Cap.Rise	[127]
283.15,395.15 K	Negative		[226]
293.15,303.15,323.15 K	Negative		[231]
C_6H_5Cl(Chlorobenzene)$+C_8H_{10}$(o-Xylene)			
			[14]
293.15,303.15 K	Negative	Cap.Rise	[101]
C_6H_5Cl(Chlorobenzene)$+C_8H_{10}$(m-Xylene)			
			[14]
293.15,303.15 K	Negative	Cap.Rise	[101]

Table 1 (*cont.*)
SYSTEM

Temperature	Deviation	Method	Ref.

C_6H_5Cl(Chlorobenzene)+C_8H_{10}(p-Xylene)

			[14]
293.15,303.15 K	Negative	Cap.Rise	[101]

C_6H_5Cl(Chlorobenzene)+$C_{13}H_{12}$(Diphenylmethane)

303.15 K		Max.Bub.Pres.	[82]

$C_6H_5NO_2$(Nitrobenzene)+C_6H_6(Benzene)

298.15 K	Negative	Max.Bub.Pres.	[16]
293.15 to 303.15 K	Negative	Max.Bub.Pres.	[38]
293.15 K			[10]
328.15 K	Negative		[238]
298.15 K	Negative		[240]

$C_6H_5NO_2$(Nitrobenzene)+C_6H_7N(Aniline)

288.15 K			[234]

$C_6H_5NO_2$(Nitrobenzene)+C_6H_{12}(Cyclohexane)

293.15 K			[10]
293.15,303.15 K	Negative	Max.Bub.Pres.	[38]
293.15 K	Negative		[195]

$C_6H_5NO_2$(Nitrobenzene)+C_6H_{14}(Hexane)

295.15,308.15 K	Neg.Infl.	Cap.Rise	[58]
293.15 K			[10]
303.15,308.15,313.15 K	Negative	Max.Bub.Pres.	[27]
295.15 to 318.15 K (6)	Neg.Infl.	Max.Bub.Pres.	[68]
273.15 to 323.15 K (10)	Neg.Infl.		[87]

$C_6H_5NO_2$(Nitrobenzene)+C_7H_8(Toluene)

293.15 K			[10]
323.15 K	Negative		[238]
298.15 K	Negative		[240]

$C_6H_5NO_2$(Nitrobenzene)+C_7H_9N(N-Methylaniline)

285.15 to 343.15 K (7)	Negative		[225]

$C_6H_5NO_2$(Nitrobenzene)+C_7H_9N(o-Toluidine)

288.15 to 343.15 K (6)			[225]
288.15 to 343.15 K (6)			[228]

$C_6H_5NO_2$(Nitrobenzene)+C_7H_{14}(Methylcyclohexane)

293.15 K			[10]

Table 1 (*cont.*)
SYSTEM

Temperature	Deviation	Method	Ref.

$C_6H_5NO_2$ (Nitrobenzene)+C_7H_{16} (Heptane)

303.15 K	Negative	Max.Bub.Pres.	[27]
293.15 K			[10]

$C_6H_5NO_2$ (Nitrobenzene)+C_8H_{10} (p-Xylene)

293.15 K			[10]

$C_6H_5NO_2$ (Nitrobenzene)+$C_8H_{11}N$ (N,N-Dimethylaniline)

283.15 to 343.15 K (9)	Negative		[225]

$C_6H_5NO_2$ (Nitrobenzene)+$C_8H_{11}N$ (N-Ethylaniline)

289.15 to 343.15 K (6)	Negative		[225]

$C_6H_5NO_2$ (Nitrobenzene)+C_8H_{18} (Octane)

293.15 K			[10]

$C_6H_5NO_2$ (Nitrobenzene)+$C_{10}H_{15}N$ (N,N-Diethylaniline)

288.15 to 343.15 K (6)	Negative		[225]

$C_6H_5NO_2$ (Nitrobenzene)+$C_{10}H_{22}$ (1,8-Dimethyloctane)

293.15 K			[10]

$C_6H_5NO_2$ (Nitrobenzene)+H_2O_4S (Sulphuric acid)

300.15 K	Neg.Min.	Drop Weight	[59]

$C_6H_5NO_3$ (p-Nitrophenol)+$C_{10}H_8$ (Naphthalene)

394.15 K	Negative	Cap.Rise	[90]

C_6H_6 (Benzene)+C_6H_6O (Phenol)

308.15 K	Negative	Drop Weight	[1]
298.15 K	Neg.Min.	Ring	[15]

C_6H_6 (Benzene)+C_6H_7N (Aniline)

298.15 K	Negative	Tensiometer	[105]
288.15 K	Negative		[234]
293.15 K	Negative		[238]

C_6H_6 (Benzene)+$C_6H_{10}O$ (Cyclohexane)

291.15 K	Negative		[191]

Table 1 (*cont.*)
SYSTEM

	Temperature	Deviation	Method	Ref.

C_6H_6(Benzene)+C_6H_{12}(Cyclohexane)

	Temperature	Deviation	Method	Ref.
	293.15,303.15 K	Negative	Max.Bub.Pres.	[9]
	298.15 K	Negative	Max.Bub.Pres.	[16]
	293.15,303.15 K	Negative	Max.Bub.Pres.	[33]
	293.15,303.15 K	Negative	Max.Bub.Pres.	[38]
	291.15 K	Negative		[48]
				[73]
	298.15 K	Negative	Drop Weight	[89]
	278.15 to 333.15 K			[147]
	293.15 K	Negative		[195]
	293.15 to 318.15 K (6)	Negative	Max.Bub.Pres.	[279]

C_6H_6(Benzene)+$C_6H_{12}O$(Cyclohexanol)

	Temperature	Deviation	Method	Ref.
	293.15 K	Negative		[192]

C_6H_6(Benzene)+C_6H_{14}(Hexane)

	Temperature	Deviation	Method	Ref.
	298.15 to 313.15 K (4)	Negative	Max.Bub.Pres.	[29]
				[73]
	278.15 to 333.15 K			[145]
	293.15 K	Negative		[195]

C_6H_6(Benzene)+$C_6H_{14}O$(Hexanol)

	Temperature	Deviation	Method	Ref.
				[160]
	293.15 K	Negative		[195]
	295.15 K	Negative	Max.Bub.Pres.	[197]

C_6H_6(Benzene)+$C_7H_6O_2$(Salicylaldehyde)

	Temperature	Deviation	Method	Ref.
	293.15 K			[250]

C_6H_6(Benzene)+C_7H_8(Toluene)

	Temperature	Deviation	Method	Ref.
	293.15,298.15,303.15 K	Negative	Cap.Rise	[47]
				[118]
	323.15 to 443.15 K	Zero	Cap.Rise	[127]
	278.15 to 333.15 K			[145]
	291.35 K	Negative	Cap.Rise	[188]
	298.15 K			[224]
	283.15 to 343.15 K	Negative		[225]
	283.15 to 315.15 K	Negative		[226]

C_6H_6(Benzene)+C_7H_8O(Benzyl alcohol)

	Temperature	Deviation	Method	Ref.
	288.15 to 323.15 K			[106]

C_6H_6(Benzene)+C_7H_8O(o-Cresol)

	Temperature	Deviation	Method	Ref.
	293.15 K	Positive		[237]

Table 1 (*cont.*)

SYSTEM

Temperature	Deviation	Method	Ref.

C_6H_6(Benzene)+C_7H_8O(m-Cresol)

283.15 to 333.15 K	Positive		[225]
293.15 K	Negative		[237]

C_6H_6(Benzene)+C_7H_8O(p-Cresol)

293.15 K	Positive		[237]

C_6H_6(Benzene)+$C_7H_8O_2$(o-Methoxyphenol)

293.15 K			[250]

C_6H_6(Benzene)+C_7H_9N(N-Methylaniline)

298.15 K	Negative	Tensiometer	[105]

C_6H_6(Benzene)+$C_7H_{14}O$(o-Methylcyclohexanol)

293.15 K	Negative		[192]

C_6H_6(Benzene)+$C_7H_{14}O$(m-Methylcyclohexanol)

293.15 K	Negative		[192]

C_6H_6(Benzene)+$C_7H_{14}O$(p-Methylcyclohexanol)

293.15 K	Negative		[192]

C_6H_6(Benzene)+C_7H_{16}(Heptane)

278.15 to 333.15 K			[145]

C_6H_6(Benzene)+C_8H_8O(Acetophenone)

298.15 K			[16]

C_6H_6(Benzene)+C_8H_{10}(o-Xylene)

298.15 K	Negative	Max.Bub.Pres.	[33]

C_6H_6(Benzene)+C_8H_{10}(m-Xylene)

298.15 K	Negative	Max.Bub.Pres.	[16]
283.15 to 343.15 K	Negative		[225]

C_6H_6(Benzene)+$C_8H_{11}N$(Dimethylaniline)

298.15 K	Negative	Pendant Drop	[239]

C_6H_6(Benzene)+C_8H_{18}(2,2,4-Trimethylpentane)

303.15 K	Negative	Max.Bub.Pres.	[13]
303.15 K	Negative	Max.Bub.Pres.	[52]
293.15,313.15,333.15 K	Negative	Max.Bub.Pres.	[53]

Table 1 (*cont.*)
SYSTEM

Temperature	Deviation	Method	Ref.

C_6H_6(Benzene)+C_9H_7N(Quinoline)

| 293.15,313.15,333.15 K | Negative | Max.Bub.Pres. | [53] |

C_6H_6(Benzene)+C_9H_{12}(Mesitylene)

| 283.15,305.95,328.15 K | Negative | | [227] |

C_6H_6(Benzene)+C_9H_{13}N(Dimethyl-o-toluidine)

| 327.75 K | Negative | | [238] |

C_6H_6(Benzene)+$C_{10}H_7$Br(1-Bromonaphthalene)

| 293.15,313.15,333.15 K | Negative | Max.Bub.Pres. | [53] |

C_6H_6(Benzene)+$C_{10}H_8$(Naphthalene)

| 352.65 K | Positive | Cap.Rise | [91] |

C_6H_6(Benzene)+$C_{12}H_{10}$(Diphenyl)

| 343.15 K | Negative | Cap.Rise | [4] |

C_6H_6(Benzene)+$C_{12}H_{26}$(Dodecane)

| 298.15 to 313.15 K (4) | Negative | Max.Bub.Pres. | [29] |

C_6H_6(Benzene)+$C_{13}H_{12}$(Diphenylmethane)

| 303.05 K | Pos.Neg. | Cap.Rise | [20] |
| 303.15 K | | Max.Bub.Pres. | [82] |

C_6H_6(Benzene)+$C_{14}H_{10}$(Diphenylethyne)

| 333.15 K | Positive | Cap.Rise | [4] |
| 333.15 K | Positive | Cap.Rise | [31] |

C_6H_6(Benzene)+$C_{14}H_{12}O_2$(Benzyl benzoate)

| 293.15,313.15,333.15 K | Negative | Max.Bub.Pres. | [53] |

C_6H_6(Benzene)+$C_{14}H_{14}$(1,2-Diphenylethane)

333.15 K	Positive	Cap.Rise	[4]
333.15 K	Positive	Cap.Rise	[31]
343.15 K	Negative	Cap.Rise	[37]

C_6H_6(Benzene)+$C_{19}H_{24}$(4,4-Dicumylmethane)

| 293.15,313.15,333.15 K | Positive | Max.Bub.Pres. | [56] |

Table 1 (*cont.*)
SYSTEM

Temperature	Deviation	Method	Ref.

C_6H_6(Benzene)+$C_{20}H_{26}$(1,8-Diphenyloctane)

298.15 K	Positive	Cap.Rise	[4]
298.15 K	Positive	Cap.Rise	[31]

C_6H_6(Benzene)+ICl(Iodine monochloride)

293.15 to 343.15 K		Cap.Rise&Bub.Pres.	[150]

C_6H_6O(Phenol)+H_4N_2(Hydrazine)

338.15,348.15,388.15 K	Pos.Neg.	Max.Bub.Pres.	[143]

C_6H_6O(Phenol)+C_6H_7N(Aniline)

292.65,313.15 K	Neg.Min.		[254]

C_6H_6O(Phenol)+C_7H_9N(N-Methylaniline)

292.65,318.15 K	Neg.Min.		[254]

C_6H_6O(Phenol)+$C_8H_{11}N$(N,N-Dimethylaniline)

292.65,318.15 K	Negative		[254]

C_6H_6O(Phenol)+$C_8H_{11}N$(N-Ethylaniline)

292.65,318.15 K	Negative		[254]

C_6H_6O(Phenol)+$C_{10}H_{15}N$(N,N-Diethylaniline)

292.65,318.15 K	Negative		[254]

C_6H_6O(Phenol)+H_2O(Water)

273.15 to 338.15 K	Neg.Infl.		[206]
342.15 K			[219]
288.15 to 373.15 K		Cap.Rise	[286]

$C_6H_6O_2$(Catechol)+$C_6H_6O_2$(Resorcinol)

383.15,388.15,393.15 K	Positive	Dif.Cap.Rise	[133]

$C_6H_6O_2$(Catechol)+$C_6H_8N_2$(o-Phenylenediamine)

383.15,388.15,393.15 K	Neg.Min.	Dif.Cap.Rise	[133]

C_6H_7N(Aniline)+C_6H_{12}(Cyclohexane)

305.15,333.15 K	Neg.Infl.	Max.Bub.Pres.	[63]
308.15 K	Negative	Max.Bub.Pres.	[27]
283.15,304.45,343.15 K	Neg.Infl.		[102]

Table 1 (*cont.*)
SYSTEM

	Temperature	Deviation	Method	Ref.
C_6H_7N(Aniline)+C_6H_{14}(Hexane)				
	298.15 to 341.15 K (11)	Neg.Infl.	Dif.Cap.Rise	[41]
	298.15 K	Negative		[236]
C_6H_7N(Aniline)+C_7H_5NS(Phenyl thiocyanate)				
	287.15,288.15 K	Pos.Max.		[246]
C_6H_7N(Aniline)+C_7H_8(Toluene)				
	293.15 K	Negative		[238]
C_6H_7N(2-Methylpyridine)+C_7H_8(Toluene)				
	293.15,313.15,333.15 K			[200]
C_6H_7N(Aniline)+C_7H_9N(o-Toluidine)				
	327.15 K	Negative		[238]
C_6H_7N(Aniline)+C_7H_9N(p-Toluidine)				
				[39]
C_6H_7N(Aniline)+$C_8H_{11}N$(N,N-Dimethylaniline)				
				[111]
	327.85 K	Positive		[238]
C_6H_7N(Aniline)+C_9H_7N(Quinoline)				
				[111]
C_6H_7N(Aniline)+$C_9H_{10}O_2$(Benzyl acetate)				
				[74]
C_6H_7N(Aniline)+$C_{10}H_{18}$(trans-Decalin)				
	313.15 K	Negative	Max.Bub.Pres.	[27]
C_6H_7N(Aniline)+H_2O(Water)				
	288.15 to 373.15 K		Cap.Rise	[286]
$C_6H_8N_2$(m-Phenylenediamine)+H_2O(Water)				
				[137]
C_6H_{10}(Cyclohexene)+C_6H_{12}(Cyclohexane)				
	273.15 to 350.15 K			[280]

Table 1 (*cont.*)

SYSTEM

Temperature	Deviation	Method	Ref.

$C_6H_{10}O$(Cyclohexanone)$+C_{12}H_{11}N$(Diphenylamine)

333.15 K	Positive		[245]

$C_6H_{10}O$(2-Methylpent-2-ene-4-one)$+H_2O$(Water)

			[187]

C_6H_{12}(Cyclohexane)$+C_6H_{14}$(Hexane)

298.15 K	Negative	Drop Weight	[34]
			[73]
			[75]
298.16 K	Negative	Drop Weight	[196]

C_6H_{12}(Methylcyclopentane)$+C_6H_{14}$(Hexane)

293.15 K	Positive		[195]

C_6H_{12}(Cyclohexane)$+C_6H_{14}O$(Hexanol)

295.15 K	Negative	Max.Bub.Pres.	[197]

C_6H_{12}(Cyclohexane)$+C_7F_{14}$(Tetradecafluoromethylcyclohexane)

329 K	Neg.Min.	Dif.Cap.Rise	[297]

C_6H_{12}(Cyclohexane)$+C_7H_8$(Toluene)

293.15 to 313.15 K	Negative	Cap.Rise&Drop Wt.	[23]
298.15 K	Negative	Max.Bub.Pres.	[33]

C_6H_{12}(Cyclohexane)$+C_7H_8O$(Benzyl alcohol)

288.15 to 333.15 K			[106]

C_6H_{12}(Cyclohexane)$+C_7H_{16}$(Heptane)

298.15 K	Negative	Drop Weight	[34]
298.15 K	Negative	Drop Weight	[196]

C_6H_{12}(Cyclohexane)$+C_8H_{10}$(Ethylbenzene)

298.15,308.15 K	Negative	Drop Weight	[32]

C_6H_{12}(Cyclohexane)$+C_8H_{10}$(o-Xylene)

298.15,308.15 K	Negative	Drop Weight	[32]

C_6H_{12}(Cyclohexane)$+C_8H_{10}$(m-Xylene)

298.15,308.15 K	Negative	Drop Weight	[32]

Table 1 (*cont.*)
SYSTEM

Temperature	Deviation	Method	Ref.

C_6H_{12}(Cyclohexane)+C_8H_{10}(p-Xylene)

298.15, 308.15 K	Negative	Drop Weight	[32]

C_6H_{12}(Cyclohexane)+C_8H_{18}(Octane)

298.15 K	Negative	Drop Weight	[34]
298.16 K	Negative	Drop Weight	[196]

C_6H_{12}(Cyclohexane)+C_8H_{18}(2,2,4-Trimethylpentane)

303.17 K	Negative	Max.Bub.Pres.	[13]
303.15 K	Negative	Max.Bub.Pres.	[52]

C_6H_{12}(Cyclohexane)+$C_{10}F_{18}$(Octadecafluorodecalin)

336 K		Dif.Cap.Rise	[297]

C_6H_{12}(Cyclohexane)+$C_{10}H_{18}$(cis-Decalin)

298.15 K	Positive	Max.Bub.Pres.	[33]

C_6H_{12}(Cyclohexane)+$C_{10}H_{18}$(trans-Decalin)

298.15 K	Positive	Max.Bub.Pres.	[33]

C_6H_{12}(Cyclohexane)+$C_{12}H_{22}$(Dicyclohexyl)

298.15, 313.15 K	Negative	Dif.Cap.Rise	[299]

C_6H_{12}(Cyclohexane)+$C_{12}H_{26}$(Dodecane)

298.16 K	Pos.Neg.	Drop Weight	[196]

C_6H_{12}(Cyclohexane)+$C_{13}H_{24}$(Dicyclohexylmethane)

298.15, 313.15 K	Negative	Dif.Cap.Rise	[299]

C_6H_{12}(Cyclohexane)+$C_{14}H_{26}$(1,2-Dicyclohexylethane)

298.15, 313.15 K	Positive	Dif.Cap.Rise	[299]

C_6H_{12}(Cyclohexane)+$C_{15}H_{28}$(1,3-Dicyclohexylpropane)

298.15, 313.15 K	Positive	Dif.Cap.Rise	[299]

C_6H_{12}(Cyclohexane)+$C_{16}H_{30}$(1,4-Dicyclohexylbutane)

298.15, 313.15 K	Positive	Dif.Cap.Rise	[299]

C_6H_{12}(Cyclohexane)+$C_{16}H_{34}$(Hexadecane)

298.16 K	Pos.Neg.	Drop Weight	[196]

Table 1 (*cont.*)
SYSTEM

	Temperature	Deviation	Method	Ref.

$C_6H_{12}O$(1-Methylpentan-2-one)+$C_6H_{14}O$(Diisopropyl ether)

			[276]

$C_6H_{12}O$(Cyclohexanol)+$C_6H_{14}O$(Hexanol)

| 295.15 K | Negative | | [197] |

$C_6H_{12}O_2$(4-Methylpentanoic acid)+$C_{12}H_{11}N$(Diphenylamine)

| 324.15 K | | | [254] |

$C_6H_{12}O_2$(Butyl acetate)+H_2O(Water)

| 298.15 K | | Pendant Drop | [167] |

$C_6H_{12}O_2$(2-Hydroxy-2-methylpentan-4-one)+H_2O(Water)

| | | | [187] |

$C_6H_{12}O_2$(Butyl acetate)+H_3O_4P(Phosphoric acid)

| 298.15,313.15,333.15 K | Negative | | [2] |

C_6H_{14}(Hexane)+$C_6H_{14}O$(Hexanol)

| 295.15 K | Negative | Max.Bub.Pres. | [197] |

C_6H_{14}(Hexane)+C_7H_9N(o-Toluidine)

| 303.15 K | Negative | Max.Bub.Pres. | [27] |

C_6H_{14}(Hexane)+C_8H_{18}(Octane)

| 303.15 to 353.15 K | | Max.Bub.Pres. | [83] |

C_6H_{14}(Hexane)+$C_{10}H_{12}$(Tetralin)

| 291.15 K | | | [191] |

C_6H_{14}(Hexane)+$C_{10}H_{22}$(Decane)

| | | | [80] |

C_6H_{14}(Hexane)+$C_{12}H_{26}$(Dodecane)

| 293.15 K | Positive | Drop Volume | [30] |
| 298.15 K | Pos.Neg. | Drop Weight | [34] |

C_6H_{14}(Hexane)+$C_{14}H_{30}$(Tetradecane)

| 293.15 K | Positive | Drop Volume | [30] |
| 307.15 to 353.15 K | | Max.Bub.Pres. | [83] |

Table 1 (*cont.*)
SYSTEM

Temperature	Deviation	Method	Ref.

C_6H_{14}(Hexane)+$C_{16}H_{34}$(Hexadecane)

293.15 K	Positive	Cap.Rise&Drop Vol.	[30]
298.15 K	Zero	Drop Weight	[34]

$C_6H_{14}O$(Hexanol)+$C_{10}H_{22}O$(Decanol)

298.15 K	Pos.Neg.	Max.Bub.Pres.	[35]

$C_6H_{14}O_2$(2-Butoxyethanol)+H_2O(Water)

313.15 to 327.15 K	Neg.Infl.	Dif.Cap.Rise	[300]

$C_6H_{14}O_4$(Triethylene glycol)+$C_7H_{13}NO$(N-Methylcaprolactam)

293.15 to 333.15 K			[109]

$C_6H_{15}N$(Triethylamine)+H_2O(Water)

313.15 to 333.15 K	Negative	Max.Bub.Pres.	[100]
273.15 to 288.15 K			[136]
273.15,292.35,303.15 K	Negative		[206]

$C_6H_{16}N_2$(1,6-Hexanediamine)+H_2O(Water)

			[137]

$C_6H_{18}OSi_2$(Hexamethyldisiloxane)+C_7H_8(Toluene)

300.15 K	Negative	Max.Bub.Pres.	[24]

$C_6H_{18}OSi_2$(Hexamethyldisiloxane)+C_7H_{16}(Heptane)

303.15 K	Negative	Dif.Cap.Rise	[290]

$C_6H_{18}OSi_2$(Hexamethyldisiloxane)+$C_{10}H_{22}$(Decane)

303.15 K	Negative	Dif.Cap.Rise	[290]

$C_6H_{18}OSi_2$(Hexamethyldisiloxane)+$C_{16}H_{34}$(Hexadecane)

303.15 K	Negative	Dif.Cap.Rise	[290]

C_7H_5NS(Phenyl isothiocyanate)+$C_{12}H_{11}N$(Diphenylamine)

323.15,348.15 K	Positive		[247]

C_7H_8(Toluene)+C_7H_8O(Benzyl alcohol)

288.15 to 333.15 K (9)			[106]

C_7H_8(Toluene)+C_7H_8O(o-Cresol)

288.15 K			[251]

Table 1 (*cont.*)
SYSTEM

Temperature	Deviation	Method	Ref.

C_7H_8(Toluene)+C_7H_8O(m-Cresol)

288.15 to 333.15 K (16) Negative			[225]
288.15 K			[251]

C_7H_8(Toluene)+C_7H_8O(p-Cresol)

288.15 K			[251]

C_7H_8(Toluene)+C_7H_9O(o-Toluidine)

327.65 K	Negative		[238]

C_7H_8(Toluene)+$C_8H_{11}N$(N,N-Dimethylaniline)

293.15 K	Negative		[238]

C_7H_8(Toluene)+$C_9H_{10}O_2$(Ethyl benzoate)

298.15 K	Negative		[224]

C_7H_8(Toluene)+$C_9H_{13}N$(Dimethyl-o-toluidine)

327.15 K	Negative		[238]

C_7H_8(Toluene)+$C_{10}H_{16}$(Turpentine)

298.15 K	Negative		[224]

C_7H_8O(m-Cresol)+C_7H_9N(o-Toluidine)

290.15 to 343.15 K (16)			[225]

C_7H_8O(m-Cresol)+$C_8H_{11}N$(N,N-Dimethylaniline)

283.15 to 343.15 K (18)			[225]

C_7H_8O(m-Cresol)+C_9H_7N(Quinoline)

298.15 K	Positive		[129]

C_7H_8O(m-Cresol)+$C_9H_{10}O_2$(Benzyl acetate)

			[74]

C_7H_8O(Benzyl alcohol)+$C_{12}H_{11}N$(Diphenylamine)

333.15 K	Negative		[245]

C_7H_8O(Benzyl alcohol)+H_2O(Water)

293.15 to 333.15 K (7)			[106]

Table 1 (*cont.*)
SYSTEM

	Temperature	Deviation	Method	Ref.

C_7H_8O(o-Cresol)+H_6N_2O(Hydrazine hydrate)

| | | | | [157] |

$C_7H_8O_2$(Phenyl acetate)+H_2O(Water)

| | 298.15 K | | Pendant Drop | [167] |

C_7H_9N(p-Toluidine)+H_2O(Water)

| | | | | [39] |

$C_7H_{13}NO$(N-Methylcaprolactam)+$C_8H_{18}O_5$(Tetraethylene glycol)

| | 293.15 to 333.15 K | | | [109] |

$C_7H_{14}O_2$(n-Pentyl acetate)+H_3O_4P(Phosphoric acid)

| | 298.15,313.15,333.15 K | Negative | | [2] |

C_7H_{16}(Heptane)+$C_7H_{16}O$(Heptanol)

| | | Positive | | [107] |

C_7H_{16}(Heptane)+$C_8H_{11}N$(N,N-Dimethylaniline)

| | 298.15 K | Negative | Pendant Drop | [239] |

C_7H_{16}(Heptane)+C_8H_{18}(2,2,4-Trimethylpentane)

| | 293.15,313.15,333.15 K | | | [200] |

C_7H_{16}(Heptane)+$C_8H_{18}O$(Octanol)

| | | Pos.Max. | | [107] |

C_7H_{16}(Heptane)+$C_8H_{24}O_2Si_3$(Octamethyltrisiloxane)

| | 303.15 K | Negative | Dif.Cap.Rise | [290] |

C_7H_{16}(Heptane)+$C_9H_{20}O$(Nonanol)

| | | Pos.Max. | | [107] |

C_7H_{16}(Heptane)+$C_{10}H_{22}$(Decane)

| | | | | [80] |

C_7H_{16}(Heptane)+$C_{10}H_{22}O$(Decanol)

| | | Pos.Max. | | [107] |

C_7H_{16}(Heptane)+$C_{10}H_{30}O_3Si_4$(Decamethyltetrasiloxane)

| | 303.15 K | Negative | Dif.Cap.Rise | [290] |

Table 1 (*cont.*)
SYSTEM

Temperature	Deviation	Method	Ref.

C_7H_{16}(Heptane)+$C_{12}H_{36}O_4Si_5$(Dodecamethylpentasiloxane)

303.15 K	Negative	Dif.Cap.Rise	[290]

C_7H_{16}(Heptane)+$C_{11}H_{24}$(Undecane)

303.15 to 373.15 K		Max.Bub.Pres.	[272]

C_7H_{16}(Heptane)+$C_{16}H_{34}$(Hexadecane)

293.15 K	Positive	Cap.Rise&Drop Vol.	[30]
303.15 to 373.15 K		Max.Bub.Pres.	[131]
293.15 to 305.05 K	Positive	Cap.Rise	[20]
			[272]

$C_8H_7O_2Cl$(Methyl-o-chlorobenzoate)+$C_{10}H_{18}$(Decalin)

291.15 K	Negative		[191]

C_8H_9NO(Acetanilide)+$C_{14}H_{10}O_2$(Benzil)

393.15 K			[11]

C_8H_{10}(o-Xylene)+C_8H_{10}(p-Xylene)

284.15 to 318.15 K (5)	Negative		[225]

C_8H_{10}(m-Xylene)+C_8H_{10}(p-Xylene)

285.15,373.15 K			[228]

C_8H_{10}(o-Xylene)+$C_{10}H_8$(Naphthalene)

290.15 to 343.15 K (8)			[225]

C_8H_{10}(o-Xylene)+$C_{19}H_{24}$(4,4'-Dicumylmethane)

293.15,313.15,333.15 K	Positive	Max.Bub.Pres.	[56]

C_8H_{10}(m-Xylene)+$C_{19}H_{24}$(4,4'-Dicumylmethane)

293.15,313.15,333.15 K	Positive	Max.Bub.Pres.	[56]

C_8H_{10}(p-Xylene)+$C_{19}H_{24}$(4,4'-Dicumylmethane)

293.15,313.15,3343.15 K	Positive	Max.Bub.Pres.	[56]

$C_8H_{11}N$(N,N-Dimethylaniline)+C_9H_{12}(Cumene)

293.15,313.15,333.15 K			[200]

$C_8H_{11}N$(2,3,6-Trimethylpyridine)+H_2O(Water)

273.15 to 353.15 K (4)	Negative		[136]

Table 1 (*cont.*)

SYSTEM

	Temperature	*Deviation*	*Method*	*Ref.*

$C_8H_{12}N_2$(m-Xylylenediamine)+H_2O(Water)

| | | | | [137] |

$C_8H_{12}N_2$(p-Xylylenediamine)+H_2O(Water)

| | | | | [137] |

C_8H_{18}(2,2,4-Trimethylpentane)+C_9H_7N(Quinoline)

| 293.15,313.15,333.15 K | Negative | Max.Bub.Pres. | [53] |

C_8H_{18}(2,2,4-Trimethylpentane)+$C_{12}H_{26}$(Dodecane)

| 303.15 K | Zero | Max.Bub.Pres. | [13] |
| 303.15 K | Zero | Max.Bub.Pres. | [52] |

C_8H_{18}(2,2,4-Trimethylpentane)+$C_{14}H_{12}O_2$(Benzyl benzoate)

| 293.15,313.15,333.15 K | Negative | Max.Bub.Pres. | [53] |

C_8H_{18}(Octane)+$C_{14}H_{30}$(Tetradecane)

| 303.15 to 353.15 K | | Max.Bub.Pres. | [83] |

C_8H_{18}(Octane)+$C_{16}H_{34}$(Hexadecane)

| 293.15 K | Positive | Cap.Rise&Drop Vol. | [30] |

C_8H_{18}(2,2,4-Trimethylpentane)+H_2O(Water)

| 293.15,313.15,333.15 K | Negative | Max.Bub.Pres. | [53] |

$C_8H_{24}O_2Si_3$(Octamethyltrisiloxane)+$C_{10}H_{22}$(Decane)

| 303.15 K | Negative | Dif.Cap.Rise | [290] |

$C_8H_{24}O_2Si_3$(Octamethyltrisiloxane)+$C_{14}H_{30}$(Tetradecane)

| 303.15 K | Negative | Dif.Cap.Rise | [290] |

C_9H_7N(Quinoline)+$C_{12}H_{11}N$(Diphenylamine)

| 333.15,363.15 K | Positive | | [227] |

$C_9H_{10}O_2$(Ethyl benzoate)+H_2O(Water)

| 298.15 K | | Pendant Drop | [167] |

C_9H_{12}(Cumene)+$C_{19}H_{24}$(4,4'-Dicumylmethane)

| 293.15,313.15,333.15 K | Positive | Max.Bub.Pres. | [56] |

Table 1 (*cont.*)
SYSTEM

Temperature	Deviation	Method	Ref.

$C_9H_{22}N_2$(2,2,4-Trimethylhexane-1,6-diamine)+H_2O(Water)

			[37]

$C_{10}H_{12}$(Tetralin)+$C_{19}H_{24}$(4,4'-Dicumylmethane)

293.15,313.15,333.15 K	Negative	Max.Bub.Pres.	[56]

$C_{10}H_{14}N_2$(Nicotine)+H_2O(Water)

273.15 to 368.15 K			[208]

$C_{10}H_{22}$(Decane)+$C_{10}H_{30}O_3Si_4$(Decamethyltetrasiloxane)

303.15 K	Negative	Dif.Cap.Rise	[290]

$C_{10}H_{22}$(Decane)+$C_{12}H_{36}O_4Si_5$(Dodecamethylpentasiloxane)

303.15 K	Negative	Dif.Cap.Rise	[290]

$C_{10}H_{22}$(Decane)+$C_{14}H_{30}$(Tetradecane)

298.15 K	Pos.Zero	Drop Vol.	[28]

$C_{10}H_{22}$(Decane)+$C_{16}H_{34}$(Hexadecane)

298.15 K	Positive	Drop Vol.	[28]

$C_{10}H_{30}O_3Si_4$(Decamethyltetrasiloxane)+$C_{14}H_{30}$(Tetradecane)

303.15 K	Negative	Dif.Cap.Rise	[290]

$C_{11}H_{24}$(Undecane)+$C_{16}H_{34}$(Hexadecane)

303.15 to 373.15 K		Max.Bub.Pres.	[131] [272]

$C_{12}H_{36}O_4Si_5$(Dodecamethylpentasiloxane)+$C_{14}H_{30}$(Tetradecane)

303.15 K	Negative	Dif.Cap.Rise	[290]

$ClHO_4$(Perchloric acid)+H_2O(Water)

288.15,298.15,323.15 K			[283]

$ClHO_4$(Perchloric acid)+H_2O_4S(Sulphuric acid)

273.15,283.15,298.15 K	Negative		[18]

DH(Deuterium hydride)+D_2(Deuterium)

293.15 K	Negative	Cap.Rise	[110]

Table 1 (*cont.*)
SYSTEM

Temperature	Deviation	Method	Ref.

DH(Deuterium hydride)+H$_2$(Hydrogen)

293.15 K	Negative	Cap.Rise	[110]

D$_2$(Deuterium)+H$_2$(Hydrogen)

293.15 K	Negative	Cap.Rise	[110]

D$_2$O(Deuterium oxide)+D$_2$O$_2$(Deuterium peroxide)

273.15 to 293.15 K	Pos.Neg.	Cap.Rise	[40]

D$_2$O(Deuterium oxide)+H$_2$O(Water)

298.15 K		Cap.Rise	[190]
293.15 K			[285]

F$_2$(Fluorine)+O$_3$(Ozone)

77.35,90.15 K			[128]

HNO$_3$(Nitric acid)+H$_2$O(Water)

291.15 K		Cap.Rise	[188]

H$_2$(Hydrogen)+Ne(Neon)

24.59 to 29 K		Dif.Cap.Rise	[178]

H$_2$O(Water)+H$_2$O$_2$(Hydrogen peroxide)

273.15,291.15 K			[281]

H$_2$O(Water)+H$_2$O$_4$S(Sulphuric acid)

283.15 to 323.15 K (5)	Positive	Cap.Rise	[62]
291.35 K	Pos.Max.	Cap.Rise	[188]
273.15,303.15,323.15 K	Pos.Max.		[282]
273.15 to 343.15 K (8)	Pos.Max.		[262]

H$_2$O(Water)+H$_4$N$_2$(Hydrazine)

298.15 K	Pos.Max.	Max.Bub.Pres.	[60]
288.15,298.15,323.15 K	Maximum		[273]

He(Helium-3)+He(Helium-4)

1.0 to 4.5 K		Cap.Rise	[174]

In(Indium)+Pb(Lead)

F.pt. to 550 K	Negative	Dif.Cap.Rise	[26]

Table 1 (*cont.*)
SYSTEM

Temperature	Deviation	Method	Ref.
N_2(Nitrogen)+O_2(Oxygen)			
83.85 K	Negative		[142]
Pb(Lead)+Sn(Tin)			
F.pt. to 550 K	Negative	Dif.Cap.Rise	[26]

3 **References**

1.   J.L.R.Morgan and A.J.Scarlett,J.Am.Chem.Soc.,1917,**39**,2280.

2.   I.L.Krupatkin and A.N.Bratushchak,Russ.J.Phys.Chem.(Engl.Transl.), 1972,**46**,1532.

3.   M.N.Chaudhri, P.K.Katti, and M.N.Baliga,Trans.Faraday Soc.,1959, **55**,2013.

4.   J.Marechal,Bull.Soc.Chim.Belg.,1952,**61**,149.

5.   K.M.S.Sundaram,Z.Phys.Chem.(Leipzig),1969,**241**,107.

6.   K.M.S.Sundaram,Z.Phys.Chem.(Leipzig),1966,**233**,85.

7.   K.M.S.Sundaram,Z.Phys.Chem.(Leipzig),1972,**249**,285.

8.   F.Goelles, O.Wolfbauer, and F.Still,Monatsh.Chem.,1965,**106**,1437.

9.   S.K.Suri and V.Ramakrishna,J.Phys.Chem.,1968,**72**,1555.

10.   A.Taubman,Acta Physicochim.URSS,1936,**5**,355.

11.   S.Baykut and O.Ozer,Chim.Acta Turc.,1976,**4**,253.

12.   R.S.Myers, G.P.Angel, and H.L.Clever,Colloid Interface Sci., (Proc.Int.Conf.)50th,1976,**3**,443(ed.by M.Kerker).

13.   H.B.Evans,jun., and H.L.Clever,J.Phys.Chem.,1964,**68**,3433.

14.   R.K.Nigam and N.N.Maini,Indian J.Chem.A,1976,**14**a,605.

15.   L.E.Swearingen,J.Phys.Chem.,1928,**32**,1346.

16.   D.L.Hammick and L.W.Andrew,J.Chem.Soc.,1929,759.

17.   J.W.Belton,Trans.Faraday Soc.,1935,**31**,1642.

18.   M.Usanovich, T.Sumarokova, and V.Udovenko,J.Gen.Chem.USSR,1939,**9**, 1967.

19.   I.A.McLure, B.Edmonds, and M.Lal,Nature Phys.Sci.,1973,**241**,71.

20.   J.Koefoed and J.V.Villadsen,Acta Chem.Scand.,1958,**12**,1124.

21.   S.Fuks and A.Bellemans,Physica,1966,**32**,594.

22.   F.B.Sprow and J.H.Prausnitz,Trans.Faraday Soc.,1966,**62**,1105.

23.   M.Siskova and V.Secova,Coll.Czech.Chem.Commun.,1970,**35**,2702.

24.   E.E.Romagosa and G.L.Gaines,jun.,J.Phys.Chem.,1969,**73**,3150.

25.   S.T.Bowden and E.T.Butler,J.Chem.Soc.,1939,79.

26.   T.D.Hoar and D.A.Melford,Trans.Faraday.Soc.,1963,**59**,2476.

[27] V.Ramakrishna and H.Patel,Indian J.Chem.,1970,**8**,256.

[28] R.Aveyard,Unpublished data.

[29] R.L.Schmidt, J.C.Randall, and H.L.Clever,J.Phys.Chem.,1966,**70**,3912.

[30] R.Aveyard,Trans.Faraday Soc.,1968,**63**,2778.

[31] J.Marechal,Trans.Faraday Soc.,1952,**48**,601.

[32] D.V.S.Jain and S.Singh,Indian J.Chem.,1974,**12**,714.

[33] V.T.Lam and G.C.Benson,Can.J.Chem.,1970,**48**,3773.

[34] V.S.Y.Dharan, S.Singh, and K.K.Wadi,J.Chem.Soc.,Faraday Trans.1, 1974,**70**,961.

[35] G.C.Benson and V.T.Lam,J.Colloid Interface Sci.,1972,**38**,259.

[36] R.S.Myers and H.L.Clever,J.Chem.Thermodyn.,1979,**6**,949.

[37] H.L.Clever and C.C.Snead,J.Phys.Chem.,1963,**67**,918.

[38] S.K.Suri and V.Ramakrishna,J.Phys.Chem.,1968,**72**,3073.

[39] A.B.Taubman,Dokl.Akad.Nauk SSSR,1950,**74**,521.

[40] D.A.Guiguine,Can.J.Chem.,1975,**29**,173.

[41] A.N.Campbell, E.M.Katzmark, S.C.Anand, Y.Cheng, H.P.Zihowski, and S.M.Skrynyk,Can.J.Chem,1968,**46**,2399.

[42] A.N.Campbell, E.M.Katzmark, and S.C.Anand,Can.J.Chem.,1971,**49**,2183.

[43] A.N.Campbell and S.C.Anand,Can.J.Chem.,1972,**50**,1109.

[44] J.E.Lane,Faraday Discuss.Chem.Soc.,1975,**59**,55.

[45] K.Chand and V.Ramakrishna,Indian J.Chem.,1969,**7**,698.

[46] S.K.Suri and V.Ramakrishna,Indian J.Chem.,1969,**7**,243.

[47] R.K.Nigam and N.N.Maini,Indian J.Chem.,1971,**9**,855.

[48] H.B.Donald and K.Ridgeway,J.Appl.Chem.,1958,**8**,403.

[49] V.T.Lam, H.D.Pflug, S.Hinakani, and G.C.Benson,J.Chem.Eng.Data,1973, **18**,63.

[50] P.K.Katti and M.M.Chandry,J.Chem.Phys.,1961,**35**,756.

[51] A.N.Campbell and E.M.Katzmark,J.Chem.Thermodyn.,1973,**5**,163.

[52] H.L.Clever,J.Phys.Chem.,1969,**68**,3433.

53 A.P.Toropov and V.M.Kabanova,Uzb.Khim.Zh.,1961,**1**,23.

54 B.S.Mahl, S.L.Chapra, and P.P.Singh,Z.Phys.Chem.(Leipzig),1972,
    249,337.

55 M.S.Dhillon,Z.Phys.Chem.(Leipzig),1979,**260**,497.

56 A.P.Toropov and U.T.Mat'yakubova,Uzb.Khim.Zh.,1963,**7**,92.

57 Kh.R.Pakhimov and R.Gafanova,Nauch.Tr.Tashkent.Gos.Univ.,1968,**323**,
    174.

58 G.B.Samsonov,Kolloidn.Zh.,1957,**13**,46.

59 K.C.Bailey,J.Chem.Soc.,1936,684.

60 N.B.Baker and E.G.Gilbert,J.Am.Chem.Soc.,1940,**62**,2479.

61 F.Hovorka, R.A.Schaefer, and D.Dreisbach,J.Am.Chem.Soc.,1936,**58**,
    2264.

62 L.Sabinina and L.Terpugov,Z.Phys.Chem.,1935,**A173**,237.

63 J.Wellni,Z.Phys.Chem.,1935,**B28**,119.

64 H.S.Dhillon and H.S.Chugh,Thermochim. Acta,1976,**16**,185.

65 H.B.Friedman, A.Barnard, W.B.Doc, and C.L.Fox,J.Am.Chem.Soc.,1940,
    62,2366.

66 J.A.V.Butler, A.Wightman, and W.H.Maclennan,J.Chem.Soc.,1934,528.

67 K.A.Aronovich, L.P.Karstororskii, and K.F.Fedorava,Russ.J.Phys.Chem.
    (Engl.Transl),1967,**41**,9.

68 A.I.Rusanov, S.A.Levichev, and O.N.Khalenko,Russ.J.Phys.Chem.,1969,
    43,1481;Fiz.Khim.Poverkh.Yavlenii Vys Temp.,1977,64.

69 Yu.V.Efremov and Yu.M.Krylov,Russ.J.Phys.Chem.(Engl.Transl.),1970,
    44,805.

70 I.M.Bokhovkin,J.Gen.Chem.USSR(Engl.Transl.),1957,**27**,933.

71 P.K.Migal and D.P.Belotskii,J.Gen.Chem.USSR(Engl.Transl.),1955,**25**,
    1849.

72 W.E.Shipp,J.Chem.Eng.Data,1970,**15**,308.

73 K.Ridgway and D.A.Butler,J.Chem.Eng.Data,1967,**12**,509.

74 P.K.Katti and Chaudri,J.Chem.Eng.Data,1964,**9**,128.

75 H.L.Clever and W.E.Chase,J.Chem.Eng.Data,1963,**8**,291.

76 Y.Saji and T.O.Kuda,Adv.Cryog.Eng.,1964,**10**,209.

[77] D.Volyak,Kolloidn.Zh.,1950,**12**,248.

[78] K.N.Kovalenko, N.A.Trifonov, and D.S.Tissen,Zh.Obshch.Khim.,1956, **26**,403,2404.

[79] N.L.Yarym-Agaev,Dometsk.Ind.Inst.,1960,**23**,129.

[80] P.P.Pugachevich and E.M.Beglyarov,Kolloidn.Zh.,1970,**32**,895.

[81] K.N.Kovalenko and N.A.Trifonov,Sbornik Statei Obshch.Khim.,Akad. Nauk SSSR,1953,**7**,229.

[82] I.A.Zrereva, I.M.Balaskova, and N.A.Smirnova,Vestn.Leningrad Univ. Fiz.Khim.,1977,**4**,107.

[83] P.P.Pugachevich and A.I.Cherkasskaya,Russ.J.Phys.Chem.,1976,**50**, 1303.

[84] V.A.Granzhan and S.K.Laktionova,Russ.J.Phys.Chem.(Engl.Transl.), 1976,**50**,324.

[85] I.M.Bokhovkin,J.Gen.Chem.USSR(Engl.Transl.),1959,**29**,1766.

[86] Y.I.Dutchak and H.F.Panchenko,J.Gen.Chem.USSR,(Engl.Transl.),1973, **43**,481.

[87] B.Y.Teitelbaum and O.A.Osipov,Colloid J.(USSR),1955,**17**,51.

[88] B.Y.Teitelbaum,Dokl.Akad.Nauk SSSR,1949,**65**,303.

[89] C.Saceanu,C.R.Acad.Sci.Paris,Ser.B,1966,**B263**,724.

[90] A.N.Campbell and A.J.R.Campbell,Can.J.Res.,1941,**19B**,73.

[91] A.N.Campbell,Can.J.Res.,1941,**19B**,143.

[92] R.C.Brown,Philos.Mag.,1932,**13**,578.

[93] B.Soucek,Coll.Czech.Chem.Commun.,1938,**10**,459.

[94] D.K.Agarwal and S.Agarwal,Indian J.Chem.,1978,**12**,22.

[95] V.Udovenko, E.V.Sichkova, and A.P.Toropov,J.Gen.Chem.USSR,1939,**9**, 2048.

[96] D.C.Jones and L.Saunders,J.Chem.Soc.,1951,2944.

[97] I.M.Bokhovkin and E.O.Vitman,J.Gen.Chem.USSR(Engl.Transl.),1963, **33**,2030.

[98] I.M.Bokhovkin and E.G.Veselkova,J.Gen.Chem.USSR(Engl.Transl.),1958, **28**,819.

[99] S.S.Vrazovskii and I.A.Shcherbalzov,J.Phys.Chem.(USSR),1948,**22**,417.

[100] F.Schnell,Z.Phys.Chem.,1927,**127**,121.

[101] M.S.Dhillon and B.S.Mahl,Z.Phys.Chem.(Leipzig),1978,**259**,249.

[102] H.Schlegel,J.Chim.Phys.,1934,**31**,668.

[103] S.A.Levichev,Fiz.Khim.,1964,219.

[104] M.V.Khlebaikova,Dokl.Tsuha,1972,**176**,263.

[105] G.Pannetier and L.Abello,Bull.Soc.Chim.France,1965,**7**,2048.

[106] W.Huckel, M.Niesel, and L.Bucks,Ber.Deut.Chem.Ges.,1944,**77B**,334.

[107] Yu.V.Efremov,Russ.J.Phys.Chem.(Engl.Transl.),1971,**45**,1349.

[108] P.K.Migal and V.S.Starchevskii,Uchenye Zapiski Kishinev.Univ.,1957,
27,135;Referat.Zhur.Khim.,1958,Abstr.No.7180.

[109] E.I.Shcherbina, A.E.Tenenbaum, and T.V.Bashun,Dokl.Akad.Nauk,1978,
22,440.

[110] V.N.Grigor'ev and N.S.Rudeuka,JETP,1965,**20**,63.

[111] K.N.Kovalenko and N.I.Balandina,Uchenye Zapiski Kishinev.Univ.,
1955, **25**,13.

[112] V.I.Syuzyaev and A.A.Berdyev,Izv.Akad.Nauk Turkmen.SSR,1957,No.6,9.

[113] G.A.Vetrova and E.N.Vasenko,Nauch.Zapiski,Lvov Politekh.,1956,
No.22,3.

[114] I.M.Bokhovkin,J.Gen.Chem.USSR(Engl.Transl.),1959,**29**,2448.

[115] K.N.Kovalenko and N.A.Trifonov,Uchenye Zapiski Rostovna Donu Univ.,
1958,**41**,45.

[116] E.Z.Zhuravlev, S.Y.Serukhina, L.I.Krupovna, A.V.Zhettova, and
I.I.Koustantinov,Zh.Prikl.Khim.,1967,**40**,1854.

[117] I.M.Bokhovkin and E.O.Vitman,Izvest.Vysshikh Ucheb.Zavednii,Lesnoi
Zh.,1960,**3**,159.

[118] I.M.Bokhovkin,Izvest.Vysshikh Ucheb.Zavednii,Lesnoi.Zh.,1962,**5**,136.

[119] D.Neagu and R.G.Salukvadze,Tr.Inst.Fiz.Akad.Nauk Gruz.SSR,1962,**8**,
183.

[120] E.Tommila, E.Lindell, M.L.Virtalaine, and R.Laakso,Suom.
Kemistilehti B,1969,**42**,95.

[121] Y.Morino,Bull.Inst.Phys.Chem.Research (Tokyo),1932,**11**,1018;
Sci.Papers.Inst.Phys.-Chem.Research(Tokyo),**19**,380.

[122] V.A.Granzhan and O.G.Kirillova,Tr.Gos.Nauch.-Issled.Prockt.Inst.
Azotn.Prom.Prod.Org.Sin.,1973,**13**,5.

123 H.Iwasaki and K.Oate,Koatsu Gasu,1975,**12**,379.

124 A.I.Toryanik and V.G.Pagrebnik,Zh.Strukt.Khim.,1976,**17**,536; J.Struct.Chem.,1976,**17**,464.

125 N.A.Trifonov and K.N.Kovaleuka,Bull.Acad.Sci.URSS,Classe Sci.Chim.,1947,153

126 Y.P.Blagoi,Ukr.Fiz.Zh.,1960,**5**,109.

127 E.Tommila and R.Yrjovnori,Suom.Kemistilehti B,1969,**42**,90.

128 A.J.Gaynor and C.K.Hersh,Adv.Chem.Ser.,1965,**54**,270;Am.Chem.Soc. Div.Fuel Chem.,preprints,1965,**9**,128.

129 H.Tschamler and H.Krischai,Monatsh.Chem.,1951,**82**,259.

130 C.Wenang and D.L.Katz,Ind.Eng.Chem.,1943,**35**,239.

131 P.P.Pugachevich and Y.A.Khuotov,Zh.Fiz.Khim.,1978,**52**,1559.

132 W.S.Bonnell, L.Byman, and D.B.Keyes,Ind.Eng.Chem.,1940,**32**,532.

133 R.K.Nigam and M.S.Dhillon,J.Chem.Thermodyn.,1977,**3**,819; Res.J.Sci.,1974,**1**,28.

134 G.L.Starobinets and K.S.Starobinets,Zh.Fiz.Khim.,1951,**25**,753.

135 G.L.Starobinets and K.S.Starobinets,Zh.Fiz.Khim.,1951,**25**,759.

136 R.V.Mertzin,J.Gen.Chem.USSR,1935,**5**,886.

137 V.B.Igovin, V.Z.Nikonov, and L.B.Sokolov,Zh.Fiz.Khim.,1978,**52**,2341.

138 E.Angelescu and C.Holszky,Acad.Rep.Pop.Romane,Bull.Stiint.Ser.Mat. Fiz.Chim.,1950,**2**,241.

139 A.V.Pamfilov, G.G.Devyatyhh, and L.V.Shirshova,Zh.Fiz.Khim.,1950, **24**,832.

140 M.Hirata, T.Kanai, and H.Ishida,Mem.Fac.Technol,Tokyo Metropolitan Univ.,1958,605;Kagaku Kogaku,1958,**22**,154.

141 S.A.Balyan,Trudy Seminara po fiz. i primenen Vl'trazuka Posvyashchem.pamyati Prof. S.A.Sokolova,Leningrad,1958,77.

142 Y.P.Blagoi and N.S.Rudenov,Izvest.Vysshikh Ucheb.Zavednii,Fiz.,1959, No.2,22.

143 I.M.Bokhovkin,Zh.Obshch.Khim.,1959,**29**,3531.

144 E.Tommila and T.Autio,Suom.Kemistilehti,B,1969,**42**,107.

145 B.I.Kondbeev and V.V.Lyapin,Zh.Prikl.Khim.(Leningrad),1970,**43**,803.

146 G.L.Starobinets, K.S.Starobinets, and L.A.Ryzhikova,Zh.Fiz.Khim., 1951,**25**,1186.

147 A.V.Pamfilov, G.G.Devyatyhh, and L.V.Shirshova,Zh.Fiz.Khim.,1950, **24**,292.

148 F.Blankenship and B.Clampitt,Proc.Okla.Acad.Sci.,1950,**31**,106.

149 A.I.Rusanov and S.A.Levichev,Vestn.Leningrad Univ.,1967,22; Fiz.Khim.No.3,124.

150 G.Junghaehnel and E.Hannemann,Wiss.Z.Tech.Hochsch.Karl-Marx-Stadt., 1964,**6**,11.

151 I.M.Bokhovkin,J.Gen.Chem.USSR(Engl.Transl.),1956,**26**,3281.

152 R.G.Larson,Proc.Indiana Acad.Sci.,1954,**64**,94.

153 V.V.Pashkov, Y.P.Blagoi, and M.P.Lobko,Fiz.Khim.Poverkh.Yavlenii Vys.Temp.,1971,129.

154 K.U.Usananov, U.T.Mat'yakubova, and A.P.Toropov,VINITI,1972,4253.

155 V.I.Obraztsov and A.A.Khrustaleva,VINITI,1973,6043.

156 H.Wagenbreth,PTB-Milt.,1972,**82**,299.

157 I.L.Krupatkin and T.K.Balina,Izv.Vyssh.Uchebn.Zaved.,Khim.Khim. Tekhnol.,1972,**15**,563.

158 V.A.Granzhan and S.K.Laktionova,Zh.Prikl.Khim.(Leningrad),1977,**50**, 1182.

159 C.Sandonini,Atti.Acad.Naz.Lincei,1925,**vi1**,448.

160 A.Giaccelone,Gazz.Chem.Ital.,1942,**72**,378.

161 A.Giaccelone,Gazz.Chem.Ital.,1942,**72**,429.

162 L.G.Nagy,Periodica Polytech.,1963,**7**,75.

163 K.S.Howard and R.A.McAllister,Am.Inst.Chem.Eng.J.,1957,**3**,325.

164 M.B.Donald and K.Ridgeway,Chem.Eng.Sci.,1956,**5**,188.

165 B.L.Dunicz,(Naval Radiol.Defense Lab.,San Francisco,Cal.)AD 628351, 1965.

166 V.A.Granzhan, V.I.Seraya, R.Ya.Muskin, and V.V.Kvasova,Khim.Tekhnol. (Kiev),1975,23.

167 H.Riede, S.Vohland, and H.Schuberth,Z.Phys.Chem.(Leipzig),1976,**257**, 529.

168 B.A.Akhverdien, S.A.Binanda, and B.G.Gasanov,Uch.Zap.Azerb.Gos. Univ.Ser.Khim.Nauk,1973,**2**,62.

169 K.Nikotovski and P.Stefanov,Chem.Ind.,1978,**32**,548.

170 G.Sembrand and G.Fabbrani,Atti.Acad.Veneto-Trentino-Istriana,1932, **23**,33.

171 V.I.Syzaev,Primenie Ul'traakust.Issled.Veshchestva,1961,No.13,199.

172 I.Prigogine and J.Narbond,Trans.Faraday Soc.,1948,**44**,628.

173 W.Herz and E.Lorentz,Z.Phys.Chem.,1929,**A140**,406.

174 B.N.Esel'son and N.G.Bereznyak,Zh.Eksp.Teor.Fiz.,1963,**44**,483; Dokl.Akad.Nauk ,1954,**98**,365,569.

175 Y.P.Blagoi, G.P.Kropachov, and V.V.Pashkov,Ukr.Fiz.Zh.,1967,**12**, 1338.

176 A.Iguchi,Kagaku Sochi(Tokyo),1968,**10**,62.

177 C.R.Gunta, R.D.Madding,jun., T.E.Hanson, and B.Musulin,Trans.Ill. State Acad.Sci.,1971,**64**,55.

178 Yu.P.Blagoi and V.V.Pashkov,Zh.Eksp.Teor.Fiz.,1968,**55**,59.

179 A.I.Rusanov, S.A.Levichev, and V.Ya.Tyushin,Vestn.Leningrad Univ., 1966,**21**,Ser.Fiz.Khim.No.4,121.

180 A.P.Kapustim,J.Exp.Theor.Phys.USSR,1947,**17**,30.

181 B.Staneic and I.Berberovic,Arh.Rudaustvo Tekhnol.,1969,**7**,55.

182 V.E.Azeu, V.A.Granzhan, and O.G.Kinillova,Zh.Prikl.Khim. (Leningrad),1971,**44**,462.

183 J.R.Dean, S.Bennett, and R.N.Maddox,Proc.Ann.Conv.Natur.Gas Processors Ass.,Tech.Paper,1967,**46**,26 and 1968,**47**,30.

184 M.F.Pachenko,Fiz.Khim.Rastorov,1972,250(Ed.Samoilov,O.Y.,"Nauka", Moscow,USSR).

185 S.Valentiner and H.W.Hohls,Z.Phys.,1937,**108**,101.

186 V.A.Granzhan and O.G.Kirillova,Zh.Prikl.Khim.(Lenigrad),1970,**43**, 1875.

187 K.Malysa and J.Pawlikowska-Czubak,Bull.Acad.Pol.Sci.,Ser.Sci.Chim., 1975,**23**,423.

188 W.H.Whatmough,Z.Phys.Chem.,1902,**39**,129.

189 B.Y.Teitelbaum, T.A.Gortalova, and E.E.Sidorova,Zh.Fiz.Khim.,1951, **25**,911

190 G.Jones and W.A.Ray,J.Chem.Phys.,1937,**5**,505.

191 G.Weissenberger, F.Schuster, and N.Mayer,Monatsh.Chem.,1924,**45**,449.

192 G.Weissenberger and F.Schuster,Monatsh.Chem.,1925,**45**,413.

193 P.P.Kosakewitsch and N.S.Kosakewitsch,Z.Phys.Chem.,1933,**A166**,113.

194 D.Faust,Z.Anorg.Chem.,1926,**154**,61.

195 K.L.Wolf,Theor.Chem.,Leipzig,1943,**499**,642(2nd edition 1948)

196 D.V.S.Jain, S.Singh, and R.K.Wadi,J.Chem.Soc.,Faraday Trans.1,1974, **70**,961.

197 H.G.Trieschmann,Z.Phys.Chem.,1935,**B29**,328.

198 S.Uchida and K.Matsumoto,Kagaku Kogaku,1958,**22**,750.

199 S.Constantin and E.Margaretta,C.R.Hebd.Seances Acad.Sci.,Ser.B, 1976,**283**,347.

200 A.P.Toropov and Y.V.Rashkes,Dokl.Akad.Nauk Uzbek.SSR,1958,No.10,27.

201 V.I.Syuzyaev, I.Redzhepov, and L.S.Serukhova,Izvest.Akad.Nauk Turkmen.SSR,1958,No.3,70.

202 G.W.Bennett,J.Chem.Educ.,1929,**6**,1544.

203 R.C.Erust, E.E.Litkenhous, and J.W.Spanyer,jun.,J.Phys.Chem.,1932, **36**,842.

204 B.Y.Teitelbaum, S.G.Gauelina, and T.A.Gortalova,Zh.Fiz.Khim.,1951, **25**,1044.

205 A.L.Vierk,Z.Anorg.Chem.,1950,**261**,283.

206 J.L.R.Morgan and G.Egloff,J.Am.Chem.Soc.,1916,**38**,844.

207 H.Hartley, N.G.Thomas, and M.P.Appleby,J.Chem.Soc.,1908,**93**,538.

208 W.E.Seyer and A.F.Gallaugher,J.Am.Chem.Soc,1930,**52**,1448.

209 J.Traube,J.Prakt.Chem.,1885,**37**,177.

210 J.L.R.Morgan and M.Neidle,J.Am.Chem.Soc,1913,**35**,1856.

211 B.Weinstein,Metronomische Beitrage,No.6,N.Eich.Konn,1889,889.

212 A.Sohet,Mem.Soc.Roy.Sci.Liege,II,20,1.

213 M.Descude,J.Phys.,1903,**2**,348.

214 R.Furth,Ann.Phys.IV,1923,**70**,63.

215 L.L.Bircumshaw,J.Chem.Soc.,1922,**121**,887.

216   J.A.V.Butler and S.Wightman,J.Chem.Soc.,1932,2089.

217   R.C.Ernst, C.H.Watkins, and H.H.Ruwe,J.Phys.Chem.,1936,**40**,627.

218   K.Drucker and E.Moles,Z.Phys.Chem.A,1910,**75**,405.

219   A.Schukarev,Z.Phys.Chem.A,1970,**71**,90.

220   K.Drucker,Z.Phys.Chem.,1905,**52**,641.

221   L.Grunmach,Drude's Ann.IV,1909,**28**,217.

222   A.A.Glagoleva, N.A.Pushin, and M.S.Vrevskii,Zh.Obshch.Khim.,
      1947,**17**,1044.

223   G.N.Antonov,J.Chim.Phys.,1907,5,363;J.Russ.Phys.-Chem.Soc.,
      1907,**39**,342.

224   C.E.Linebarger,Z.Phys.Chem.A,1896,**20**,131;Phys.Rev.,1896,**3**,418;
      J.Am.Chem.Soc,1896,**18**,429;Am.J.Sci.,1896,2,227,331.

225   R.Kremann and R.Meingast,Monatsh.Chem.,1914,**35**,1323.

226   J.L.R.Morgan and M.A.Griggs,J.Am.Chem.Soc,1917,**39**,2261.

227   K.N.Kovalenko and N.A.Trifonov,Russ.J.Phys.Chem.,1953,**27**,527.

228   R.Kremann, F.Gugl, and R.Mengast,Monatsh.Chem.,1914,**35**,1365.

229   W.Ramsay and E.Aston,Proc.Roy.Soc.London,1894,**56**,182;
      Z.Phys.Chem.A,1894,**15**,89.

230   R.P.Worley,J.Chem.Soc.,1914,**105**,273.

231   N.A.Yasnik, B.K.Sharma, and M.C.Bharadwaj,J.Indian Chem.Soc.,
      1926,3,63.

232   D.L.Hammick and H.F.Wilmut,J.Chem.Soc.,1934,32.

233   W.Ramsay and E.Aston,Trans.Roy.Irish Acad.,1902,**32A**,93.

234   W.Sutherland,Philos.Mag.,1894,**38**,V,188.

235   D.Pekar,Z.Phys.Chem.A,1902,**39**,433.

236   O.B.Keyes and J.H.Hildebrand,J.Am.Chem.Soc.,1917,**39**,2110;
      1917,**39**,2126.

237   G.Weissenberger and L.Piatti,Monatsh.Chem.,1924,**45**,187,281;
      Sitzungsb.Wien,1924,**116**,133.

238   E.Herzen,Helv.Chim.Acta,1902,**14**,232.

239   J.H.Mathews and A.J.Stamm,J.Am.Chem.Soc,1924,**46**,1071.

240 A.Ritzel,Z.Phys.Chem.A,1907,**60**,319.

241 N.J.Geschhus,J.Russ.Phys.Chem.Soc.,1900,**32**,97.

242 G.S.Pavlov,J.Russ.Phys.Chem.Soc.,1927,**58**,1302,1309.

243 G.Weissenberger, R.Henke, and F.Schuster,Monatsh.Chem.,1925,**46**, 44,57.

244 G.Weissenberger, F.Schuster, and K.Schuler,Monatsh.Chem.,1924, **45**,425.

245 G.G.Devyatikh, A.V.Pamilov, and G.Starobinets,Russ.J.Phys.Chem., 1948,**22**,1072.

246 O.A.Osipov and N.A.Trifonov,Russ.J.Gen.Chem.,1951,**21**,811.

247 O.A.Osipov and N.A.Trifonov,Russ.J.Gen.Chem.,1949,**19**,1822.

248 T.C.Sutton and H.L.Harden,J.Phys.Chem.,1934,**38**,779.

249 R.Kremann and F.W.Kuster,Monatsh.Chem.,1909,**29**,863.

250 G.Weissenberger, R.Henke, and L.Bergmann,Monatsh.Chem.,1925,**46**,471.

251 G.Weissenberger, F.Schuster, and K.Wojnoff,Monatsh.Chem.,1925, **46**,471.

252 R.M.Conrad and J.L.Hall,J.Am.Chem.Soc.,1935,**57**,861.

253 W.Herz and M.Levi,Z.Anorg.Chem.,1929,**183**,340.

254 N.N.Efremov, A.D.Vinogradova, and A.M.Tikhomirova,Bull.Acad.Sci. USSR,1937,**443**.

255 R.N.Coleman and B.R.Prideaux,J.Chem.Soc.,1937,1022.

256 E.Angelescu and C.Eustatin,Z.Phys.Chem.A,1936,**177**,263.

257 G.L.Starobinets, A.V.Pamfilov, G.G.Devyatikh, and G.A.Lazerko, Russ.J.Phys.Chem.,1951,**25**,1186.

258 E.B.Prideaux and R.N.Coleman,J.Chem.Soc.,1936,1346.

259 K.N.Kozlenko and S.P.Miskidzhyan,Russ.J.Gen.Chem.,1955,**25**,87,35.

260 Y.I.Bokhovkina and I.M.Bokhovkin,J.Gen.Chem.USSR(Engl.Transl.), 1956,**26**,2399,2255.

261 I.M.Bokhovkin and ·Y.I.Bokhovkina,J.Gen.Chem.USSR(Engl.Transl.), 1956,**26**,1316,1319.

262 C.E.Linebarger,J.Am.Chem.Soc,1900,**22**,5.

263 D.G.LeGrand and G.L.Gaines,jun.,J.Polym.Sci.,Part C,1977,**34**,45; Am.Chem.Soc.Div.Org.Coatings Plast.Chem.,Pap.1970,**30**,459.

264 I.A.McLure and B.Edmonds,J.Chem.Phys.,1979,**70**,3999.

265 J.C.G.Calado, I.A.McLure, and V.A.M.Soares,Fluid Phase Equilibria, 1979,**2**,199.

266 P.P.Pugachevich, A.L.Martirosyan, R.M.Kamalayan, and I.A.Lavygin, Russ.J.Phys.Chem.,1974,**48**,566.

267 M.P.Khosla and B.Widom,J.Colloid Interface Sci.,1980,**76**,375.

268 G.L.Gaines,jun.,J.Polym.Sci.Part A2,1969,**7**,1379.

269 A.B.Ponter and V.Peier,Int.J.Heat Mass Transfer,1978,**21**,1025.

270 V.I.Syuzyaev,Uchenye Zapiski Turkmen Gosudarst Univ.im. A.M.Gor'Koyo,1959,No.15,113.

271 W.Warudzin and P.Skubla,Petrochemia,1974,**14**,141.

272 P.P.Pugachevich, A.Y.Khyorov, and E.M.Beglyarov,Zh.Fiz.Khim., 1979,**53**,429.

273 H.D.Oerf and J.O.Birzer,Colloid Polym.Sci.,1978,**256**,1034.

274 D.K.Agarwal, R.Gopal, and S.Agarwal,J.Chem.Eng.Data,1979,**24**,181.

275 Nai-Fu Zhou and Ti-Ren Gu,Sci.Sin.(English Ed),1979,**22**,1033.

276 T.Nishida, T.Shimiza, and H.Jido,Hakodate Kogyo Koto Senmon Gakko Kiyo,1979,**13**,73.

277 H.D.Doerfler,Colloid Polym.Sci.,1979,**257**,882.

278 M.S.Dhillon,Monatsh.Chem.,1979,**110**,847.

279 M.Patel and V.Ramakrishra,J.Colloid Interface Sci.,1980,**76**,166.

280 W.Huckel and H.Harder,Ber.Deut.Chem.Ges.,1947,**80**,357.

281 O.Maas and W.H.Hatcher,J.Am.Chem.Soc.,1920,**42**,2548.

282 J.L.R.Morgan and C.E.Davies,J.Am.Chem.Soc.,1916,**38**,555.

283 C.A.Meros and W.G.Eversole,J.Phys.Chem.,1941,**45**,388.

284 E.N.Butskaya,Uchenye Zapiski,Leningrad.Gosudarst.Pedagog.Inst.im. A.I.Gertsena,1959,**60**,141.

285 P.W.Selwood and A.A.Frost,J.Am.Chem.Soc,1933,**55**,4335.

286 R.P.Worley,J.Chem.Soc.,1914,**105**,260.

287 S.P.Miskidzhyan and N.A.Trifonov,Zh.Obshch.Khim.,1947,**17**,1038,1234.

288 E.Moles, E.Cabrera, K.Drucker, N.M.Retortillo, and J.Robles, An.Fis.Quim.Madrid,1911,**9**,156.

289 I.A.McLure, J.T.Sipowska, and I.L.Pegg,J.Chem.Thermodyn.,1982, 14,733.

290 B.Edmonds and I.A.McLure,J.Chem.Soc.,Faraday Trans.1,1982,in press.

291 I.A.McLure and J.L.Arriaga Colina,unpublished work.

292 V.A.M.Soares and I.A.McLure,Rev.Port.Quim.,1982,in press.

293 I.A.McLure, B.E.Edmonds, and M.Lal,J.Colloid Interface Sci.,1982, in press.

294 V.A.M.Soares and I.A.McLure,J.Chem.Soc.,Faraday Trans.1,1982, in press.

295 B.Edmonds,Ph.D. Thesis (University of Sheffield,1972).

296 J.F.Neville,Ph.D. Thesis (University of Sheffield,1979).

297 I.A.McLure and W.P.Edwards,unpublished work.

298 I.A.McLure and S.Lee,unpublished work.

299 A.Al-Nakash,M.Sc. Thesis (University of Sheffield,1980).

300 I.L.Pegg,Ph.D. Thesis (University of Sheffield,1982).

4 Compound Index

Column 2 in Table 2 shows all the locations where a particular component shown in Column 1 is in admixture. Figures in parentheses denote the number of entries in which the chosen formula appears as the second component of a mixture.

Table 2

Ar	see Ar.
Br_2	see Br_2.
CBr_4	see CBr_4.
CCl_4	see Br_2, CBr_4, CCl_4.
$CHCl_3$	see CCl_4, $CHCl_3$.
CH_2O_2	see CH_2O_2.
CH_3I	see CH_3I.
CH_3NO	see CH_3NO.
CH_3NO_2	see CCl_4, CH_3NO_2.
CH_4	see Ar, CH_4.
CH_4N_2O	see CH_4N_2O.
CH_4O	see $CHCl_3$, CH_4O.
CO	see CH_4, CO.
CS_2	see $CHCl_3$, CH_3I, CH_4, CS_2.
C_2Cl_4	see C_2Cl_4.
C_2HCl_3	see C_2HCl_3.
C_2HCl_3O	see CH_2O_2, CH_4O, C_2HCl_3O.
$C_2HCl_3O_2$	see $C_2HCl_3O_2$.
C_2H_2	see C_2H_2.
$C_2H_2Br_4$	see $C_2H_2Br_4$.
$C_2H_2Cl_2$	see CS_2, $C_2H_2Cl_2$.
$C_2H_2Cl_2O_2$	see $C_2H_2Cl_2O_2$.
$C_2H_3ClO_2$	see $C_2H_3ClO_2$.

Table 2 (*cont.*)

C_2H_3N	see C_2H_3N.
$C_2H_4Br_2$	see $C_2H_4Br_2$.
$C_2H_4Cl_2$	see CCl_4, $CHCl_3$, $C_2H_4Cl_2$.
C_2H_4O	see C_2H_4O.
$C_2H_4O_2$	see CCl_4, $CHCl_3$, CH_2O_2, CH_3I, CH_3NO_2, C_2HCl_3O, $C_2HCl_3O_2$, $C_2H_2Cl_2O_2$, $C_2H_3ClO_2$, $C_2H_4Br_2$, C_2H_4O, $C_2H_4O_2$.
C_2H_5I	see CH_4O, $C_2H_4O_2$, C_2H_5I.
C_2H_5NO	see $C_2HCl_3O_2$, $C_2H_3ClO_2$, $C_2H_4O_2$, C_2H_5NO.
$C_2H_5NO_2$	see CCl_4, $C_2H_5NO_2$.
C_2H_6O	see CCl_4, $CHCl_3$, CH_4O, C_2HCl_3O, C_2H_3N, $C_2H_4O_2$, C_2H_6O.
C_2H_6OS	see CCl_4, $CHCl_3$, CH_4O, C_2H_6O, C_2H_6OS.
C_2H_6OSi	see C_2H_6OSi.
$C_2H_6O_2$	see $C_2H_6O_2$.
C_2H_7N	see C_2H_7N.
C_2H_7NO	see CH_4O.
$C_2H_8N_2$	see $C_2H_8N_2$.
C_3F_8	see C_3F_8.
$C_3H_4O_3$	see $C_3H_4O_3$.
C_3H_5N	see C_3H_5N.
$C_3H_5N_3O_9$	see $C_3H_5N_3O_9$.
$C_3H_6Br_2$	see $C_2H_4Br_2$.
C_3H_6O	see CCl_4, $CHCl_3$, CH_2O_2, CH_4O, CS_2, C_2HCl_3O, $C_2HCl_3O_2$, $C_2H_3ClO_2$, C_2H_4O, $C_2H_4O_2$, C_2H_6O, C_2H_6OS, C_3H_6O.
$C_3H_6O_2$	see CH_3I, CH_3NO_2, $C_2H_4O_2$, $C_3H_6O_2$.
$C_3H_6O_3$	see $CHCl_3$, $C_3H_6O_3$.
C_3H_7NO	see C_3H_7NO.
C_3H_8	see CH_3I, CH_4, C_3H_8.
C_3H_8O	see $CH_4O(2)$, C_3H_6O, C_3H_8O.

Table 2 (*cont.*)

$C_3H_8O_2$ see $C_3H_8O_2$.

$C_3H_8O_3$ see CH_3NO, C_2H_6O, $C_3H_8O_3$.

C_4F_8 see C_4F_8.

C_4F_{10} see C_4F_{10}.

$C_4H_5Cl_3O_2$ see $C_4H_5Cl_3O_2$.

C_4H_5NS see C_2H_6O, C_4H_5NS.

C_4H_6O see C_2H_4O, $C_2H_4O_2$, C_3H_6O, C_4H_6O.

$C_4H_6O_2$ see C_2H_4O, $C_2H_4O_2$, C_3H_6O, C_4H_6O.

$C_4H_6O_3$ see CH_3I, CS_2, $C_2H_4O_2$, C_2H_6O, C_3H_6O, $C_3H_6O_2$, $C_4H_6O_3$.

$C_4H_7O_2$ see $C_4H_7O_2$.

C_4H_8 see C_4H_8.

C_4H_8O see $C_2H_4O_2$, C_3H_6O, C_4H_8O.

$C_4H_8O_2$ see $CCl_4(2)$, $CHCl_3$, $CH_3NO_2(2)$, $CH_4O(2)$, CS_2, C_2HCl_3O, $C_2H_4O_2(2)$
                    C_2H_5I, C_2H_6O, C_2H_6OS, $C_2H_6O_2$, C_3H_6O, $C_3H_6O_2$, $C_4H_5Cl_3O_2$,
                    $C_4H_6O_3$, $C_4H_8O_2$.

$C_4H_8O_2S$ see CH_4O, C_2H_6O, $C_4H_8O_2S$.

C_4H_9I see $C_2H_4Br_2$.

C_4H_9NO see C_4H_9NO.

C_4H_{10} see C_3F_8, C_4F_8, $C_4F_{10}(2)$.

$C_4H_{10}N_2$ see $C_4H_{10}N_2$.

$C_4H_{10}O$ see $CHCl_3$, CH_3I, $CH_4O(4)$, CS_2, C_2HCl_3O, $C_2H_2Br_4$, C_2H_5I, $C_3H_6O(2$
                    C_3H_8O, $C_4H_{10}O$.

$C_4H_{10}O_3$ see $C_4H_{10}O_3$.

$C_4H_{11}N$ see $C_3H_6O_2$, $C_4H_{11}N$.

$C_4H_{11}NO_2$ see $C_4H_{11}NO_2$.

$C_4H_{12}Si$ see $C_4H_{12}Si$.

$C_5H_4O_2$ see $C_2HCl_3O_2$, $C_2H_3ClO_2$, $C_2H_4O_2$, C_2H_5NO, C_3H_6O, $C_5H_4O_2$.

C_5H_5N see CH_2O_2, $C_2H_4O_2$, $C_3H_4O_3$, $C_3H_6O_2$, $C_4H_8O_2$, C_5H_5N.

Table 2 (*cont.*)

$C_5H_8N_2$	see $C_5H_8N_2$.
$C_5H_8O_2$	see C_2HCl_3O.
C_5H_{10}	see CCl_4, C_2Cl_4, C_5H_{10}.
$C_5H_{10}O$	see CH_2O_2, $C_2H_4O_2$, $C_4H_8O_2$.
$C_5H_{10}O_2$	see CH_3NO_2, C_3H_6O, $C_5H_{10}O_2$.
$C_5H_{10}O_3$	see C_2H_3N, $C_5H_{10}O_2$, $C_5H_{10}O_3$.
$C_5H_{11}N$	see CH_4O, $C_3H_6O_2$, $C_4H_8O_2$, $C_5H_{10}O_2$, $C_5H_{11}N$.
$C_5H_{11}NO$	see CH_3NO, $C_5H_{11}NO$.
$C_5H_{11}NO_2$	see C_2H_2.
C_5H_{12}	see C_3F_8, C_3H_5N, $C_4H_{12}Si$, C_5H_{12}.
$C_5H_{12}N_2O$	see $C_5H_{12}N_2O$.
$C_5H_{12}O$	see C_2HCl_3O, C_2H_6O, C_3H_8O, $C_4H_8O_2$, $C_4H_{10}O$, $C_5H_{12}O$.
C_6F_6	see CCl_4, CH_3NO_2, C_2H_6OS, C_3H_7NO, $C_5H_{12}N_2O$, C_6F_6.
C_6F_{12}	see C_6F_{12}.
C_6F_{14}	see C_5H_{12}, C_6F_{14}.
C_6H_4BrCl	see C_6H_4BrCl.
$C_6H_4BrNO_2$	see $C_6H_4BrNO_2$.
$C_6H_4Br_2$	see C_6H_4BrCl, $C_6H_4Br_2$.
$C_6H_4ClNO_2$	see C_6H_4BrCl, $C_6H_4BrNO_2$, $C_6H_4Br_2$.
$C_6H_4Cl_2$	see C_2H_6O, $C_3H_8O(2)$, $C_4H_{10}O$, $C_5H_{12}O$, $C_6H_4Cl_2$.
C_6H_5Br	see C_2H_6OS, C_5H_{12}, C_6H_5Br.
C_6H_5Cl	see CCl_4, C_2H_6OS, C_3H_6O, $C_4H_8O_2$, $C_5H_{10}O_2(2)$, C_5H_{12}, C_6H_5Br, C_6H_5Cl.
$C_6H_5NO_2$	see CCl_4, $CHCl_3$, CS_2, C_2H_6O, C_2H_6OS, $C_3H_5N_3O_9$, $C_3H_6O_3$, C_3H_8O, $C_4H_8O_2(3)$, $C_4H_{10}O(3)$, $C_5H_{10}O_2$, $C_5H_{10}O_3$, $C_5H_{12}O$, C_6F_6, $C_6H_5NO_2$.
$C_6H_5NO_3$	see $C_6H_5NO_3$.

Table 2 *(cont.)*

C_6H_6 see Br_2, CCl_4, $CHCl_3$, CH_2O_2, CH_3NO_2, CH_4O, CS_2, C_2HCl_3O, C_2H_2Br, $C_2H_2Cl_2$, $C_2H_4Br_2$, $C_2H_4Cl_2$, C_2H_4O, C_2H_5I, $C_2H_5NO_2$, C_2H_6O, C_2H_6OS, C_3H_6O, $C_3H_6O_2$, $C_3H_6O_3$, $C_3H_8O(2)$, C_4H_8O, $C_4H_8O_2(3)$, $C_4H_{10}O(3)$, C_5H_5N, $C_5H_{10}O_2(3)$, $C_5H_{10}O_3$, C_5H_{12}, $C_5H_{12}O$, C_6F_6, $C_6H_4Cl_2$, C_6H_5Br, C_6H_5Cl, $C_6H_5NO_2$, C_6H_6.

C_6H_6ClN see $C_6H_4Br_2$, C_6H_5Cl.

C_6H_6O see CH_4N_2O, C_2HCl_3O, C_2H_5NO, C_2H_6O, C_3H_6O, C_5H_5N, C_6H_6, C_6H_6O.

$C_6H_6O_2$ see $C_6H_6O_2$.

C_6H_7N see $CHCl_3$, CS_2, $C_2H_4O_2$, C_2H_6O, C_3H_8O, $C_4H_8O_2(2)$, $C_4H_{10}O$, C_5H_5N, C_5H_{10}, C_6H_5Cl, $C_6H_5NO_2$, C_6H_6, C_6H_6O, C_6H_7N.

$C_6H_8N_2$ see $C_6H_6O_2$, $C_6H_8N_2$.

C_6H_{10} see C_6F_6, C_6H_{10}.

$C_6H_{10}O$ see $CHCl_3$, C_2HCl_3, C_2H_6O, C_3H_6O, $C_4H_8O_2$, C_6H_6, $C_6H_{10}O$.

$C_6H_{10}O_3$ see $CHCl_3$, CS_2.

C_6H_{12} see CCl_4, CH_4O, C_2Cl_4, $C_2H_4Br_2$, $C_2H_5NO_2$, C_2H_6O, C_4H_8O, $C_4H_8O_2$, $C_4H_{10}O(2)$, C_5H_{12}, C_6F_6, C_6F_{12}, $C_6H_4Cl_2$, C_6H_5Br, C_6H_5Cl, $C_6H_5NO_2$, C_6H_6, C_6H_7N, C_6H_{10}, C_6H_{12}.

$C_6H_{12}O$ see $CHCl_3$, $CH_4O(2)$, C_2HCl_3O, $C_3H_6O(2)$, C_3H_8O, $C_4H_8O_2$, C_6H_6, $C_6H_{12}O$.

$C_6H_{12}O_2$ see CH_3NO_2, C_2HCl_3O, $C_2H_6O_2$, $C_3H_8O_2$, $C_5H_{10}O_2(2)$, $C_5H_{11}N$, $C_6H_{12}O_2$

C_6H_{14} see CCl_4, CH_4O, $C_2H_5NO_2(2)$, C_2H_6O, C_3H_6O, C_3H_8O, $C_4H_8O_2$, $C_4H_{10}O$, $C_5H_{12}O$, $C_6F_{14}(2)$, $C_6H_5NO_2$, C_6H_6, C_6H_7N, $C_6H_{12}(2)$, C_6H_{14}.

$C_6H_{14}O$ see CH_4O, C_3H_6O, C_3H_8O, C_6H_6, C_6H_{12}, $C_6H_{12}O(2)$, C_6H_{14}, $C_6H_{14}O$.

$C_6H_{14}O_2$ see $C_6H_{14}O_2$.

$C_6H_{14}O_4$ see $C_6H_{14}O_4$.

$C_6H_{15}N$ see $C_6H_{15}N$.

$C_6H_{16}N_2$ see $C_6H_{16}N_2$.

$C_6H_{18}OSi_2$ see C_2H_6OSi, $C_4H_{12}Si$, C_5H_{12}, C_6F_{14}, $C_6H_{18}OSi_2$.

Table 2 (*cont.*)

C_7F_{14} see C_6H_{12}.

C_7H_5NS see $C_4H_{11}N$, C_6H_7N, C_7H_5NS.

$C_7H_6O_2$ see C_2H_6O, C_3H_6O, $C_4H_{10}O$, C_6H_6.

C_7H_8 see CCl_4, CS_2, $C_2H_4Br_2$, $C_2H_5NO_2$, C_2H_6O, C_2H_6OS, C_2H_6OSi, $C_3H_6O_2$, $C_4H_8O_2(2)$, $C_4H_{10}O$, C_5H_{10}, $C_5H_{10}O_2$, $C_5H_{11}N$, C_5H_{12}, C_6H_5Br, C_6H_5Cl, $C_6H_5NO_2$, C_6H_6, $C_6H_7N(2)$, C_6H_{12}, $C_6H_{18}OSi_2$, C_7H_8.

C_7H_8O see $CHCl_3(3)$, $CH_4O(3)$, $CS_2(4)$, $C_2HCl_3O(2)$, $C_2H_6O(3)$, $C_3H_6O(4)$, $C_4H_8O_2(3)$, $C_4H_{10}O(3)$, $C_5H_4O_2(3)$, $C_6H_6(4)$, C_6H_{12}, $C_7H_8(4)$, C_7H_8O.

$C_7H_8O_2$ see C_2H_6O, $C_2H_6O_2$, C_3H_6O, $C_4H_{10}O$, C_6H_6, $C_7H_8O_2$.

C_7H_9N see $CH_2O_2(2)$, $C_2H_4O_2(2)$, $C_3H_6O_2(2)$, C_4H_5NS, $C_4H_8O_2(3)$, C_5H_5N, $C_6H_5NO_2(2)$, C_6H_6, C_6H_6O, $C_6H_7N(2)$, C_6H_{14}, C_7H_8, C_7H_8O, C_7H_9N.

$C_7H_{12}O_4$ see C_2HCl_3O.

$C_7H_{13}NO$ see $C_2H_6O_2$, $C_4H_{10}O_3$, $C_6H_{14}O_4$, $C_7H_{13}NO$.

C_7H_{14} see $C_2H_5NO_2$, C_2H_6O, $C_4H_{10}O$, $C_6H_5NO_2$.

$C_7H_{14}O$ see $CHCl_3(3)$, $CH_4O(3)$, $C_4H_8O_2(3)$, $C_6H_6(3)$.

$C_7H_{14}O_2$ see CH_3NO_2, CS_2, $C_4H_{10}O$, $C_5H_{11}N$, $C_7H_{14}O_2$.

C_7H_{16} see C_3H_5N, C_3H_8O, C_4H_8, C_6F_{14}, $C_6H_5NO_2$, C_6H_6, C_6H_{12}, $C_6H_{18}OSi_2$, C_7H_{16}.

$C_7H_{16}O$ see $C_6H_4Cl_2$, C_7H_{16}.

C_8F_{18} see C_5H_{12}.

$C_8H_7O_2Cl$ see $C_8H_7O_2Cl$.

C_8H_8O see C_6H_6.

C_8H_9NO see C_8H_9NO.

C_8H_{10} see $C_2H_4Br_2(3)$, $C_4H_8O_2$, $C_5H_{12}(2)$, $C_6H_4Cl_2(3)$, $C_6H_5Br(3)$, $C_6H_5Cl(3)$, $C_6H_5NO_2$, $C_6H_6(2)$, $C_6H_{12}(4)$, C_8H_{10}.

$C_8H_{11}N$ see CS_2, C_4H_5NS, $C_4H_8O_2(2)$, $C_4H_{10}O$, C_5H_5N, $C_6H_5NO_2(2)$, C_6H_6, $C_6H_6O(2)$, C_6H_7N, C_7H_8, C_7H_8O, C_7H_{16}, $C_8H_{11}N$.

Table 2 (*cont.*)

$C_8H_{12}N_2$ see $C_8H_{12}N_2$.

$C_8H_{14}O_4$ see $C_4H_8O_2$.

$C_8H_{16}O_2$ see CH_3NO_2, C_4H_7N, $C_5H_{11}N$.

C_8H_{18} see $C_2H_2Br_4$, C_4H_8, C_6F_{14}, $C_6H_5NO_2$, C_6H_6, $C_6H_{12}(2)$, C_6H_{14}, C_7H_1
               C_8H_{18}.

$C_8H_{18}O$ see $C_6H_4Cl_2$, C_7H_{16}.

$C_8H_{18}O_5$ see $C_7H_{13}NO$.

$C_8H_{24}O_2Si_3$ see C_2H_6OSi, C_5H_{12}, C_6F_{14}, C_7H_{16}, $C_8H_{24}O_2Si_3$.

C_9H_7N see $C_2H_4O_2$, C_6H_6, C_6H_7N, C_7H_8O, C_9H_7N.

$C_9H_{10}O_2$ see $C_2H_6O_2$, $C_4H_8O_2$, C_6H_7N, C_7H_8, C_7H_8O, $C_9H_{10}O_2$.

C_9H_{12} see CH_4O, C_6H_6, $C_8H_{11}N$, C_9H_{12}.

$C_9H_{13}N$ see C_6H_6, C_7H_8.

C_9H_{20} see CH_4.

$C_9H_{20}O$ see C_7H_{16}.

$C_9H_{22}N_2$ see $C_9H_{22}N_2$.

$C_{10}F_{18}$ see C_6H_{12}.

$C_{10}H_7Br$ see C_6H_6.

$C_{10}H_8$ see $C_4H_8O_2$, $C_4H_{10}O(2)$, $C_5H_{10}O_2$, $C_5H_{12}O$, $C_6H_5NO_3$, C_6H_6, C_8H_{10}.

$C_{10}H_{12}$ see $CHCl_3$, C_2H_6O, C_3H_6O, C_4H_8, C_6H_{14}, $C_{10}H_{12}$.

$C_{10}H_{14}N_2$ see CH_4O, $C_{10}H_{14}N_2$.

$C_{10}H_{15}N$ see $C_4H_8O_2$, $C_6H_5NO_2$, C_6H_6O.

$C_{10}H_{16}$ see C_7H_8.

$C_{10}H_{18}$ see C_6H_7N, $C_6H_{12}(2)$, $C_8H_7O_2Cl$.

$C_{10}H_{22}$ see $C_6H_5NO_2$, C_6H_{14}, $C_6H_{18}OSi_2$, C_7H_{16}, $C_8H_{24}O_2Si_3$, $C_{10}H_{22}$.

$C_{10}H_{22}O$ see CH_4O, C_2H_6O, $C_3H_8O(2)$, $C_4H_8O_2$, $C_4H_{10}O(4)$, $C_6H_{14}O$, C_7H_{16}.

$C_{10}H_{30}O_3Si_4$ see C_5H_{12}, C_7H_{16}, $C_{10}H_{22}$, $C_{10}H_{30}O_3Si_4$.

$C_{11}H_{24}$ see C_7H_{16}, $C_{11}H_{24}$.

$C_{12}F_{27}N$ see C_5H_{12}.

Table 2 (*cont.*)

Table 2 (*cont.*)

Author Index